OPERATION OF DISTRIBUTED ENERGY RESOURCES IN SMART DISTRIBUTION NETWORKS

OPERATION OF DISTRIBUTED ENERGY RESOURCES IN SMART DISTRIBUTION NETWORKS

Edited by

DR. KAZEM ZARE

University of Tabriz, Tabriz, Iran

DR. SAYYAD NOJAVAN

University of Tabriz, Tabriz, Iran

Academic Press is an imprint of Elsevier
125 London Wall, London EC2Y 5AS, United Kingdom
525 B Street, Suite 1650, San Diego, CA 92101, USA
50 Hampshire Street, 5th Floor, Cambridge, MA 02139, United States
The Boulevard, Langford Lane, Kidlington, Oxford OX5 1GB, United Kingdom

Notices
Knowledge and best practice in this field are constantly changing. As new research and experience broaden our understanding, changes in research methods, professional practices, or medical treatment may become necessary.

Practitioners and researchers must always rely on their own experience and knowledge in evaluating and using any information, methods, compounds, or experiments described herein. In using such information or methods they should be mindful of their own safety and the safety of others, including parties for whom they have a professional responsibility.

To the fullest extent of the law, neither the Publisher nor the authors, contributors, or editors, assume any liability for any injury and/or damage to persons or property as a matter of products liability, negligence or otherwise, or from any use or operation of any methods, products, instructions, or ideas contained in the material herein.

British Library Cataloguing-in-Publication Data
A catalogue record for this book is available from the British Library

Library of Congress Cataloging-in-Publication Data
A catalog record for this book is available from the Library of Congress

ISBN: 978-0-12-814891-4

For Information on all Academic Press publications
visit our website at https://www.elsevier.com/books-and-journals

Publisher: Joe Hayton
Acquisition Editor: Lisa Reading
Editorial Project Manager: Jennifer Pierce
Production Project Manager: Kamesh Ramajogi
Cover Designer: Mark Rogers

Typeset by MPS Limited, Chennai, India

DEDICATION

To:

My wife, Maryam, and my daughter, Panisa.

—Kazem

To:

My mother's soul.

—Sayyad

CONTENTS

LIST OF CONTRIBUTORS

Saeed Abapour
University of Tabriz, Tabriz, Iran

Gianni Celli
University of Cagliari, Cagliari, Italy

Mohammadreza Daneshvar
University of Tabriz, Tabriz, Iran

Naser Nourani Esfetanaj
Sahand University of Technology, Tabriz, Iran

Sina Ghaemi
Azarbaijan Shahid Madani University, Tabriz, Iran

Mehrdad Ghahramani
University of Tabriz, Tabriz, Iran

Emilio Ghiani
University of Cagliari, Cagliari, Italy

Farkhondeh Jabari
University of Tabriz, Tabriz, Iran

Moein Manbachi
The University of British Columbia, Vancouver, BC, Canada

Susanna Mocci
University of Cagliari, Cagliari, Italy

Behnam Mohammadi-ivatloo
University of Tabriz, Tabriz, Iran

Solmaz Moradi-Moghadam
Department of Electrical Engineering and Computer Science, University of Wisconsin-Milwaukee, Milwaukee, WI, United States

Sajad Najafi-Ravadanegh
Smart Distribution Grid Research Lab, Department of Electrical Engineering, Azarbaijan Shahid Madani University, Tabriz, Iran

Nima Nikmehr
Department of Electrical Engineering and Computer Science, University of Wisconsin-Milwaukee, Milwaukee, WI, United States

Sayyad Nojavan
University of Tabriz, Tabriz, Iran

Fabrizio Pilo
University of Cagliari, Cagliari, Italy

Giuditta Pisano
University of Cagliari, Cagliari, Italy

Sadjad Sarkhani
University of Tabriz, Tabriz, Iran

Gian Giuseppe Soma
University of Cagliari, Cagliari, Italy

Lingfeng Wang
Department of Electrical Engineering and Computer Science, University of Wisconsin-Milwaukee, Milwaukee, WI, United States

Kazem Zare
University of Tabriz, Tabriz, Iran

PREFACE

Distribution networks are deeply involved in the challenge for the integration of renewable energy sources (RES), in which their electricity production depends upon meteorological conditions and/or the time of the day. Therefore, the persisting growth of variable renewable generation—predominantly on distribution level—is gradually altering the operating condition of distribution networks.

From the distribution system operator (DSO) point of view, the new electricity generation mix is causing a dramatic revolution of their distribution systems. Furthermore, the presence of bi-directional power flows and the increasingly occurrence of reverse power flow on distribution transformers are compromising the traditional passive management of distribution networks, particularly for voltage regulation and protection systems. At the same time, DSO needs to be ready to deal with external inputs coming from the "smart world" that will transform the loads from simple points of electricity absorption to effective players in the distribution system, with new requirements but also with a new capability of interaction. The flexibility of loads can be regarded as one of the distributed energy resources (DER) issues.

Also, energy storage systems (ESS) are valuable components of the future distribution system, thanks to their ability to increase the flexibility of the overall system and to provide a wide range of services to the DSOs and to the customers with the final goal of mitigating the negative effects of RES. Services provided by ESS may be such as deferment of network investments, energy losses reduction, voltage regulation, reactive power balance, congestion overcoming, etc. ESS will allow both load profile flattening and solving extra production conditions, especially if they are connected to critical nodes of the network.

The electricity distribution is relying more and more on the intelligent active management of the system instead of traditional absence of operational actions. This is named smart distribution network (SDN) or Smart Grid. This concept is proposed to improve the local flexibility and reliability of electrical systems, which is defined as a radial network with a cluster of DER, various types of loads, and ESS. Indeed, the term Smart Grid refers to a modernized electricity grid that uses information and communications technology (ICT) to gather and act on data about the behavior of suppliers and consumers, in an automated fashion to improve reliability, efficiency, economics, and sustainability of the use of electricity. In addition, SDN can be a proper solution to the raised energy crisis and environmental problems. However, huge penetration of DER and the intermittent nature of renewable generations may put the system operation at risk. According to the importance of continuous and secure operation of the electricity distribution networks, further and complete studies are required. In other words, the distribution system complexity is increasing mainly due to technological innovation, renewable distributed generation, and responsive loads. This complexity makes the monitoring, control, and operation of distribution networks for operators, decision-makers, investors, etc., difficult. Therefore, this book seeks to find, analyze, and introduce features and problems of operation of DER in SDNs from different aspects. The following topics are comprehensively discussed throughout the book:

1. integrating different types of elements including electrical vehicles, demand response programs, and various RES in distribution networks,
2. proposing optimal operational models for short term performance and scheduling of distribution networks,
3. considering and modeling the uncertainties of renewable resources and intermittent load in decision making of distribution networks.

Finally, it was a great pleasure to deal with Elsevier; its encouragement in the early stages of the project was important in motivating us to write this book.

Kazem Zare and Sayyad Nojavan
University of Tabriz, Tabriz, Iran

CHAPTER 1

Definition of Smart Distribution Networks

Emilio Ghiani, Fabrizio Pilo and Gianni Celli
University of Cagliari, Cagliari, Italy

1.1 INTRODUCTION TO SMART GRID PARADIGM

The persisting growth of variable renewable generation—predominantly on distribution level—is gradually altering the operating conditions of both electricity transmission and distribution networks. The transmission system operators (TSO) are concerned about the uncertain dynamic behavior of the system with a huge amount of nonprogrammable generators connected. Moreover, since distributed generation (DG) is typically interfaced through static conversion systems, the share of production from conventional generators (e.g., synchronous generators driven by steam turbines in thermoelectric power plants) is reducing. This fact causes a decrement of units able to provide ancillary services, especially to the transmission system with a considerable reduction of inertia. From the distribution system operator (DSO) point of view, the new electricity generation mix is causing a dramatic revolution of their distribution systems. Indeed, a high penetration of photovoltaic (PV) and wind plants increases the uncertainty of the system state, raises the risk of contingencies and lowers the overall service quality. Furthermore, the presence of bi-directional power flows and the increasingly occurrence of reverse power flow on distribution transformers are compromising the correct traditional passive management of the distribution networks, particularly for voltage regulation and protection systems. At the same time, DSO needs to be ready to deal with external inputs coming from the "smart world" (smart cities, smart transports, smart industries, and smart customers) that will transform the loads from simple points of electricity absorption into effective actors of the distribution system, with new requirements but also with new capability of interaction.

The electricity distribution is relying more and more on the intelligent active management of the system in lieu of the traditional absence of

Operation of Distributed Energy Resources in Smart Distribution Networks
DOI: https://doi.org/10.1016/B978-0-12-814891-4.00001-1

operational actions. This is named smart distribution network (SDN) or Smart Grid. Indeed, the term Smart Grid refers to a modernized electricity grid that uses information and communications technology (ICT) to gather and act on data, such as statistics about the behaviors of suppliers and consumers, in an automated fashion to improve reliability, efficiency, economics, and sustainability of the use of electricity [1].

The *reliability and resiliency* of power systems may be enhanced with the use of modern technologies oriented to the self-healing of the network, obtained by increasing its flexibility by means of automatic network reconfiguration and the ability to operate as intentional islanding portions of the main grid isolated by faults. Improved techniques for fault detection and selective protection can be also implemented thanks to the use of data from advanced metering infrastructure systems and real-time state estimators.

The *efficiency* of the distribution and utilization of electricity may be improved with smart grid functionalities like the energy losses reduction through Volt/VAR optimization, the demand-side management, the optimization of power consumption, the advanced intelligent building automation for controlling all aspects of the building's mechanical, electrical and electronic systems, such as air conditioners activation/deactivation, temperature control, renewable thermal/electrical energy production optimization and storage.

In the search for options for allowing distributors to actively manage the varying generation and demand, the ICT solutions that are being introduced in the smart distribution grids, adding communication, sensors and automation, could be less costly than extending/reinforcing physical infrastructure [2]. The *economics* of residential and businesses consumption of energy may benefit from the participation to new electricity market frameworks in which the customers can participate to demand side response (DSR) programs, real time energy pricing, providing ancillary services of peak curtailment or peak leveling to distributors along with voltage support and regulation.

The *sustainability* and decarbonization of the power sector has to deal with the management of intermittent energy sources such as wind and solar energy and their integration in the power system. The improved flexibility of the smart grid permits increasing the distribution network hosting capacity of highly variable renewable energy sources (RES) like PV plants and wind turbines, eventually with the addition of energy storage systems (ESS).

The Smart Grid paradigm encompasses all the traditional sectors of transmission, distribution, and utilization of electricity towards the power system, including the functions of markets and network management, planning and operation. In particular, the power distribution sector is foreseen to be the more involved in the evolution of the power system toward future smart grid scenarios, in which active networks will be more intelligently capable to integrate RES, DSR and ESS. The transition toward the SDN will require the implementation of new concepts with novel electrical system structure as well as the addition of novel types of control systems and equipment able to manage bidirectional power flows.

The massive penetration of DG has been leading toward a new approach in the distribution system. Until now, distribution networks are regarded as a passive termination of the transmission network, having the goal of supplying reliably and efficiently the end users. According to this scheme, distribution networks are radial, with unidirectional power flows and with a simple and efficient protection scheme. A greater penetration of DG will change completely this well consolidated environment: definitely, distribution networks will be no longer passive and it has been foreseen a gradual, but ineluctable, changing towards a new kind of active networks. Active networks could be strongly interconnected (no radial scheme, no unidirectional power flow) and could be subdivided into small cells or portions as microgrids, locally managed by a power controller, which controls power flow among local generators, loads and adjacent cells or network portions. Automation, control systems and ICT will be widely used in active networks to manage local cells, to control generated and absorbed power, to manage congestions and prevent propagation of overloads and faults by isolating the affected part of the network. Intentional islanding, the ability of managing a portion of the network by using DG to enhance the service provided thereby, plays a key role in active networks and requires the development of new control algorithms for a rapid network reconfiguration, of new communication protocols, for data exchange between generators and loads, and of simple and reliable communications systems. Also post fault actions can benefit from DER and flexibility with a contextual release of power capacity currently locked for managing post fault reconfiguration. People involvement is a crucial part of mart Grid programs, and specific projects on such a nontechnical aspect must be realized as a follow up of the following high level political and regulatory drivers:

- liberalization;
- public concern over the environment;

- ensure diverse, secure, sustainable supplies of energy at competitive prices;
- integration of renewables;
- new technologies which closely interact with power systems:
  - power electronic converters;
  - electronic control systems;
  - communications.

A major change is also requested to distribution network operators (DNO), which should evolve toward system operators responsible for the operation of the system. Indeed, distribution networks have been worldwide traditionally managed by the geographically located distribution companies, which operate typically as a utility with distribution network operator duties. DNO still are companies (or regulated bodies) licensed to distribute and measure electricity from the transmission grid to residential customers and commercial/industrial businesses. In the past, and in some places till now, DNOs were also committed to supply electricity to final customers; with the liberalization of the electricity sector this task is assigned to separate supply companies or energy service companies (ESCO) who buy the energy in the wholesale power market and, make use of the distribution network, and sell the energy to final consumers who have the power to chose their energy provider. DNOs are also responsible for the planning and maintenance of the network under their jurisdiction, as well as to guarantee a specified service quality and continuity to the users of the network.

The traditional DNO distribution model focused on the one-way delivery of electricity, with clear and univocal direction of the flow of electricity. With the introduction of DG, renewable sources, microgrids, electric vehicles and energy storage, the classic DNO model is mandatory to change, as well as the design, operation, and management of distribution networks.

Besides the traditional mission of DNOs to operate, maintain, and develop an efficient electricity distribution system, the DSOs are asked to facilitate effective and well-functioning retail markets. Effective retail markets are markets which should give options to the customers allowing them to choose the best supplier and should allow suppliers to offer options and services best tailored to customer needs [3]. In this new scenario customers can be influenced to change their consumption behaviors through some form of variable pricing. It will be possible to have time-of-use pricing offering low rates during periods when renewable sources are available and wholesale prices are expected to be high. Retailers will propose load curtailment and price changes according to customer/prosumer features and preferences and their energy use and generation behaviors, and how those behaviors are likely to respond to retailer actions.

This may be either when demand is normally high, such as late afternoon in areas with high air-conditioning loads, or when renewable sources are unavailable, such as hours of darkness in areas with significant solar production.

According to the foreseen scenario in the medium, and also in the long term, DNOs/DSOs will have to radically change their mission and will have a new role as wiring companies. In the following the most important activities for a DNO evolving to a DSO are listed:

- system access facilitation for integrating distributed energy resources (DERs);
- real-time studies and state estimation;
- voltage and power flow control;
- security assessment and stability management;
- generator interaction management;
- outage coordination, system recovery, and restoration;
- constraint/contingency management;
- island operation;
- power and service quality assessment;
- distribution retail market organization and management.

Many of these activities can be regarded as system services and they will be charged to customers and DNOs/DSOs will require new guidelines to drive network development toward the future smart distribution grids.

1.1.1 The Transition From Passive to Active Networks

The future structure of the distribution systems has been envisaged by many authors, research institutions, regulators, and distribution operators. The concept of Smart Grid is now commonly used in the literature. The internet-like model with distributed decision-making and bidirectional power flows constitute the final goal to reach in the long term. The control will be distributed across nodes spread throughout the system. Not only the supplier of power for a given consumer could vary from one time period to the next, but also the network use could vary as the network self-determines its configuration [1].

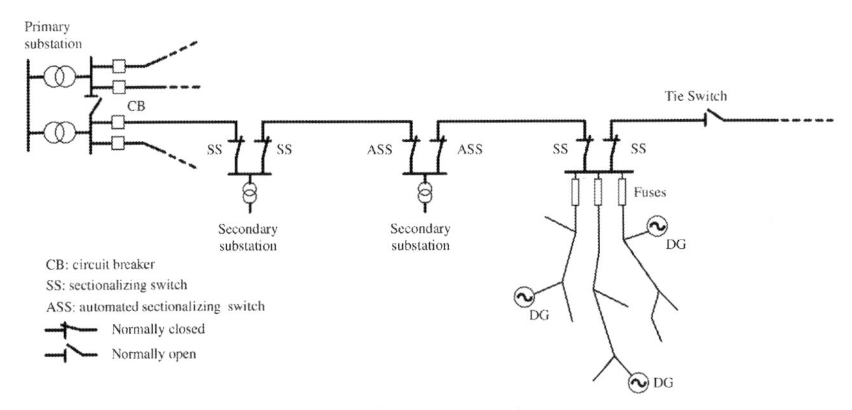

Figure 1.1 Conventional power distribution network scenario.

The contemporary political drives and environmental concerns determined a situation in which the electricity distribution industry is facing the problem of the connection and integration of renewable and other generation plants to networks, which have traditionally carried electricity from transmission systems to final consumers in one direction only, according to the conventional distribution network scheme illustrated in Fig. 1.1.

Traditional distribution schemes are radial and the distribution network operator in case of connection of DG applies the "fit and forget" approach, signifying the design of the distribution system so as to meet technical constraints in the most onerous conditions (e.g., full generation/no load or no generation/full load) even if such situations have a small probability of occurrence; no possibility of controlling the DG is assumed over the possibility of disconnect the DG in case of faults or contingencies in the distribution network.

The evolution toward SDN requires an extensive use of ICT and innovative control systems, enabling various intelligent and automated applications, such as building automation, distribution automation, outage and contingency management, and integration of electric vehicles.

In an SDN network, the distribution network operator can control loads, generators, node voltages, and power flows and the transition toward SDN involves software, automation, and controls, to ensure that the power distribution network not only remains within its operating limits (e.g., node voltages and branch currents within the acceptable thresholds), but it is also operated in an optimal way.

The inherent variability and unpredictability of the power generation from renewable sources, the expected growing number of electric vehicles

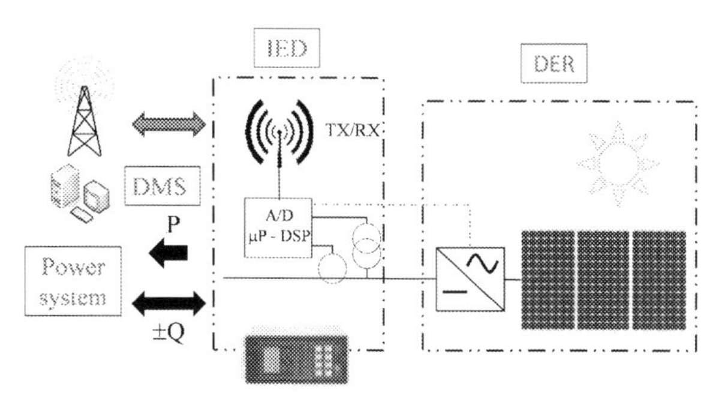

Figure 1.2 IED associated to a photovoltaic plant in a smart distribution grid.

and, in general, the DER management require a continuous flow of information in order to verify the compliance with network constraints and optimize the energy transfers within the network. For this reason, in the SDN each power network element: generators, switches, electric vehicles, etc., will be equipped with intelligent electronic devices (IEDs) able to transmit/receive information by means of a communication channel and control the element to which is associated (Fig. 1.2).

Waiting to the enhancement and smartization of the network, at the moment at the distribution level, the use of supervisory control and data acquisition (SCADA) are typically not cost effective at the substation level and rarely at the feeder level, and the majority of the distribution networks is not extensively monitored or controlled. Traditional secondary substations are organized according to the scheme in Fig. 1.3. Each power line is equipped with automatic sectionalizer (AS) used in conjunction with source-side protection devices, such as reclosers or circuit breakers (CBs), positioned along the medium voltage (MV) distribution line, to automatically isolate faulted sections of electrical distribution systems with support of SCADA systems. Power to operate the control circuitry and the mechanism is obtained from the line through the sensing-current/voltage transformers. No auxiliary power supply, external connections, or external equipment is required. The AS, when the source-side protective device opens to deenergize the circuit, permits disconnecting a portion of the distribution system or a single MV user (typically passive). Sectionalizers are the most economical method of improving service continuity on distribution lines equipped with reclosers or reclosing CBs.

In the future, the Smart Grid applications will need even more pervasive and real-time control of each network component, that will have to be equipped with smart meters and communication devices as well as new

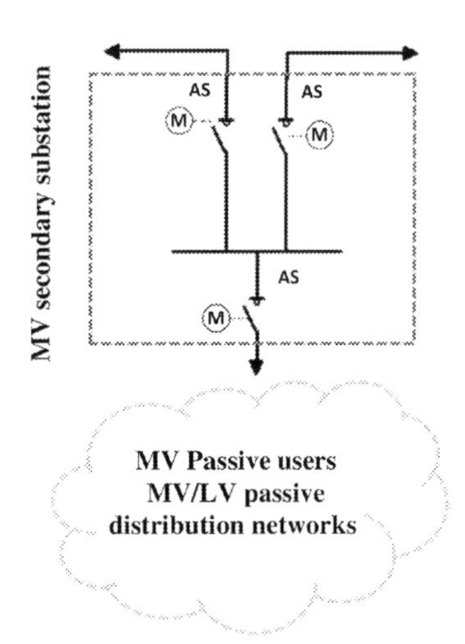

Figure 1.3 Schematic representation of a traditional secondary substation with automatic sectionalizers.

sophisticated control and protection devices. For instance, secondary substations are getting more and more strategic in MV and low voltage (LV) networks in order to achieve service quality improvement due to their key role in power flow monitoring, DG management, and network automation and control. Some DSO have already started developing a significant activity of refurbishment of secondary substations with new solutions for technological improvement of MV and LV equipment, MV/LV transformers, protection systems, remote control devices, and auxiliary components. The goals of distributors involve the exploration of existing functionalities of apparatus, to create a new smart secondary substation (SSS) model with reliable power components, high performance protection schemes, efficient flow monitoring systems and trusty communication infrastructures, thus optimizing the overall control of distribution network in order to:

- manage energy flows and voltage profiles according to load and DER needs,
- ensure fast reconfiguration after a failure,
- identify and pursue efficiency opportunities.

All devices and their features must be integrated into a global architecture of MV and LV network control, communicating according to a centralized or decentralized logic that relies on information exchange and allows a reliable operation on them. According to this view a possible configuration of an SSS as the one depicted in Fig. 1.4.

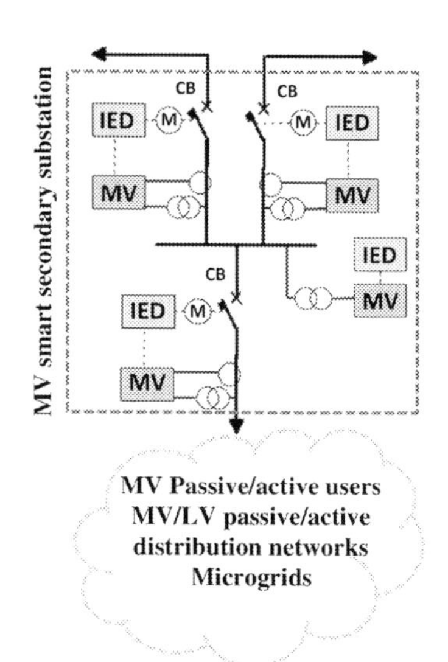

Figure 1.4 Schematic representation of smart secondary substation.

Communication technology is seen as an essential enabling component of future smart grids. In particular, the smart metering, protection, and control communication is the major component of the overall smart grid communication architecture and consists of smart meters, protection, and control systems which are two-way communicating devices with the central SDN controller (Fig. 1.5). The whole system could be organized according to hierarchical structure with a home area network (HAN) formed by appliances and devices to support different distributed applications, neighborhood area network (NAN) that collects data from multiple HANs and delivers the data to data concentrator, wide area network (WAN) which carries metering data to central control centers and, finally, a gateway which is in charge of collecting or directly measuring energy usage information from the HAN members and that transmits this data to interested parties. All the above network scenarios have different communication requirements and several communication technologies could be employed to fulfill them. In [4] a deep analysis of communication requirements and capabilities of the different type of network is presented. A review of the main current challenges and possible solutions to open problems in the development of smart distribution grids is offered in [5].

The SDN will consist then of an interconnected MV and LV network with mixed communication infrastructure between the distribution/energy management system (DMS/EMS) of the network communicating

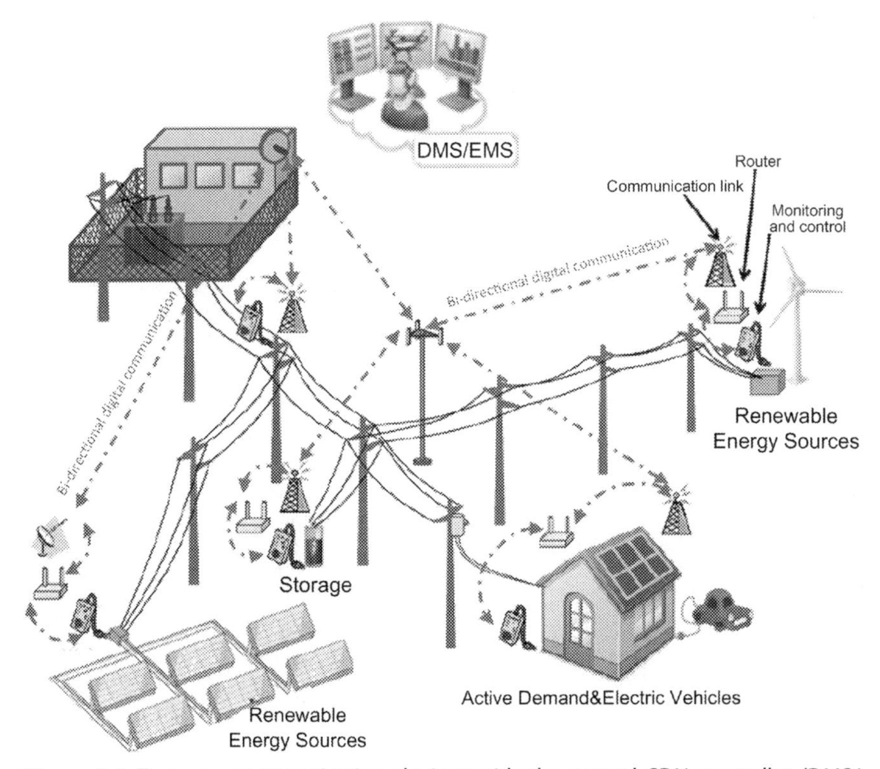

Figure 1.5 Two-way communicating devices with the central SDN controller (DMS/EMS).

with the SCADA and the active resources in the network by means of IEDs as schematized in Fig. 1.6.

In the SDN a centralized/decentralized DMS/EMS supervises the operation of the electric distribution network gathering measures of the main electric variables from IEDs and, according to the control scheme, modifies the set point of DERs directly connected to MV distribution network or interacst with the energy management of LV DERs and/or local controllers of microgrids.

Many experimental projects in the European Union are focused on finding new solutions for the operation of the distribution network with reconfiguration possibility in order to deal with the augmented complexity that comes from the increased generation that is injected on the distribution network. For instance, the Italian DNO Enel Distribuzione is actively involved in the implementation of the Smart Grid concept with several ongoing experimental projects. One of them, called ScheMa

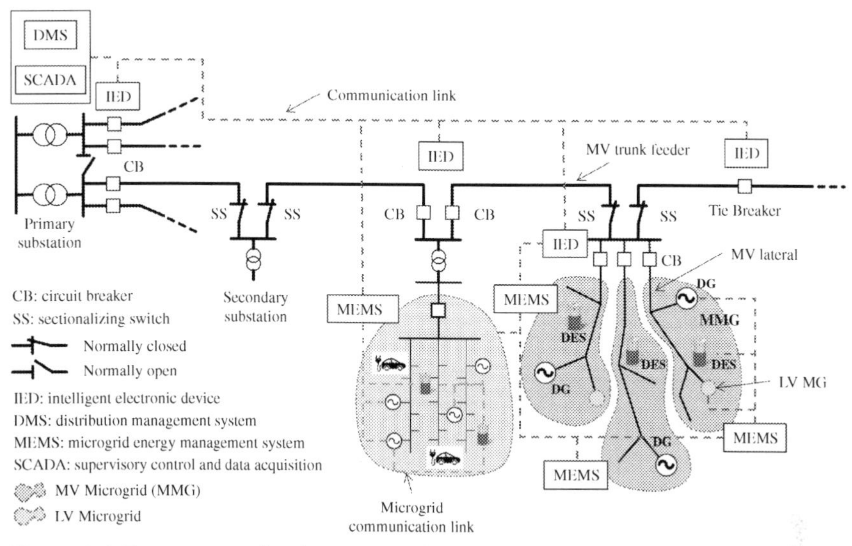

Figure 1.6 Future smart distribution network scenario.

Project, tests the performance of MV network portions in closed ring operation using an innovative control and fault detection system. This control system exploits a communication network realized with optic fiber that allows a data exchange of measurements and control signals between automated substations [6]. The automated substations along the loop line are equipped with CBs or reclosers instead of the usual sectionalizer: each feeder between adjacent automated SS is individually protected and switched if needed [7].

1.1.2 Challenges in Smart Distribution Network Implementation

The development of effective SDNs needs to control electricity productions, consumptions and storage, as well as power flow by means of a DMS/EMS, and to implement features in order to manage a more flexible and reconfigurable network structure. The possibility of having a large number of controllable DER, DG units and loads under demand side management (DSM) control, electrical vehicles (EVs) and distributed ESSs require the use of a hierarchical control scheme, which enables an effective and efficient control of this kind of system, in which the DMS operates at the higher level and is able to exchange information with local control systems. A multidisciplinary approach will be necessary and focused on the development of new communication systems and their interfacing with the power system elements and the development of DMS/EMS able to manage in a smart way the distributed system for energy production, consumption, and storage [5].

Communication technology is seen as an essential enabling component of future smart grids. In particular, the smart metering, protection and control communication is the major component of the overall smart grid communication architecture and consists of smart meters, protection and control systems which are two-way communicating devices with the central SG controller.

The whole system could be organized according to hierarchical structure with a HAN formed by appliances and devices within a home to support different distributed applications, NAN that collects data from multiple HANs and delivers the data to data concentrator, WAN which carries metering data to central control centers and, finally, a hateway which is in charge of collecting or directly measuring energy usage information from the HAN members and that transmits this data to interested parties.

Wireless architecture can be either an option for HANs, NANs and WANs, or mandatory in case of vehicle-to-grid (V2G) communications.

Wired communication will offer a valid counterpart to wireless connection in the SG. Different technologies are available for usage depending on the desired coverage area. Fiber optic links may be adopted for WAN coverage of transmission lines (tens of kilometers, and more). Power line communications (PLCs) may instead provide a good compromise for NANs and HANs coverage of local/micro SG portions (up to hundreds of meters).

Advances over the state-of-the-art are needed since the amount of information that will be exchanged in the SG will be much higher than in current systems, which are mostly limited to over relaxed time scales (days). Also, optimization necessarily needs to be matched to specific application requirements in a world in which the target application is still unclear, the interest being today focused on advanced metering infrastructures (AMI) aspects (centralized/distributed micro grid control), but being likely to converge in a longer time scale to a fully distributed management of distributed resources (the so-called energy Internet scenario).

Communication technology needs to accomplish the IEC 61850 standard to be used with DERs and power distribution networks. IEDs with security features and control are becoming standard equipment for power system components, new substations, and are also used to improve the systems of protection and control of existing substations. These smart devices are required for power flow congestion, voltage regulation, DG, load control, and fast reconfiguration. Microprocessor based protection relays provide, in addition to the basic function of protection, different functions, such as measurement, data acquisition, recording events and disturbances, tools for failure analysis, and control.

EMS/DMS and SCADA systems will be extensively used to manage data from distributed elements of the transmission and distribution system. The SCADA system transmits the measurement data, provided by an AMI and by a set of remote collecting data devices (RTUs) placed in strategic positions along the smart grid, to the EMS/DMS.

New ICT infrastructure supporting a more efficient smart grid will offer new challenges for DSM allowing to communicate frequent price updates to follow the evolution of the balance between supply and demand in near real-time. Thanks to this ICT infrastructure, much more dynamic, reactive pricing mechanisms required to take into account real-time availability of fluctuating renewable sources will be carried out.

Electric energy storage will play a key role in smart grids because it enhances flexibility of renewable DGs and of loads. Most of the problems in power quality, distribution reliability, and power flow management can be solved with the widespread roll-out of energy storage devices intelligently controlled.

Flexibility of loads can be regarded as a significant DER thanks to the large number of loads that are connected to distribution systems. The actual capabilities and potential of load flexibility are still to be investigated in the real applications. Flexibility can be introduced in two ways. In the first case, loads can be shifted in time, i.e., it can be modified for the starting time and, partially, the duration of the consumption cycle; these loads absorb a fixed consumption of energy. In the second case, loads can be curtailed for a time interval, i.e., they are temporarily disconnected or reduced; the energy that is not absorbed is not recovered in another time interval. Typical examples are washing machines as shiftable loads and lights as curtailable loads. In both the cases, smart control and communication capabilities are required at load level.

The complexity of the several aspects underlined requires to rethink the planning of the power distribution networks. Novel planning techniques should integrate operation models within planning as well as the models of the communication system. The simulation planning tools have to be capable of reproducing fluctuating renewable generation caused by the moving of cloud patterns and/or to simulate local meteorological conditions, and the cosimulation of both power and communication systems will be an essential part of the planning process.

1.1.3 Observability and Controllability of Smart Distribution Networks

In the described context of electricity distribution network evolution toward SDN, it is becoming even more important is the system observability and the suitable sharing of relevant information among all system operators and market participants. Indeed, without the necessary observability it will not be possible to implement a sufficient level of controllability on forthcoming SDN (active power management of DER and DSR used to solve contingencies at distribution level and to provide ancillary services to TSO). Moreover, improved observability not only helps network operators to maintain a good level of service continuity but it also reduces errors in forecasting demand and limits the growth of reserves caused by the uncertainty of RES (predictability).

Therefore, the three features of observability, predictability, and controllability are distinctive aspects of a SDN and they are essential for:

- improving the prediction of load and generation at all electricity system levels,
- improving the hosting capacity of the distribution system for additional RES while maintaining adequate standards of reliability and system quality,
- increasing communication between automation systems of TSO and DSO.

At present, in opposition to the quite high level of observability of the transmission system, the MV and LV systems are barely monitored by the DSO. Indeed, the distribution grids are characterized by a huge number of feeders and nodes with limited metering points. Consequently, observability and knowledge of these distribution networks are not sufficient for DSOs to establish the complex interactions among the increasing intermittent renewable generation units and flexible loads. Additionally, as the large-scale integration of RES units and other DERs occur in the distribution networks, the TSO needs to acquire their active contribution in ancillary services for ensuring power system stability and security of supply. Lastly, in order to economically profit from their small generating and flexible demand units through the participation in electricity market, the "prosumers" (customers with simultaneous capability of electricity production and consumption) at the local grids need better visibility and information exchange with the network operators and electricity market entities.

Thus, the initial major effort for the SDN implementation is the increment of the distribution system observability. In this regard, DSOs need to utilize ICT, automation and smart metering infrastructure so that abundant real-time and accurate information is made available to all system actors. The installation of smart metering and intelligent monitoring at each nodes of the network for the grid operation and control requires extensive communication infrastructure and huge data processing. This is extremely complex and expensive for the utilities. Accordingly, for implementing cost-effective solutions of accurate real-time representation of the grid and advanced grid management, a trade-off between the use of existing monitoring/control infrastructure and minimal possible instrumentation is essential. One of the viable solutions for the real-time optimal observability of the grid is to use scalable sistribution state estimator (DSE) algorithms [8], which use a minimum amount of static and dynamic information of the grid assets for providing robust and accurate results with low-computational burden. The DSE tools use grid topology as static data and pseudo and real measurements, and forecasted information as dynamic data for estimating the network conditions at all nodes [9].

1.2 MICROGRIDS, NANOGRIDS, AND VIRTUAL POWER PLANTS IN DISTRIBUTION NETWORKS

When a small part of the power system, with a group of interconnected loads and DERs, within a clearly defined electrical and geographical position, can act as a single controllable entity with respect to the main grid, it is possible to talk about virtual power plants (VPPs), microgrids, and nanogrids.

1.2.1 Microgrids

Microgrids are small-scale distribution networks containing renewable generation, load, ESS equipped with a communication and control system capable to coordinate all energy resources and the power exchange with the network (if any).

MGs can act as energy-balanced cells within existing power distribution grids or stand-alone power networks within small communities and/or remote sites, and can be operated autonomously and disconnected from the main power grid (MGs) [10,11].

The MG EMS maximizes the system efficiency and improves the quality of supply for the customers supplied. Suitable communication systems allow the EMS to send set-point and control signals to generators, loads and storage devices and to receive the necessary inputs from measurement devices and form other external sources (e.g., weather forecasts, energy prices, distribution network signals coming from sensors, actuators and protection devices).

A possible realization of hybrid solutions arranged as microgrid, with a combination of generators, both fossil fuel and RES, and storage is depicted in Fig. 1.7. Hybrid systems are a cost-effective solution able to reach higher reliability level than systems that use only one energy source.

A storage system is typically included to mitigate the intermittency of renewable generation and the demand variations. Furthermore, batteries are fundamental to smooth frequency and voltage fluctuations caused by renewable's production. A diesel generator is included in the MG as back-up source, useful in case of unavailability of other components.

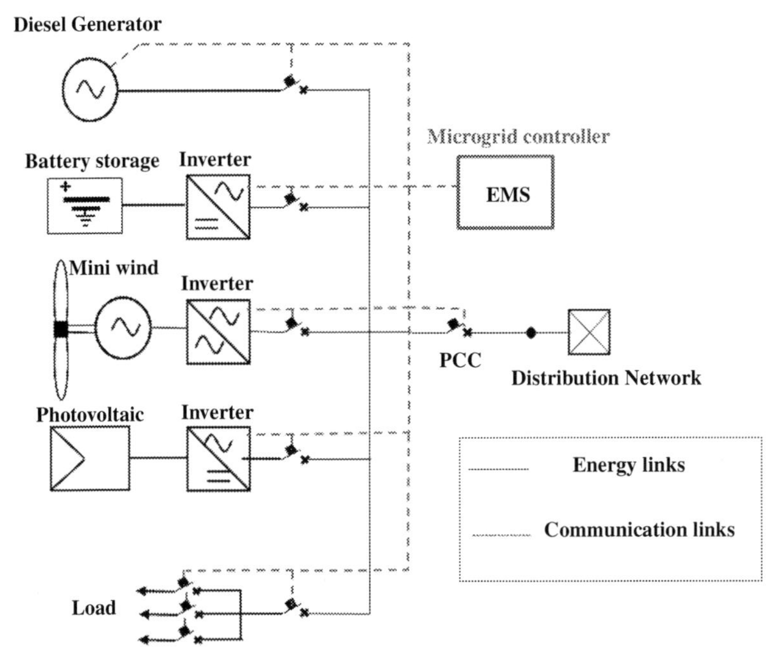

Figure 1.7 Hybrid microgrid architecture.

In this way the diesel generator does not run continuously, so its lifespan will be increased and maintenance costs will be reduced, the fuel consumption will also be reduced. Otherwise by the use of RES a significant reduction in greenhouse gases emitted in the atmosphere would be achieved.

MGs ensure several benefits. The proximity to the served loads permits reducing the losses in transmission and distribution. In the distribution networks of developed countries MG can help increase the hosting capacity and release capacity locked by the need of guaranteeing backup paths for network reconfiguration during faults. Furthermore, MG can be a crucial resource to make the power system more resilient against low probability-high risk events (e.g., fires, snowfall, icing, floods, hurricanes, and typhons that are getting more frequent due to climate changes).

Electrification programs with development plans based on traditional centralized production and networked electrical system have often caused the slow-down of rural electrification programs. As a consequence of the prohibitive cost of grid extensions, villages located in remote areas usually have autonomous power supply systems; the most common solutions are based on small diesel generators running continuously [10,11]. These units have low initial capital cost and can generate electricity on demand, but their operation and maintenance costs are high. The fuel supply is affected by fluctuations and could be very expensive due to the cost of transportation in remote areas. Moreover, the generator operation is inefficient at low load factors and the continuous operation reduces the generator lifespan, so the life-cycle cost of the system usually results very high. Furthermore, the combustion of fossil fuel has a detrimental effect on the environment due to the emissions of greenhouse gases and other pollutants which involves global warming, cancer or mutations, depletion of ozone layer, and acidification problems [12]. In this context, MGs can be adopted where the rough topography and the long distance from the nearest main grid makes the extension of the grid extremely costly and technically difficult if not even unfeasible. Thus, the electrification of remote areas, particularly in developing countries, can be fostered by MGs, which are scalable systems with dimensions that can be adapted to growing energy needs.

1.2.2 Nanogrids

Within the Smart Grid framework applied to the LV system, the nanogrid can be considered a step ahead of the "prosumer" concept, an energy consumer who moves beyond passive consumption to become active producer.

Nanogrid refers to a single small power system with the following features:

- residential, small industrial or commercial site,
- small in size (<50 kW),
- behind a single meter,
- includes the generating source, in-house distribution, and energy storage functions (thermal/electric), also obtained with electric vehicles.

The nanogrid incorporates smart appliances, lighting and heating/ventilation/air conditioning with the on-site power generation coordinated by an efficient electric EMS as depicted in Fig. 1.5. The EMS can also communicate with the power system operator or with an aggregator entity for energy exchange purposes, while also acting as a data acquisition unit. The EMS collects and stores the power flow data not only from and toward the public network, but also from the power converters and smart appliances in the nanogrid. The EMS may also control unidirectional and bidirectional inverters that may operate to isolate the end user from the public grid, work in stand-alone mode, and synchronize and reconnect the nanogrid to the distribution network without power interruptions [13] (Fig. 1.8).

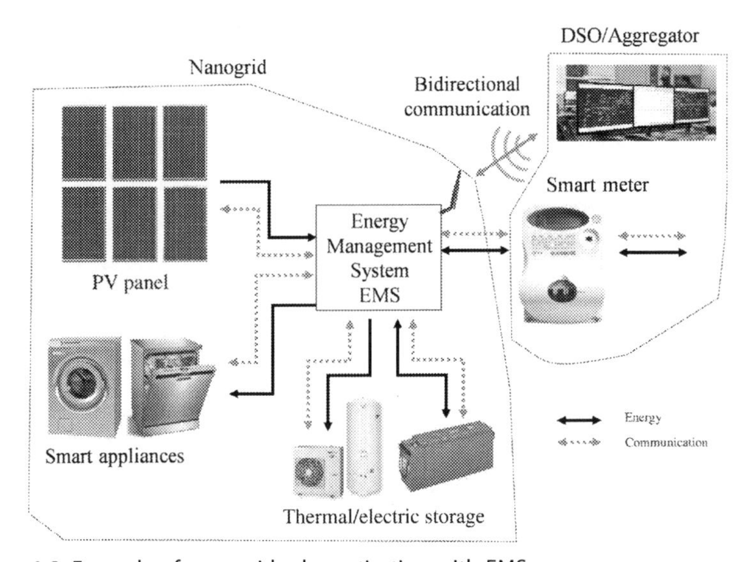

Figure 1.8 Example of nanogrid schematization with EMS.

The EMS has the objective of achieving significant benefits for the nanogrid owner:

- savings in the energy bill,
- incomes from selling energy to the public network,
- incomes from ancillary services provided to the DSO or other entities, like aggregators or ESCO.

In [14] the advantages for the nanogrid owner are shown considering the capabilities of the nanogrid of balancing the variability of the production by storing the exceeding energy during low demand/peak production and low-cost energy hours, and selling it during high cost energy hours or under the request of DSOs according to a proper ancillary service market scheme.

1.2.3 Virtual Power Plants

A probable evolution of the distribution system is represented by the definition of clusters of many different geographically dispersed energy sources, controlled by aggregators or ESCOs as VPPs and smartly coordinated by the DSO.

VPP may be considered such as a form of "Internet of energy," able to integrate diverse smaller energy sources like microgrids, nanogrids, electric vehicles, DG and ESS [15] (Fig. 1.9).

In order to exploit the potentialities of the VPP completely, a fundamental role is taken by its EMS, together with the availability of good data models of the energy resources and the existence of a defined energy market environment [15]. VPPs are obtained by clustering a group of DERs. It can deliver energy demanded at peak usage times under request

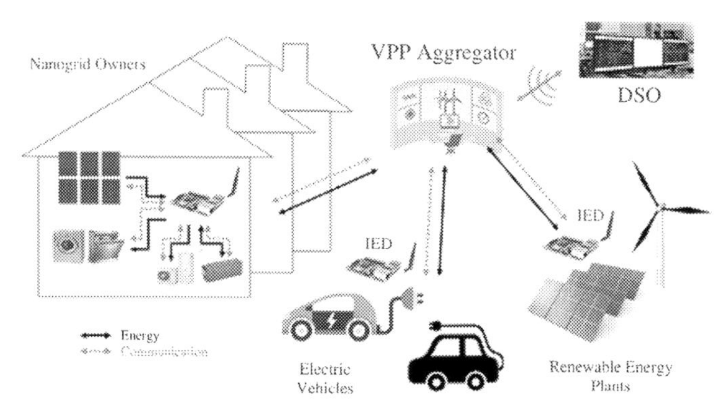

Figure 1.9 Virtual power plant schematization in a smart distribution network.

from DSO, and can store any surplus power, giving the energy aggregator more options, for instance by balancing the variability of RES. Other advantages include improved power network efficiency and security, and cost savings in distribution systems by increasing the use of existing assets.

The VPP can use IoT and cloud systems for the aggregation, coordination, and optimization of DER, and can provide dispatching services to the electric distributor. The VPP is able to monitor the power absorbed by the electrical and thermal loads and the one generated by the renewable sources. The VPP controller can determine, in compliance with the operational and network constraints, the energy flows between generation, loads and existing storage systems.

To collect and exchange information among DERs IEDs are used. Each IED can receive data from sensors and power supplies as well as issue control commands and can also transmit and receive data using existing, wireless or wired network infrastructures. Thanks to the use of IED, it is possible to monitor and collect information on the operation and expenditure of the individual DER resources (microgrids, nanogrids, electric vehicles, DG) and implement control actions.

A case study with a LV distribution network with several nanogrids managed as a VPP by an aggregator who receives power dispatching requests by DSO to improve voltage profile and reduce renewable energy curtailment is presented in [14].

1.2.4 Smart Distribution Networks With Multimicrogrids Reorganization

Innovative network schemes and operation policies have been proposed to help facing the new challenges mainly based on the exploitation of communication and information technologies to increase the observability and controllability of the distribution system. The power infrastructure might then benefit of the MGs by exploiting their typical features for reliability and resiliency with a new operational strategy that can be referred as multi-microgrid (MMG) [16,17] or networked microgrids [18] (Fig. 1.10).

According to transformative architecture for the normal operation and self-healing of networked MGs, MGs can support and interchange electricity with each other in the proposed infrastructure. The networked MGs are connected to the main power distribution backbone with a designed communication network. In each MG the EMS schedules the MG operation and all EMSs are globally optimized. In the normal operation mode, the objective is to schedule dispatchable DG, ESS, and

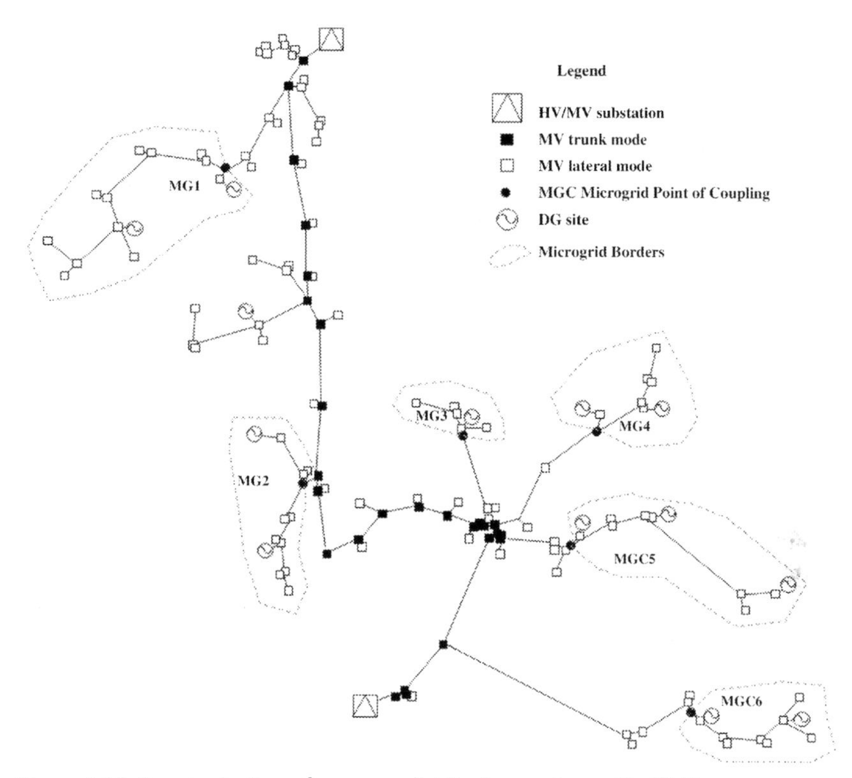

Figure 1.10 Reorganization of a power distribution system with MMGs.

controllable loads to minimize the operation costs and maximize the supply adequacy of each MG. When a generation deficiency or fault happens in a MG, the architecture can switch to the self-healing mode and the local generation capacities of other MGs can be used to support the on-emergency portion of the system [18] (Fig. 1.10).

The reorganization of the distribution system into MMGs can determine significant benefits to the customers, with the opportunity of reducing the energy bill and increasing the perceived reliability, to the DSO, that may postpone or avoid investments or activate new business, and to the environment, thanks to the exploitation of RESs.

The advantages of the reconfiguration of distribution networks into interconnected MGs, autonomously operated, are summarized in the following points [16–18].

- *Deferment of investments*: the MGs take advantage of all benefits related to the energy sources properly placed in proximity of loads. In fact, in a distribution network, the production of energy close the point of consumption can reduce power losses, relieve congestions, and defer DSO CAPEX;

- *Reliability and power quality*: the opportunity of using local generators and storage during system faults can improve the reliability. However, it should be noted that the MGs can also improve the power quality of the distribution system;
- *Economic and market outlook*: the combination of MGs with RESs can lead to savings for consumers. Moreover, the potential of a MG to sell ancillary service to the distributors, and to actively participate at the electricity market, both as purchasers and as sellers has to be taken into account;
- *Environmental impact*: the control of generation and loads in the MG increase the penetration of RES. The local management makes the customers more conscious about the importance of a wise use of energy.

1.3 CONCLUSION

The chapter introduces the general concept of Smart Grid with possible different implementations (e.g., active distribution network, microgrids, nanogrids, and VPPs), and the motivations that lead to this power systems evolution. A discussion on the transition from the traditional distribution architecture toward the innovative Smart Grid is also included, paying attention to the challenges in front of distribution planners and decision makers at any level of the chain value. Emphasis is given to the enabling technologies required for the Smart Grid implementation. Indeed, advanced metering, IEDs, and communication infrastructures assure the observability of the electric distribution system, essential prerequisite for any efficient controllability. ESS are also crucial together with the other DERs to guarantee the sufficient level of flexibility for making the control also effective.

REFERENCES

[1] Smartgrids, European Technology Platform on Vision and Strategy for Europe's Electricity Networks of the Future. s. l.: EUR 22040, 2006.
[2] Distribution System Operators' Association for Smart Grids. http://www.edsoforsmartgrids.eu/.
[3] Eureletric. The Role of Distribution System Operators (DSOs) as Information Hubs. 2010.

[4] Z. Fan, P. Kulkarni, S. Gormus, C. Efthymiou, G. Kalogridis, M. Sooriyabandara, et al., Smart grid communications: overview of research challenges, solutions, and standardization activities, IEEE Commun. Surv. Tutor. 15 (2013) 21–38.
[5] A.R. Di Fazio, T. Erseghe, E. Ghiani, M. Murroni, P. Siano, F. Silvestro, Integration of renewable energy sources, energy storage systems, and electrical vehicles with smart power distribution networks, J. Ambient Intellig. Humanized Computing (May 2013).
[6] S. Botton, L. Cavalletto, F. Marmeggi. Schema project-innovative criteria for management and operation of a closed ring MV network. 22nd International Conference and Exhibition on Electricity Distribution (CIRED 2013), 10–13 June 2013. Stockholm, Sweden.
[7] S. Lauria, A. Codino, R. Calone, Protection System Studies for ENEL Distribuzione's MV Loop Lines, IEEE Eindhoven PowerTech, Power Tech 2015, Eindhoven, Netherlands, 29 June 2015–2 July 2015.
[8] Y. Huang, S. Werner, J. Huang, N. Kashyap, V. Gupta, State estimation in electric power grids, IEEE Signal Processing Magazine (Sept. 2012).
[9] K. Kouzelis, D. Diaz, I. Mendaza, B. Bak-Jensen, J.R. Pillai, K. Katsavounis, Enhancing the Observability of Traditional Distribution Grids by Strategic Meter Allocation, IEEE Eindhoven Power Tech, Aalborg, 2015.
[10] E. Ghiani, C. Vertuccio, F. Pilo. Optimal Sizing and Management of a Smart Microgrid for Prevailing Self-Consumption: PowerTech 2015 conference. Eindhoven, Netherlands. June 29 – July 2, 2015.
[11] E. Ghiani, C. Vertuccio, F. Pilo. Optimal Sizing of Multi-Generation Set for Off-Grid Rural Electrification. Proc. 2016 IEEE PES General Meeting, July 17–21, 2016 in Boston, MA, USA.
[12] H. Louie, E. O'Grady, V. Van Acker, S. Szablya, N.P. Kumar, R. Podmore, Rural off-grid electricity service in Sub-Saharan Africa, Electrif. Mag. IEEE 3 (1) (March 2015) 7–15.
[13] A. Zipperer, et al., Electric energy management in the smart home: perspectives on enabling technologies and consumer behavior, Proc. IEEE 101 (11) (Nov. 2013) 2397–2408.
[14] G. Celli, E. Ghiani, F. Pilo, G. Marongiu, Smart Integration and Aggregation of Nanogrids: Benefits for Users and DSO, Powertech, Manchester, 2017. June 18–22.
[15] D. Pudjianto, C. Ramsay, G. Strbac, Virtual power plant and system integration of distributed energy resources, IET Renew. Power Generation 1 (1) (2007) 10–16.
[16] E. Ghiani, S. Mocci, F. Pilo, Optimal reconfiguration of distribution networks according to the microgrid paradigm 2005 IEEE International Conference on Future Power Systems, 2005. Amsterdam; Netherlands; 16–18 November 2005.
[17] G. Celli, E. Ghiani, S. Mocci, F. Pilo, G.G. Soma, C. Vertuccio. Probabilistic Planning of Multi-Microgrids with Optimal Hybrid Multi-Generation sets. 2016 CIGRE Session. Paris 21–26 August 2016.
[18] Z. Wang, B. Chen, J. Wang, C. Chen, Networked microgrids for self-healing power systemsin IEEE Trans. Smart Grid 7 (1) (Jan. 2016) 310–319.

FURTHER READING

Microgrids: Architectures and Control, First Edition. Edited by Nikos Hatziargyriou. John Wiley & Sons. 2014.
J.A. Lopes Peças, A.G. Madureira, C. Moreira, A View of Microgrids, Wiley Interdisciplinary Reviews: Energy and Environment. WENE, 2013.

CHAPTER 2

Impact of Renewable Energy Sources and Energy Storage Technologies on the Operation and Planning of Smart Distribution Networks

Emilio Ghiani and Giuditta Pisano
University of Cagliari, Cagliari, Italy

2.1 INTRODUCTION

In the last 20 years the power system experienced a tremendous change at both transmission and distribution levels due to the connection of renewable energy sources (RES). In distribution systems, in order to accommodate RES, institutional, regulatory and commercial reorganizations started a process of renovation that has not been completed yet. Such a renovation process is involving markets of energy and ancillary services in order to allow distribution system players to place offers and bids. The well-known environmental concerns, the sustainable development, and the electric market liberalization are posing to distribution engineers important challenges in order to achieve the best compromise among technical aspects, economics, and environmental needs. The current politic and economic international scenario, characterized by the petroleum and gas price volatility, by the intrinsic depletion of the primary fuels, and by an increasing electricity demand, may give an important impulse to the exploitation of RES, of demand side response actions, and, in general, to all those actions that may contribute to improve the overall energy efficiency. RES and/or newest high-efficiency technologies are destined to receive a rising interest and will play a significant role in the electric power system. But, in order to remove the barriers against the integration of distributed energy sources (DERs), many technical problems have still to be solved and adequate operating and control systems have to be developed.

Operation of Distributed Energy Resources in Smart Distribution Networks
DOI: https://doi.org/10.1016/B978-0-12-814891-4.00002-3

2.2 IMPACT OF DISTRIBUTED ENERGY RESOURCES ON DISTRIBUTION NETWORKS

2.2.1 Transition From Passive to Active Distribution Networks

Until now, distribution networks have been viewed as a passive termination of the transmission network, with the goal of supplying customers reliably and efficiently. According to this scheme, distribution networks are radial, with unidirectional power flows and with a simple and efficient protection system. The greater penetration of DERs is changing completely this well consolidated environment. Definitely, distribution networks are no longer passive, and the active operation of distribution networks started to be a reality in many countries with partial realization of smart grid that is named active distribution network. Active distribution networks may be strongly interconnected (no radial scheme, no unidirectional power flow) and subdivided into small cells, locally managed by a power controller, which superintends power flows among local generators, loads and adjacent cells. Automation, control systems and information and communication Technologies (ICT) will be widely used in active networks to govern local cells, to control generated and absorbed power, to manage congestions and prevent propagation of overloads and faults by isolating the affected part of the network.

In the existing contest, the DER connection complies with the *connect and forget policy* with distributed generation regarded as a combination of negative loads. According to such rule, only those generators that do not alter the normal operating of a network can be connected without network investments. The DG effect is assessed considering the worst-case conditions of no-load/maximum-generation and maximum-load/no-generation. Generally, the higher the voltage level, the higher the connection costs and, consequently, lower the profitability of the investments for the DG owner. It is clear that the *connect and forget policy* poses a severe limitation to the integration of DG, especially in weak distribution networks. This policy is unfair because penalizes the last DG owner asking to be connected. For these reasons, it should be recognized that the passive operation of distribution systems causes technical and economic barriers that still limit the use of DG and RES and it can no longer be accepted whether the reduction CO_2 emission becomes mandatory.

Distribution system designs and operating practices are normally based on radial power flows and this creates a special challenge to the successful integration of DERs. Few of the issues that must be considered to guarantee that DG will not degrade distribution system power quality, safety or reliability are well described in [1].

2.2.2 Modeling of Renewable Output Power in Smart Distribution Networks

In the power system, distribution networks are deeply involved in the challenge for the integration of non-dispatchable energy sources (e.g., solar, wind), in which electricity production depends upon meteorological conditions and/or the time of the day. Among the various renewable generation plants, solar PV generators and wind turbines can be deployed more flexibly, in terms of location, than other renewable resources. In particular, small scale installation of solar PV panels can be roof-mounted on commercial or residential buildings or ground-mounted close to load in proximity of commercial and industrial areas, while small wind generators are quite common in rural areas.

Typical models for representing the most frequently installed photovoltaic and wind generation system in distribution networks are described in the following.

2.2.2.1 Photovoltaic Generators Modelling

The amount of energy that can be produced in a PV system is directly dependent on the irradiance and the angle at which solar PV cells are radiated. The variability of the irradiance, and then of the production, depends on meteorological conditions (Fig. 2.1).

The power output of PV generators can be then assumed linearly dependent on the global solar irradiance. The proportionality has to take into account of:

1. a parameter related to of a superimposition of a cloud distribution model to the global solar radiation time series during clear atmospheric conditions;

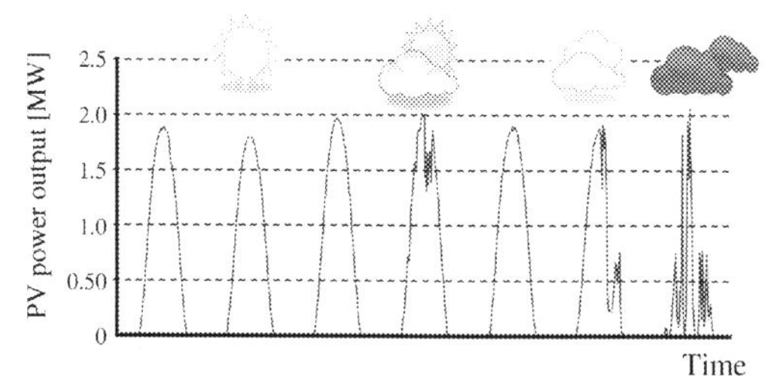

Figure 2.1 Tipical variability of the irradiance and production of PV systems.

2. a parameter that considers the environmental and electrical factors that can influence the production; and
3. the performances of the PV system, as reported in (2.1)

$$P_{PV_{hm}} = k_{cloud} \cdot k_{BOS} \cdot \frac{G_{hm}}{G_{stc}} \cdot P_{PV_r} \tag{2.1}$$

where G_{stc} is the global solar irradiance in the standard test condition (1000 W/m^2), G_{hm} is the global solar irradiance for clear atmospheric conditions in the hth hour of a generic day of the mth month of the year, P_{PV_r} is the rated power of the PV plant, k_{cloud} is a reduction factor due to the presence of clouds, and k_{BOS} is a reduction factor that takes into account the performance of the whole balance of system (BOS), that includes all components of a photovoltaic system other than the photovoltaic panels (wiring, switches, solar inverter) in real operating conditions.

Many environmental and electrical factors can influence the k_{BOS} value [2]:

- variable solar irradiation,
- temperature of the PV module,
- PV module power dissipation,
- PV module shaded or soiled,
- cabling, switchboards and circuit losses,
- efficiency of the dc/ac inverter,
- efficiency of the transformer (if needed).

The solar in-plane irradiation can be directly measured with a solarimeter/pyranometer or by using a different database available in the literature [3], and in the specific web resources [4] for each installation site.

In order to properly model the power output of a PV system would be necessary knowing the relative geometry of the cloud distribution at the site of the PV plant. In the practice, some assumption can be made to represent various conditions when the PV plant is shaded: in partially cloudy days, the global solar irradiance can be reduced to 50%, instead, for overcast days, a global irradiance reduction of 90% can be assumed ($k_{cloud} = 0.1$).

2.2.2.2 Wind Generation Modeling

In the wind generation plant the output production depends on the availability of the primary source that can fluctuate at various time scales; wind is subjected to seasonal variations, diurnal and hourly changes as

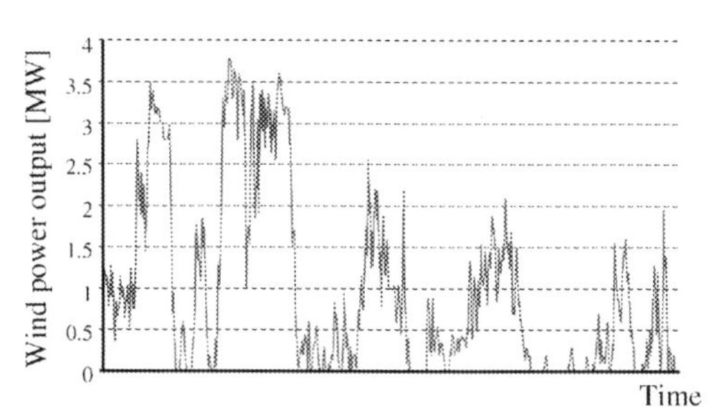

Figure 2.2 Typical variability of the production of Wind systems.

well as to very short-term fluctuations in the intra-minute and inter-minute timeframe, that can be difficult to manage in weak distribution networks (Fig. 2.2).

The output power of wind production systems can be modelized by taking into account time series of the primary source and applying them to the generator model in order to calculate the hourly power output, otherwise, suitable time series may be randomly formed in accordance with the probabilistic nature of the primary source. Starting from time series, the output of the turbine can be evaluated with the nonlinear relationship (2.2):

$$P_{WF_h} = \begin{cases} 0 & 0 \leq v_h \leq v_{ci} \\ P_{WT_r} \cdot \dfrac{v_h - v_{ci}}{v_r - v_{ci}} & v_{ci} < v_h < v_r \\ P_{WT_r} & v_r \leq v_h \leq v_{co} \\ 0 & v_h > v_{co} \end{cases} \tag{2.2}$$

where ν_{ci}, ν_r and ν_{co}, are respectively the cut-in, rated and cut-off wind speed of the generic wind turbine of a specific wind farm, ν_h is the actual wind speed at hour h, P_{WTr} is the rated power of the wind turbine.

2.2.3 Critical Operation of Distribution Networks With High Penetration of Renewable Sources

In this section a particular attention is dedicated to some common cases when power distribution networks may result inadequate to integrate RES installations, due to voltage regulation problems caused by power fluctuations of renewable energy at the point of connection or PCC (point of common coupling). This happens for instance when photovoltaic systems (PV) are installed in radial distribution networks. During the

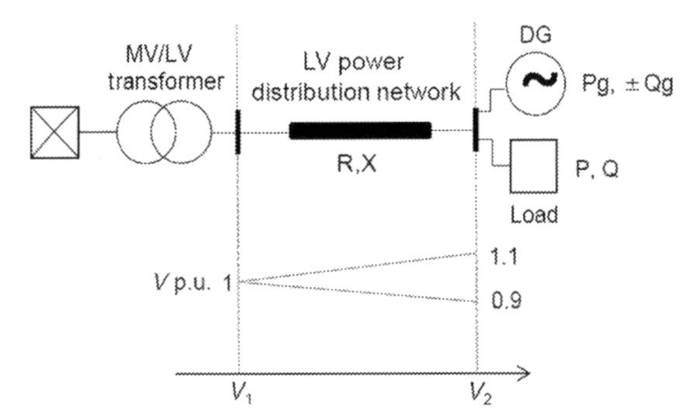

Figure 2.3 Voltage rise effects in radial weak distribution networks.

PV system operation in case of a weak power distribution network (e.g., rural or extra urban long MV or LV distribution lines with high R/X ratios), the voltage variation may fall outside the admissible regulation band as defined by the connection standard, causing the intervention of the loss of mains relay and the unintentional disconnection of the plant.

The DG effects on weak distribution networks are well-known, among all the limitation of the integration of further amounts of DG caused by voltage rise is one of the most relevant. In fact, particularly in weak radial distribution networks, characterized by long lines, with high R/X ratio, the voltage rise can be considerable.

The voltage regulation problem is described by Fig. 2.3, where one generator and one load are connected at the end of the feeder.

The voltage variation in the feeder can be expressed by (2.3).

$$\Delta V = V_1 - V_2 = \frac{R(P - P_g) + X(Q \pm Q_g)}{V} \tag{2.3}$$

where:

- V: Nominal voltage;
- V_1: Voltage at the begin of the feeder;
- V_2: Voltage at the end of the feeder;
- R: resistance of the feeder;
- X: reactance of the feeder;
- P: active power absorbed by the load at the end of the feeder;
- Q: reactive power absorbed by the load at the end of the feeder;
- P_g: active power generated by DG at the end of the feeder;
- $\pm Q_g$: reactive power absorbed/generated by DG at the end of the feeder.

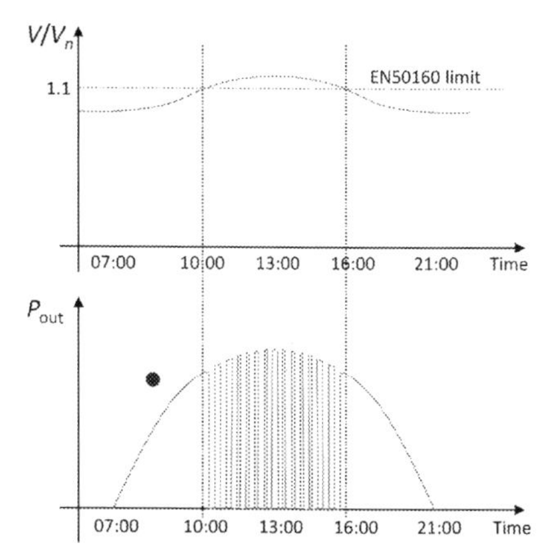

Figure 2.4 Repeated network disconnections due to overvoltage.

In case of no load condition ($P = Q = 0$), (2.3) can be rewritten as in (2.4) that shows the relationship between the injected power and the voltage at the PCC.

$$V_2 = V_1 + \frac{R\cdot(P_g) + X\cdot(\pm Q_g)}{V} \tag{2.4}$$

According to (2.4), DG can provide voltage support to raise too low voltage V_2 at the receiving-end of the feeder, but, in case of minimum/no load and high values of R, the magnitude of V_2 can result an overvoltage, limiting the DG capacity which can be installed, or, in general, the active power that can be injected at the end of the line. The PV power generation could not only offset the load, but could also cause reverse power flow through the distribution system and MV/LV transformer causing operational issues, including overvoltage and loss of voltage regulation.

An intelligent and coordinated management of the active and reactive power generated by the DG can mitigate the adverse impact on the voltage at the end of the line, and improving voltage regulation by limiting overvoltage and compensating excessive voltage drops.

A typical situation of disconnection of photovoltaic plants due to voltage regulation is showed in Fig. 2.4, which shows a cycling behavior of disconnection, automatic reclosing and further disconnection of the inverter, due to the intervention of the overvoltage protection. Voltage fluctuations may result in frequent unintentional disconnections of the

PV system, causing accelerated detriment of the apparatus and reduction of energy production with related economic loss for the PV owner.

A detailed analysis on unintentional disconnection of photovoltaic plants connected to weak low voltage distribution networks and a possible solution to the problem is available in [5] in which smart inverter capabilities have been adopted. New smart inverters are able to:

- set the automatic reduction of the active power generated by the PV in case of overvoltage condition (i.e., generation curtailment); in this way the inverter will remain connected to the network avoiding continuous connections and disconnections that cause significant production losses;
- operate the PV system at a not-unity power factor, e.g., by setting the inverter power factor as inductive (PF = 0.9); in this way the reactive power in (4) is negative (adsorbed Q) and the voltage can be reduced; this operation mode has been considered by the Italian technical standard for the connections to the distribution networks (namely CEI 0–21 [6]) but it requires the approval by the DSO;
- adopt a volt/var control method to provide a continuous voltage support for voltage variations resulting from changes in PV output; this approach could be used, for instance, during cloud passages that cause a sudden drop in active power at the PCC: the inverter volt/var control reacts to the sudden drop in voltage by injecting additional reactive power (capacitive Q, positive in (4)).

2.3 USE OF DISTRIBUTION ENERGY STORAGE FOR THE INTEGRATION OF RENEWABLE ENERGY SOURCES

2.3.1 Smart Distribution Networks and Energy Storage Systems

Distribution networks are approaching the critical point of saturating DER hosting capacity if the assessment is made with the *connect and forget policy*. Further connection of DERs inevitably requires taking an active approach in distribution operation (Smart Grid (SG) paradigm). Indeed, the intermittent nature of the wind and solar generation poses planning and operational challenges not only at distribution but also at transmission level, including additional ramping and regulation requirements and impacts on system stability. In order to mitigate the negative effects of the inherent nonprogrammability of RES, new protection and control strategies, enhanced distribution automation, Volt/VAr management should be integrated with the enforcement of distribution grid infrastructure. For avoiding as much as possible, or at least for deferring, grid investment costs, optimal and coordinated management of DERs and energy storage are solutions that could alleviate some of the cited critical issues and postpone the investments for transformers and lines.

In this context, energy storage systems (ESS) are valuable components of the future smart distribution systems, thanks to their ability to increase the flexibility of the overall system and to provide a wide range of services to the DSOs and to the customers with the final goal of mitigating the negative effects of RES. Services provided by ESS may be such as deferment of network investments, energy losses reduction, voltage regulation, reactive power balance, congestion overcoming, power quality improvement through intentional islanding, etc. [7]. ESS will allow both load profile flattening and solving extra production conditions, especially if they are connected to critical nodes of the network.

Under specific regulatory framework and strict conditions only, ESS may be also owned and managed by DSOs (see the derogation within the rules of the European Commission "winter package" [8]). In case that ESS are under DSO full control, they could be easily used for enabling services that should increase the hosting capacity for RES and electrical vehicles and consequentially reduce the operation costs, in distribution and transmission grids. Indeed, with Smart Grid controls in place, the DSO is a natural aggregator of producers and consumers, and it can offer to the TSO ancillary services that are necessary to operate the system with intermittent generation, e.g., being involved in voltage/VAR regulation. SG and ESS could enable DSO to provide several services, as producing reactive power following a pattern imposed by the TSO. The greater and well planned the storage capability, the smaller the involvement of RES in network management, which will be exploited almost exclusively for active power generation and less charged by extra costs for oversized power converters and voltage regulators. More likely, only the biggest generators will be integrated in voltage/VAR regulation, whereas ESS and plug in electric vehicles will compensate the large number of small distributed generators (DG) connected to the MV and LV networks.

In fact, as the ESS, plug-in electric vehicles (EV) can add more flexibility to load demand if smart charging or vehicle to grid (V2G) points will be increasingly integrated within the SG as expected in the near future. In fact, plug-in vehicles can effectively contribute to the grid operation thanks to their potentially rapid response [9–10].

It is worth noting that the wide deployment of ESS, private, but particularly if owned by DSO, should be analyzed and compared with other grid development strategies, taking into consideration the current regulatory frameworks and the still high investment cost for ESS, but it is a matter of fact that the willingness to reduce greenhouse gas emission poses them in a privileged position compared to the other actions that leads to the maximum exploitation of RES.

A brief review of energy storage classification and technologies that are currently or promisingly engaged for smart grid applications is presented in the following [11–12].

2.3.2 ESS Overview

The two fields more involved in the development of ESS are the power system and the transport sector. The demand for especially lithium-ion battery systems rises rapidly due to the electro-mobility promotion (i.e., plug-in hybrid and full electric vehicles). The link between this sector and the electric grid integration of the ESS is quite straightforward. Electro mobility requires more efficient batteries and large-scale production is expected to give impetus to the usage of ESS in distribution systems thanks to cost reduction. Furthermore, of the V2G option that allows using the vehicle batteries as grid storage during times when the vehicles are plugged-in for charging; secondly because the utility companies necessarily start getting involved in the upgrading of their infrastructures to integrate the EV charging stations. However, the development of batteries dedicated to grid applications goes rapidly on and several technologies can be considered quite mature for these purposes.

The ESS technologies, and consequently their cost, strongly depend on the specific services that they are called to perform. ESS services can be: ancillary services (i.e., frequency control, voltage regulation, spinning and stand reserves, black start service, etc.), peak shaving, load leveling, islanding support, or other service mainly related to private uses of the ESS (e.g., residential use for increased self-consumption of DG production, industrial applications, uninterruptable power supply etc.). Different classifications can be applied to the ESS, but one of the most effective is that one related to the duration and frequency of power supply from the ESS:

1. short-term (seconds to minutes),
2. medium-term (daily storage), and
3. long-term ESS (weekly to monthly).

The short-term ESS (<0.25 hours) can be used for primary and secondary frequency control, spinning reserve, black start, peak shaving, islanding, electro-mobility, and uninterruptable power supply (UPS).

The medium-term ESS (1−10 hours) are able to provide services of tertiary frequency control, standing reserve, load leveling, islanding, electro-mobility, residential self-consumption increase, UPS. Finally, the long-term ESS (from 50 hours and typically less than 3 weeks) can be exploited for long duration services, during periods when there is no or scarce generation of electricity from wind and solar ("dark- calm periods").

Super-capacitors, superconductive magnetic coils, or flywheels may offer short-term services. Pumped hydropower, compressed air ESS, thermoelectric storages, and electrochemical ESS, as lithium-ion, lead-acid, high temperature and flow batteries, are able to perform medium-term services. Long-term services can be offered by hydrogen or natural gas storage systems.

The main applications that can be suitably exploited by the smart distribution networks fall into the medium, or at least short-term services.

The pure electrical super-capacitors, superconductive magnetic coil, (since they have some strengths, as high efficiency, high power capability, and long life), are yet affected by the lack in the validation and experimentation for grid purposes and by their very high costs, due to the high innovation degree.

The mechanical systems can be subdivided into well-established technologies (i.e., pumped hydropower), the ones that need short time-to-market (i.e., compressed air energy storage), or those that are developed for other applications than the network operation (i.e., flywheels, that are well-established in UPS systems).

The technologies useful for distribution networks applications and that already reached a higher technical readiness level are the electrical-chemical batteries. Such technologies may have internal or external storages. Examples of the latter systems, disregarding the hydrogen or methane storages that are useful for long-term services, are the redox-flow batteries, which have the advantage that energy and power are independently scalable (energy capacity depends on the tank while the cell stack determines the power). Vanadium redox-flow batteries are commercially available with different modular scalable sizes but the still high costs of the electrolyte solution and the maintenance obstacle their large-scale diffusion. The internal storages systems, in which the energy and power

Table 2.1 Parameters for chemical storage systems with internal storage

Technology	Round-trip efficiency	Energy density	Power density	Cycle life
Lithium-Ion	83% ÷ 86%	200 Wh/L ÷ 350 Wh/l	100 W/L ÷ 3500 W/l	1000 ÷ 5000
Lead-Acid	75% ÷ 80%	50 Wh/L ÷ 100 Wh/l	10 W/L ÷ 500 W/l	500 ÷ 2000
NaS	75% ÷ 80%	150 Wh/L ÷ 250 Wh/l	—	5000 ÷ 10000

depend each other, work at low (Li-ion, lead-acid, Ni-Cd) or high temperature ($NaNiCl_2$, NaS).

In Table 2.1 the range of the most important parameters for some batteries are reported [12].

Lithium-ion batteries have become the most important storage technology in different areas (e.g., portable devices and EVs) and can be an option for stationary applications also, due to their high energy density, high efficiency, and relatively long lifetime. Despite lithium resources being limited to only few countries, the current fervent development activity related to this kind of batteries could result soon in a significant cost reduction and improved lifetime and safety. Lead—acid batteries are one of the most developed and long time installed technologies. They are mainly used for cars but are widely used also for stationary grid application, e.g., in islanded grid. Their main disadvantage is the toxicity of lead and this fact causes social acceptance problems. However, the market opportunity is the low investment costs and the existing large number of manufacturers.

Sodium-nickel-chloride batteries ($NaNiCl_2$, also called zebra-battery) and sodium-sulphur batteries (NaS) operate at high temperature, in a range from 270°C to 350°C. The temperature has to be maintained during the charging/discharging cycles, thus a suitable isolation has to be designed. Typical stationary applications can be peak shaving and load shifting. They are commercially available from few manufacturers, thus they are not much diffused around the world. For their characteristics high temperature batteries may be competitive with the lead-acid and lithium-ion batteries, but the main obstacle to their diffusion is related to safety issues (fire incident caused by NaS batteries).

2.3.3 Use of Distributed ESS in a Low Voltage Distribution Network

Many studies dealing with the integration of high share of electricity from nondispatchable RES in a power system consider the option of the installation of storage devices to balance the fluctuations in power production. It is, in fact, expected that future DSOs will be likely committed to trade energy services in local energy markets dedicated to distribution networks, such as renewable balancing and voltage regulation. The benefits from the DSO side are explained with the aid of a case study. A LV distribution network with several *nanogrids* is managed as a virtual power plant (VPP) by an aggregator that receives power dispatching requests by the DSO for improving voltage profile and reducing renewable energy curtailment. The aggregator can use an ESS [13]. The proposed case study considers a 3-phase 4-wire LV distribution network (230/400 V), which may be representative of a small smart district. The test network scheme is depicted in Fig. 2.5.

The secondary substation is equipped with a 630 kVA MV/LV transformer, that supplies an area corresponding to a radius of approximately 600 m from the substation, with 48 LV customers. Some customers are passive but some of them are equipped with PV generators with or without ESS. It is assumed that those equipped with an ESS can be operated as nanogrids.

The load's nominal power and the phases where the users are connected (L1/L2/L3) are summarized in Table 2.2, while in Table 2.3 the corresponding PV and ESS sizes related to the same end-user point are reported.

In the case study, during some critical operating conditions caused by the lack of coincidence between demand and production, the LV distribution network may result inadequate to integrate such a PV amount, due to voltage regulation problems caused by PV power fluctuations [5]. In those cases, according to current standards, PV plants have to be disconnected (e.g., each time that voltage threshold are reached), and it results in loss of energy production with economic disadvantages for the PV owners and environmental burden.

According to the Italian standard CEI 0–21 [6], the PV plant has to be disconnected from the distribution network when:

- node voltage U, measured as an average on 10 minute period, is over 253V ($U > 110\% \ U_n$) with time of intervention ≤ 3 s;

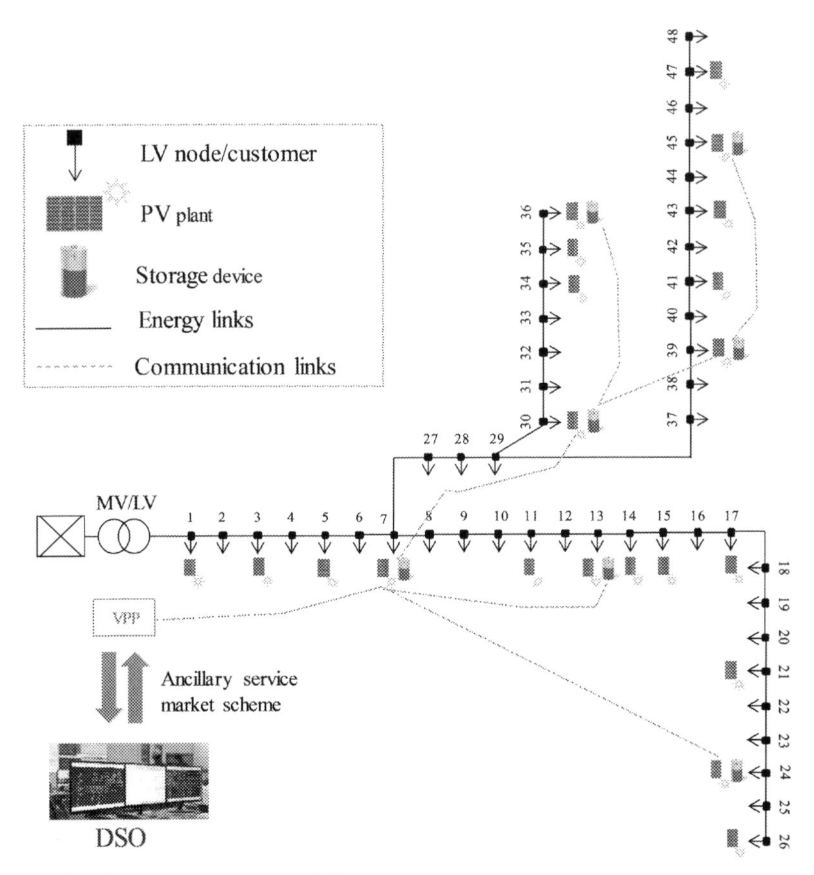

Figure 2.5 LV test network with VPP Management.

- node voltage U is over 264V (U > 115% U_n) with time of intervention <0.2 second;
- node voltage U is below 195.5V (U < 85% U_n) with time of intervention <0.4 second;
- node voltage U is below 92V (U < 40% U_n) with time of intervention <0.2 second.

The coordinated use of the ESS by means of a VPP may result successfully in solving the contingencies (i.e., avoid PV disconnection) and for improving the voltage profile in the distribution network.

Two case studies are discussed in the following related to the use of VPP for solving voltage regulation problems.

In the first case study, an overvoltage occurs on node #47 of Fig. 2.5 during the PV generation peak hours. In a passive network management

Table 2.2 End-user load data

Bus	Phase	Load (kW)	Bus	Phase	Load (kW)
1	L1	3	25	L1	6
2	L2	4.5	26	L3	6
3	L1	3	27	L2	4.5
4	L2	4.5	28	L3	3
5	L3	3	29	L2	6
6	L2	3	30	L1	3
7	L3	4.5	31	L1	3
8	L3	6	32	L2	4.5
9	L1	3	33	L1	3
10	L2	4.5	34	L3	3
11	L1	6	35	L2	3
12	L1	3	36	L3	3
13	L2	4.5	37	L1	3
14	L2	3	38	L2	4.5
15	L3	3	39	L2	6
16	L1	4.5	40	L1	3
17	L2	3	41	L2	3
18	L2	6	42	L3	3
19	L3	3	43	L3	3
20	L1	6	44	L2	3
21	L2	4.5	45	L3	4.5
22	L2	3	46	L1	3
23	L2	3	47	L3	3
24	L2	3	48	L1	3

Table 2.3 End-user PV and ESS data

Bus	PV (kW)	ESS (kW/kWh)	Bus	PV (kW)	ESS (kW/kWh)
1	6	—	26	6	—
3	3	—	30	10	6/30
5	3	—	34	3	
7	10	6/30	35	3	
11	6		36	6	6/30
13	3	6/30	39	10	6/30
14	3		41	10	
15	3		43	3	
17	6		45	6	10/50
21	6		47	10	
24	3	10/50			

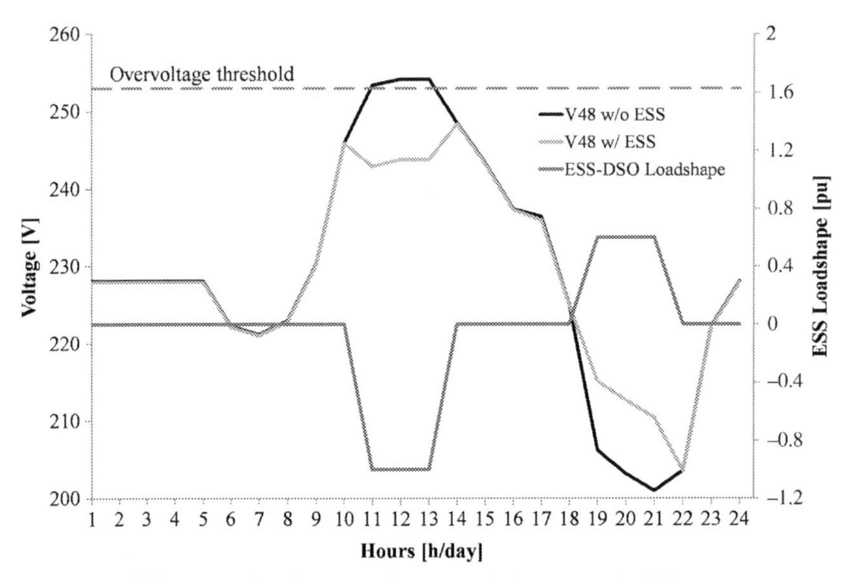

Figure 2.6 VPP intervention for overvoltage regulation on node #47.

in case of high renewable energy production, the voltage at node #47 overcomes the voltage thresholds of 1.1 p.u. and the PV generator has to be disconnected. In order to avoid the disconnection and the loss of energy production the DSO decides to enable the VPP control for the nanogrids at nodes #7, #36, #45, that are connected to the same phase as node #47, where the overvoltage is detected. The voltages of the involved node and the ESS load-shape of the charging/discharging profile of the VPP is depicted in the same in Fig. 2.6. The green line profile shows that the DSO-driven intervention over the charging/discharging profile of the ESS plants is capable of solving the overvoltage avoiding PV plants disconnection.

In the second case study an excessive voltage drop is registered during peak of demand. The DSO, as responsible for the network operation, has to guarantee the voltage profile within the range 1 p.u. $\pm$ 10%. In the test network of Fig. 2.5, due to the excessive loading condition, the voltage drop registered at node #26 in the time interval 19:00—22:00 is below the minimum value. DSO may solve this problem by purchasing a service from a VPP that with an intelligent operation of charging/discharging the ESS of the nanogrids at nodes #13 and #24 may allow maintaining the voltage at node #26 into the acceptable limits (Fig. 2.7).

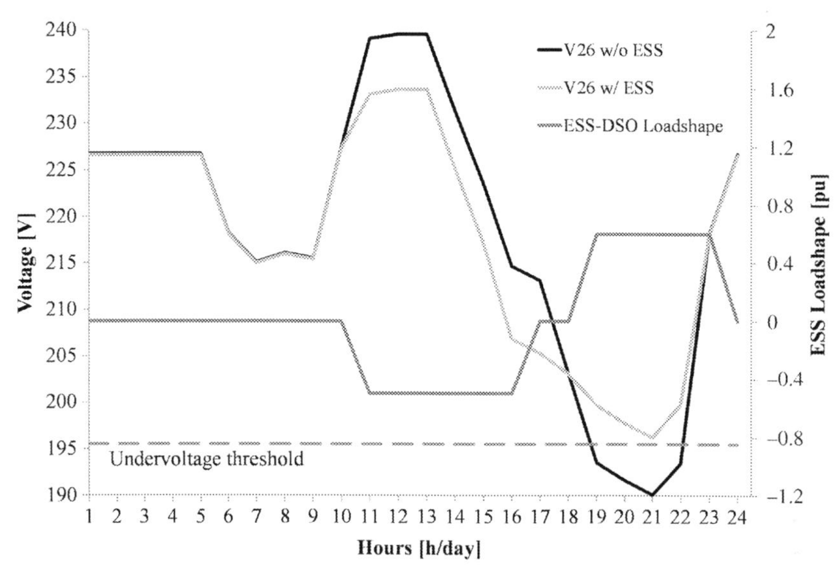

Figure 2.7 VPP intervention for undervoltage regulation on node #26.

2.4 PLANNING APPROACHES FOR INTEGRATING HIGH SHARES OF RENEWABLE ENERGY SOURCES

The availability at the distribution level of active operation systems is changing the objectives of the traditional planning and operation of the distribution networks that will be mostly oriented to the maximum exploitation of existing assets and infrastructures, by managing them much closer to their physical limits than in the past. The hosting capacity for RES can then be increased with less network investments since operational issues can be fixed with the "no-network" solutions. The novel planning methodologies, by incorporating "no-network" in the list of design options, are intrinsically capable to enhance the integration of RES [14–17].

Table 2.4 shows the most common issues in the distribution system arising from the current high integration of renewable generation.

It is interesting that each possible issue, e.g., voltage regulation, can be faced with traditional network solutions and/or with innovative no-network solutions, and that some of the most innovative solutions require that the not regulated players (e.g., the DER owners) help the DNO, the committed entity responsible for the operation of the distribution network. For this reason, some of the most innovative concepts in

Table 2.4 Network and no-network solutions in modern distribution planning with high shares of renewable energy sources installed [16]

Challenge	Current solution	Future alternatives
Voltage rise	Reinforcement Operational p.f. 0.95 lagging Generation tripping	Volt/VAR control Storage Generation Curtailment On-line reconfiguration
Voltage drop	Reinforcement Fixed capacitor banks	Volt/VAR control Storage Demand Side Response On-line reconfiguration
Network capacity	Reinforcement	Storage Generation Curtailment Demand Side Response On-line reconfiguration
Network power factor	Limits / bands for demand and generation	Storage Not-Unity power factor generation
Sources of reactive power	Transmission network Fixed capacitor banks	Storage SVC Volt/VAR control
Network asset loss of life	Strict network designs specifications based on technical and economic analyses	Dynamic protection settings Asset condition monitoring

distribution operation and, consequently, in distribution planning will require a suitable regulatory environment that allows the DNO to transform itself into a distribution system operator (DSO) that replicates at the distribution level the functions of transmission system operators (TSO). Some of the main topics related to the Smart Grid paradigm and the planning of the future active distribution network (ADN) or smart distribution networks (SDN) are briefly illustrated in the following.

2.4.1 Distribution Network Planning and Operation With Smart Grids

Planning the evolution of the system without planning the integration of the enabling technologies can lead to diseconomies or can slow down the RES development. From the analysis of Table 2.4, it emerges the recommendation that novel planning techniques should integrate operation models within planning as well as the models of the communication system. Lacking in this point leads to unreliable plans making the transition towards active distribution networks and the integration of novel technologies too expensive.

Indeed, the distribution management system (DMS), the core of active networks operation, relies heavily upon control capabilities and automation that are generally absent in current distribution networks. The lack of experience, the increase in complexity, and the use of novel communication systems are perceived by DSOs weaknesses of ADNs, but few quantitative analyses on the expected reliability of SG have been published. The expected reliability of a smart distribution network can be evaluated by applying to the distribution system techniques based on a pseudo–sequential montecarlo simulation (PSMC), typically used in the classical composite generation-transmission system. Meteorological models may be implemented with the PSMC simulation in order to reproduce fluctuating renewable generation caused by the moving of cloud patterns and/or to simulate local meteorological conditions that can degrade the performance of ICT wireless communication networks [18].

Indeed, current planning algorithms and software tools have significant shortcomings in dealing with these emerging issues. Considering that the communication facility will not be an add-on of the power system, but an essential part of active distribution, responsible for the reliability of SG, the simultaneous analysis of both power and communication systems will be useful.

In addition, the strategic system planning, in which politicians and decision makers have to be aware about the impact on the system of different options, and the related necessary investments, need to be considered. Distribution planners can provide decision makers with information on the areas where RES would be less harmful for the network or more useful. The integration of planning methodologies within geographic information systems allows finding the most promising areas for RES or the areas where RES should be promoted, according to the regulatory environment and the incentive schemes.

2.4.2 Optimal Network Topology for RES Integration—Meshed or Radial?

Active networks and RES may benefit from the adoption of meshed topology for the operation in opposition to the dominant radial scheme [19–21]. Meshed networks may have positive effects on power losses, voltage regulation, and network reliability and on the exploitation of lines and substations, particularly if high shares of DG have to be integrated. It should be recognized that the more DG installed in distribution networks, the fewer the reasons to keep using radial operated networks and the greater the motivation to adopt weakly meshed systems. Compared to radial ones, meshed networks require more complex operation, but the level of complexity is not higher than in radial ADNs or SDN. On the contrary, with the automation and communication level of ADNs, the operation of meshed systems is greatly simplified. The major concerns are related to the protection system that might become inadequate to cope with the increased short circuit level. Furthermore, the coordination of overcurrent relays for selectivity is another serious issue that could be solved with SDN [21]. Meshed operated distribution networks should be taken into account in planning for active distribution systems with efficient network automation and proactive protections. Indeed, the use of short circuit limiters in meshed networks is a valid planning alternative that allows avoiding the refurbishment of existing breakers. The optimal allocation of short circuit limiters might be solved with algorithms similar to the ones used for the optimal allocation of capacitor banks.

2.4.3 Flexible Network Reconfiguration

Network reconfiguration is a central characteristic of future smart distribution networks and can favorite the RES integration. The flexible network topology with reconfiguration allows reduction of energy losses, minimizing changes in the DG scheduled power production and achieving a more reliable and resilient system. Network portions with a good balance between production and demand may be formed for autonomous functioning according to intentional islanding operation [22–23] or as a microgrid and multimicrogrid architectures [24–26]. The reorganization of the distribution system in networked microgrids and therefore the management of energy exchanged with the public network, obtainable by the introduction of appropriate storage systems and local/virtual energy management and control systems allows avoiding the construction of new infrastructures, improving the quality of electricity supply and permitting greater penetration of non-programmable renewable sources.

The problem of on-line network reconfiguration has been studied for many years but few practical application on the field have been realized. There are several reasons for that, but the most important are the absence in some cases of adequate SCADA systems, automated nodes, and communication systems. Furthermore, without RES installed the resort to network reconfiguration is safer and less expensive if done few times per year in order to adapt the configuration to consumption. More frequent reconfiguration can lead to unreliable schemes and increase the risk of failures. Nowadays, the implementation of true on-line network reconfiguration integrated with other typical operation action such as reactive power optimization, generation curtailment, demand response, and storage operation can help find the optimal point of work regardless the non-programmability of RES. For that reason, more researches are needed in order to integrate different functionalities in reliable DMS [27].

2.4.4 Management of Uncertainties and Risk With a Probabilistic Approach

The massive penetration of RES in the distribution network needs that uncertainties related to public opinion attitude to sustainable development, political drivers (i.e., subsidies or penalties), regulatory environment, fuel cost and energy price, intermittent power production (i.e., renewables), DER availability, and SG success are properly considered in planning. Making a decision in uncertain scenarios causes risks, and planning algorithms, particularly in active distribution networks, need to explicitly deal with risks, by allowing planners to make objective and transparent decisions. Planning in the SG era is then a decision-making process applied in an uncertain scenario and decision theory based algorithms can be successfully applied.

Distribution planning relies on deterministic models. The majority of DNOs still adopts deterministic models even with high shares of RES in place, and do not take into account SG opportunities in planning studies. Probabilistic methods are better suited for planning with DG and DER and should be adopted in modern planning.

2.4.5 Energy Losses Reduction

DSO's are committed worldwide to reducing energy losses for environmental reasons and this is particularly necessary as the penetration of RES is increasing but funded with a huge amount of money paid to compensate the grid parity. SG and RES may cause an increase in losses due to the combination of investment deferment, and the non-homothetic growth of power production and load demand. DSO would like to have truly dispersed generators for loss reduction; producers strive to fully exploit the local energy resources and do not care about negative network impacts. Multicriteria programming can be used to solve the conflicts between DSO and producers.

2.4.6 Ageing

Distribution systems in developed countries are reaching the end of their life cycles, with the majority of electrical infrastructures built in the 60s. Failures of the equipment associated with the ageing of distribution infrastructures is a major issue that could negate the reliability improvements achieved in recent years with the combination of asset management, predictive and corrective maintenance, novel planning strategies, and network automation. In the long term, active management strategies could amplify ageing problems due to the deferment of investments but this is a general topic not specific related with RES integration.

2.5 CONCLUSION

This chapter describes the main technical issues arising by the intermittency and non-programmability of DER that are going to be integrated in the future distribution networks. The impacts of RES on distribution networks require to modify the traditional planning, operation and management of power distribution systems. The development of the future energy system will be based on planning and management of the distribution system in accordance with the philosophy of Smart Grid involving the extensive use of ICT and innovative control systems in order to enable the realization of smart distribution systems, the active participation of demand and energy storage.

REFERENCES

[1] P.P. Barker, R. W. De Mello, Determining the impact of distributed generation on power systems. I. Radial distribution systems, in: 2000 Power Engineering Society Summer Meeting, Seattle, WA, 2000, pp. 1645–1656, vol. 3.

[2] E. Ghiani, F. Pilo, S. Cossu, Evaluation of photovoltaic installations performances in Sardinia. Energy conversion and management 76, 1134–1142.

[3] M. Šúri, T.A. Huld, E.D. Dunlop, H.A. Ossenbrink, Potential of solar electricity generation in the European Union member states and candidate countries, Solar Energy 81 (2007) 1295–1305.

[4] European Commission. Photovoltaic Geographical Information System. JRC Solar database http://re.jrc.ec.europa.eu/pvgis/cmaps/eur.htm.

[5] E. Ghiani, F. Pilo, Smart inverter operation in distribution networks with high penetration of photovoltaic systems, J. Modern Power Systems Clean Energy - Special Issue Active Distribution Systems. (Oct. 2015).
[6] Italian Standard CEI 0–21. Technical rules for the connection of active and passive users to the LV networks of electricity distribution companies.
[7] G. Carpinelli, G. Celli, S. Mocci, F. Mottola, F. Pilo, D. Proto, Optimal integration of distributed energy storage devices in smart grids, IEEE Trans. Smart Grid 4 (2) (2013) 985–995.
[8] European Commission, Clean energy for all Europeans package, Nov 2016.
[9] M. Yilmaz, P.T. Krein, Review of the impact of vehicle-to-grid technologies on distribution systems and utility interfacesin IEEE Trans Power Electron 28 (12) (Dec. 2013) 5673–5689.
[10] F. Mwasilu, J.J. Justo, E.K. Kim, T. Duc Do, J. Jung. Electric vehicles and smart grid interaction: A review on vehicle to grid and renewable energy sources integration, Renew. Sustain. Energy Rev. 34, 2014, 501–516.
[11] G. Fuchs, B. Lunz, M. Leuthold, D.U. Sauer, Technology Overview on Electricity Storage, Smart Energy for Europe Platform GmbH (SEFEP), Aachen, 2012.
[12] D.O. Akinyele, R.K. Rayudu, Review of energy storage technologies for sustainable power networks, Sustainable Energy Technologies and Assessments, Volume 8, December 2014, Pages 74–91.
[13] G. Celli, E. Ghiani, F. Pilo, G. Marongiu, Smart Integration and Aggregation of Nanogrids: Benefits for Users and DSO, Powertech, Manchester, 2017. June 18–22.
[14] C. Abbey, F. Pilo et al., Active Distribution Networks: general features, present status of implementation and operation practices, ELECTRA, n. 2 4 6, Oct. 2009.
[15] A. Keane, F. Pilo, et al., State-of-the-art techniques and challenges ahead for distributed generation planning and optimizationin IEEE Trans. Power Systems 28 (2) (May 2013) 1493–1502.
[16] F. Pilo, G. Celli, E. Ghiani, G.G. Soma, New electricity distribution network planning approaches for integrating renewable, Wiley Interd. Rev. Energy Environ. 2 (2) (March/April 2013) 121–250.
[17] CIGRE WG SC6.11, Development and Operation of Active Distribution Networks, CIGRE Technical Brochure, April 2011978-2-85873-146-6.
[18] G. Celli, E. Ghiani, F. Pilo, G.G. Soma, Reliability assessment in smart distribution networks, Electric Power Systems Res. 104 (November 2013) 164–175. F. Pilo, G. Celli, E. Ghiani, G. G. Soma.
[19] G. Celli, F. Pilo, G. Pisano, V. Allegranza, R. Cicoria and A. Iaria. Meshed vs. radial MV distribution network in presence of large amount of DG. IEEE PES Power Systems Conference and Exposition, 2004., 2004, 709–714, vol.2.
[20] D. Wolter, M. Zdrallek, M. Stötzel, C. Schacherer, I. Mladenovic, M. Biller. Impact of meshed grid topologies on distribution grid planning and operation. CIRED 2017 24th International Conference on Electricity Distribution. 12–15 June 2017. Glasgow, UK.
[21] S. Botton, L. Cavalletto, F. Marmeggi. Schema project-innovative criteria for management and operation of a closed ring MV network. CIRED 2013 22nd International Conference and Exhibition on Electricity Distribution, 10–13 June 2013. Stockholm, Sweden.
[22] F. Pilo, G. Celli, S. Mocci. Improvement of reliability in active networks with intentional islanding. Proc. of 2004 DRPT Conference. 2004. 2: 474–479.
[23] M. Bollen, Y. Sun, G. Ault. Reliability of distribution networks with DER including intentional islanding. Proc. of 2005 International Conference on Future Power Systems. 2005. 498–503.

[24] N. Hatziargyriou, H. Asano, R. Iravani, C. Marnay, Microgrids, IEEE Power and Energy Magazine 5 (4) (2007) 78–94.
[25] R.H. Lasseter, Smart Distribution: Coupled Microgrids, Proc. IEEE 99 (6) (2011) 1074–1082.
[26] E. Ghiani, S. Mocci, F. Pilo. Optimal reconfiguration of distribution networks according to the Microgrid paradigm. Proc. of 2005 International Conference on Future Power Systems. 2005.
[27] F. Pilo, G. Pisano, G.G. Soma, Optimal coordination of energy resources with a two-stageon-line active management, IEEE Trans. Ind. Electron. 58 (2011) 10.

CHAPTER 3

Demand Response Enabled Optimal Energy Management of Networked Microgrids for Resilience Enhancement

Nima Nikmehr[1], Lingfeng Wang[1], Sajad Najafi-Ravadanegh[2] and Solmaz Moradi-Moghadam[1]
[1]Department of Electrical Engineering and Computer Science, University of Wisconsin-Milwaukee, Milwaukee, WI, United States
[2]Smart Distribution Grid Research Lab, Department of Electrical Engineering, Azarbaijan Shahid Madani University, Tabriz, Iran

3.1 INTRODUCTION

In the presence of devastating natural disasters like earthquakes, landslides, tsunamis and hurricanes, the resilience studies of power systems become extremely important. Since natural or man-made disasters cause huge damages especially to the electric power systems, further studies are required on resiliency particularly in recent years due to the tremendous impact of global climate changes [1]. Power system resilience is defined as the flexibility of the electric network that can be recovered immediately after an unpredicted, adverse event occurs such as a natural or a man-made disaster [2].

Microgrids (MGs) have a great potential to enhance important parameters of distribution grids such as reliability [3], environmental benefits [4], and techno-economic performance. The modern structure of MGs, in which MGs have a great flexibility to connect to each other and to the main grid is called networked MGs (NMGs). In this structure, MGs can operate in both interconnected and islanded modes. Therefore, the resilience of distribution grid can enhance due to sharing of power between MGs after occurring an unpredicted disaster [5]. The concept of NMGs is the main issue in mitigating the operation costs of distribution network through optimal operation of MGs [6]. In fact, designing an efficient energy management system (EMS) is a key factor to ensure a reliable

Operation of Distributed Energy Resources in Smart Distribution Networks
DOI: https://doi.org/10.1016/B978-0-12-814891-4.00003-5

cooperation between MGs and distribution network operator (DNO) to improve the resilience, reliability, and sustainability of the whole system. In [7], a stochastic bi-level optimization problem is considered for distribution network operator as an upper level and for each MG as a lower level, in which both levels have their own objective functions. A decentralized EMS is exploited to ensure the coordination between MGs and DNO in any given time. The proposed EMS structure in [8] is proper for self-healing of NMGs in which MGs are able to share their powers with each other. A two-layer cyber communication and control system is employed in a decentralized EMS to ensure the optimal performance of MGs.

In recent studies, the interconnected operation mode of MGs is considered as a popular solution for improving the resilience of MGs. In [9], the authors have tried to survive the critical loads during disconnection of MGs from radial distribution network using MGs concept. In this regard, a mixed-integer linear program (MILP) optimizes total priority weighted picked up loads. Maharjan et al. [10] have considered electric vehicles as important tools for restoring the critical loads after disasters so that by using renewable resources and optimal operation of EVs, the resilience of MGs enhances during emergencies. In [11], a two-stage stochastic programming approach is applied in day-ahead scheduling problem to improve the resilience of MGs considering demand response (DR) and energy storage systems (ESSs). In [12], a decentralized outage detection method using advanced metering infrastructure tries to amend the resilience of communication networks through picking up the emergency loads within the MGs. Nevertheless, the uncertainties of renewable sources and loads have been neglected. In [13], stochastic MILP is used to evaluate the effects of coordinated MGs after an unanticipated disaster.

With deploying smart distribution grids, power companies try to gather the power energy information from their customers through utilizing efficient EMS, in which the optimal scheduling of loads can be guaranteed [14]. In this regard, DR programs play an important role in ensuring supply-demand balance considering the economic aspects of both MG owners and costumers [15]. One of the main challenges for power companies with DR process is injecting malicious attacks during manmade disasters. In this way, authors in [16] have utilized a DR-based MG structure to enhance the resilience of distribution grids. DR programs enable the demand profile to avoid excessive use of electricity especially in peak time, which finally results in an economic advantage for the whole distribution system. In this chapter, the authors have tested the potentials of different DR programs to shift the critical loads from an interval to another when a detrimental natural disaster occurs. The current study tries to use the strong advantages of DR to shift possible loads in order to decrease the load shedding in emergency hours.

The contributions of this chapter can be summarized as follow:

- Real time pricing (RTP) is exerted to mitigate the total operation costs for MG owners and expenses of consumers.
- In proposed NMGs structure, every MG is connected with other MGs and main grid. In the presented framework, MGs can operate both in islanded mode and interconnected mode. In the case of resilience, first, MGs operate in normal operation mode meaning that MGs can provide their supply-demand balance through linking to different MGs and main distribution grid. In the second state, after occurring a fault or unanticipated natural disasters, the MGs operates in the emergency mode, in which MGs lose their connection with the main grid. What makes the proposed NMGs prominent is the direct connection between MGs which can provide them with their requirements.
- In this chapter, an efficient EMS is required to make a coordination between MGs. In proposed EMS, the optimization process will be handled in two different stages. Firstly, the optimal distribution of power will be done inside each MGs considering their own renewable sources and loads, and then in second stage, the final optimization will be handled to schedule the amount of power transaction between each MGs, as well as between MGs and main grid.
- Considering potentials of the proposed structure especially in making coordination between MGs by EMS, DR programs can decrease the load shedding through shifting critical loads when a natural disaster occurs. In the presented DR, costumers react to hourly prices.

3.2 APPLIED STRATEGIES TO ENHANCE THE RESILIENCE OF MICROGRIDS

Recently, various strategies have been introduced to address the resilience of MGs during a natural or man-made disaster. In [17], the reconfigurable MGs through switching between the existing MGs, can be reconnected to or disconnected from each other to optimally supply the critical loads. This paradigm can increase the economic benefits of the distribution network as compared to the existing power distribution system. The concept of energy hubs is considered as an important structure for analyzing the resilience of the electric network, in which the impacts of disruption in one type of energy source can affect the serving of other energy infrastructures. Authors in [18] introduce a strategy based on multiple energy carriers to enhance the resilience operation of MGs through coordinating electricity and natural gas infrastructures. Linking multiple islanded MGs is another strategy to improve the resilience of distribution system [19]. The power deficiencies of a MG can be supplied by other less loaded MG through either a centralized or decentralized self-healing agent; however, the performance of the utilized EMS in optimal operation of MGs and delivering data in a secure way is often times neglected. The hybrid MGs and ESSs can be utilized as important tools to mitigate the interruption of critical loads during a natural disaster [20]. The authors prioritize the loads within the MGs to increase the reliability of the distribution network, and also optimal charging and discharging scheduling of batteries is carried out to ensure the load survivability.

In this chapter, two important concepts are used to enhance the resilience of MGs during an unanticipated disaster. First, NMGs structure, in which MGs can operate both in islanded mode and interconnected mode through an efficient EMS. Second, RTP as a strong DR program has a great capability to minimize the interruption time.

3.2.1 Energy Management System in Proposed Networked Microgrids Structure

In this study, MGs are small scale energy zones with various types of nondispatchable generation sources like wind turbines (WT), and photovoltaic panels (PV), as well as dispatchable energy sources such as microturbines (MTs), fuel cells (FCs), and combined heat and power (CHP) units. The duty of these energy sources is supplying the load demand of each MG and even the demands of other MGs. In the proposed structure for MGs, various numbers of MGs can be considered according to the power system planning and operation benefits. The local energy sources and loads within the MGs are controlled by local controllers (LCs), in which the managing and scheduling of renewable energy sources (RESs) inside the MG is handled to guarantee the supply-demand balance. In addition to LC, MG control center (MGCC) gathers data dealing with generation and load rates from each MG to coordinate the performance of them in an optimal way. In this regard, the proposed

EMS has two main responsibilities. First, managing the local generation resources of each MG to supply their own loads. In second stage, MGs can supply their deficiencies especially their supply-demand balance through linking to other MGs and the main distribution grid. In this level, EMS coordinates the interactions of MGs with each other and with the main network so that the economical operation of multiple MGs is achieved.

In the proposed NMGs framework, MGs can operate either in islanded mode or in interconnected mode. In the isolated mode, MGs are disconnected from the main distribution grid; however, each MG can link to other MGs and share its own power with other MGs. This mode is utilized particularly during a natural disaster or a fault occurrence [20]. On the other hand, MGs have an option to connect to the main grid through distribution lines besides having links with other MGs. This kind of operation greatly improves the reliability of the network [3]. In this chapter, two different operation modes are defined to guarantee the optimal, economical, and resilient operation of MGs when a fault or natural disaster occurs in the network. First one is the normal operation mode in which MGs can operate in either islanded or interconnected mode to ensure their supply-demand balance considering the economic benefits. Actually, in normal operation there is no fault in the network. In second operation mode or emergency mode, MGs have to be disconnected from the main grid after a fault and continue their operations in the islanded mode.

The proposed EMS in this chapter tries to obviate drawbacks of the conventional EMS. In centralized EMS, all MGs are controlled and optimized by an EMS. Although this kind of EMS is effective in the islanded mode, it is not able to handle the cooperation among the MGs in interconnected mode [21]. On the other hand, the decentralized EMS is beneficial to enabling the linked operation mode of MGs. Nevertheless, it cannot impose an efficient control on the local energy sources and load demands which results in noneconomic costs for customers and MG owners [22]. The proposed EMS in this study firstly optimizes the local sources of MGs to supply the electrical demand, and then sends the information to the MGCC so that the requirements of MGs could be fulfilled through power sharing with other MGs and the main grid. Therefore, the proposed EMS fulfills the both technical and economic aspects of MGs no matter if they operate in the islanded mode or the interconnected mode. Fig. 3.1 shows the structure for NMGs.

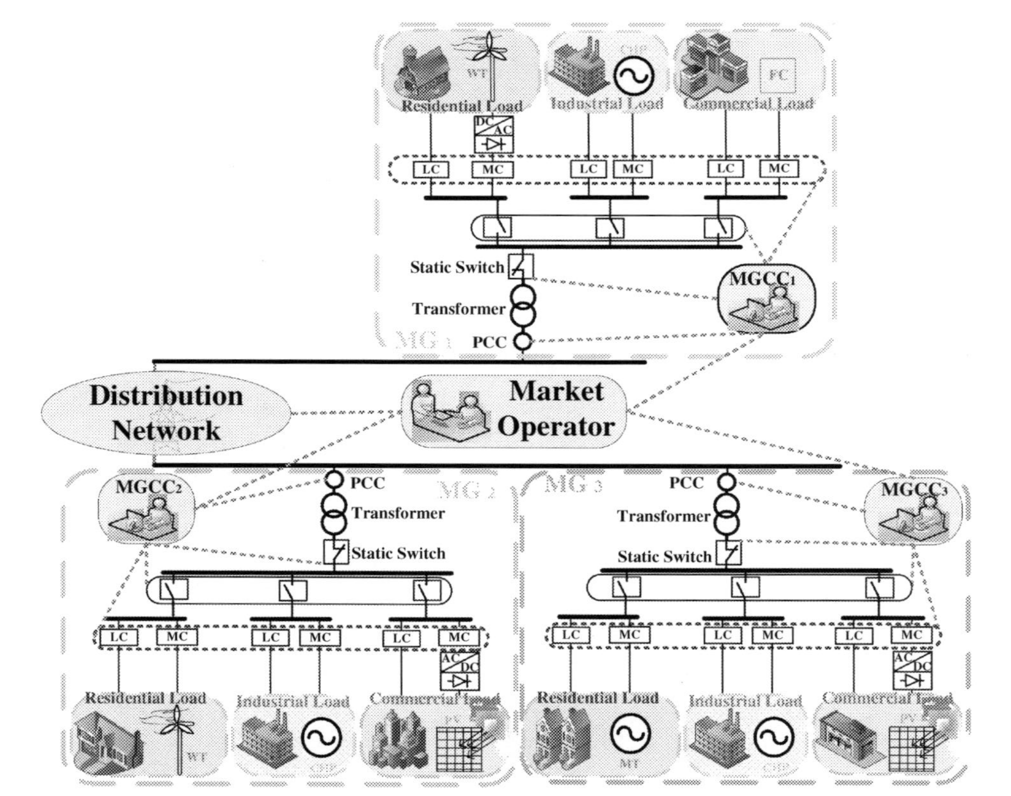

Figure 3.1 Networked MG-based structure of smart distribution grid.

3.2.2 Demand Response Programs

DR programs are potentially powerful programs which lead the electricity companies and costumers toward the economic and environmental benefits [23]. The DR can categorized into two main programs: time-based programs, and incentive-based ones [24]. Time-based DR programs include time of use (TOU), critical peak pricing (CPP), and RTP; whereas emergency DR program (EDRP), direct load control (DLC), interruptible/curtailable (I/C) service and capacity market program (CAP) fall into the category of incentive-based programs.

In this chapter, RTP is considered as a DR program to change the load demand profile within the MGs based on the electricity price changes in any given time. In the proposed RTP program, the prices are determined one hour ahead of the consumption. Therefore, consumers react to the new prices through increasing or decreasing their power consumptions. The RTP is established according to microeconomic theory, in which price-elasticity approximations are generally obtained from the model in [24]. So, elasticity is defined as the demand sensitivity with respect to the price [25] as:

$$E = \frac{C_{l0}}{P_{l0}} \times \frac{\partial P_{l,new}}{\partial C_{l,new}} \tag{3.1}$$

Self-elasticity and cross-elasticity which are negative and positive values respectively can be defined as follows:

$$E(i,j) = \frac{C_{l0}(j)}{P_{l0}(i)} \times \frac{\partial P_{l,new}(i)}{\partial C_{l,new}(j)} \quad \forall i,j \in T \tag{3.2}$$

$$\begin{cases} E(i,j) \leq 0, & \forall i = j, \quad \forall i,j \in T \\ E(i,j) \geq 0, & \forall i \neq j, \quad \forall i,j \in T \end{cases} \tag{3.3}$$

Within the MGs, some loads are fixed and they cannot be shifted from one period to another and are sensitive just in a single period which is called self-elasticity. In the proposed model for DR, loads proportionally react to the price changing. The load variations can be defined as follows:

$$\Delta P_l(i) @ P_{l,new}(i) - P_{l0}(i), \quad \forall i \in T \tag{3.4}$$

The benefits of consumers for each hour given in (3.5) can be maximized by (3.6):

$$S(P_{l,new}(i))@B(P_{l,new}(i)) - P_{l,new}(i) \times C_{l,new}(i), \quad \forall i \in T \tag{3.5}$$

$$\frac{\partial B(P_{l,new}(i))}{\partial P_{l,new}(i)} = C_{l,new}(i) \tag{3.6}$$

A quadratic function based on the second-order Taylor series expansion of $B(Pl,new(i))$ can be used to estimate the benefit function as follows:

$$\begin{aligned} &B(P_{l,new}(i)) = B(P_{l0}(i)) + C_{l0}(i) \times [P_{l,new}(i) - P_{l0}(i)] \times \\ &\left\{1 + \frac{P_{l,new}(i) - P_{l0}(i)}{2E(i,i)P_{l0}(i)}\right\}, \quad i \in T \end{aligned} \tag{3.7}$$

Finally, the single period model for load demands can be achieved by (3.8):

$$P_{l,new}(i) = P_{l0}(i) \times \left\{1 + E(i,i)\frac{C_{l,new}(i) - C_{l0}(i)}{C_{l0}(i)}\right\}, \quad i \in T \tag{3.8}$$

In the second model of DR called the multisensitivity model, loads can be shifted from an interval to others considering the price changes. This model can be calculated through cross elasticity which is mathematically defined as follows:

$$E(i,j), \frac{C_{l0}(j)}{P_{l0}(i)} \times \frac{\partial P_{l,new}(i)}{\partial C_{l,new}(j)} \quad \forall i \neq j, \quad \forall i,j \in T \tag{3.9}$$

In the presence of price variations, the loads in each MG reacts as follows:

$$P_{l,new}(i) = P_{l0}(i) + \sum_{i \neq j} E(i,i)\frac{P_{l0}(i)}{C_{l0}(i)}\left[C_{l,new}(i) - C_{l0}(i)\right], i \in T \tag{3.10}$$

At last, the summation of the single period and multiperiod models results in the final model for the time-based DR:

$$\begin{aligned} &P_{l,new}(i) = P_{l0}(i) \times \left\{1 + E(i,i)\frac{C_{l,new}(i) - C_{l0}(i)}{C_{l0}(i)}\right\} + \\ &P_{l0}(i) \times \left\{1 + \textstyle\sum_{i \neq j} E(i,i)\frac{C_{l,new}(i) - C_{l0}(i)}{C_{l0}(i)}\right\}, i \in T \end{aligned} \tag{3.11}$$

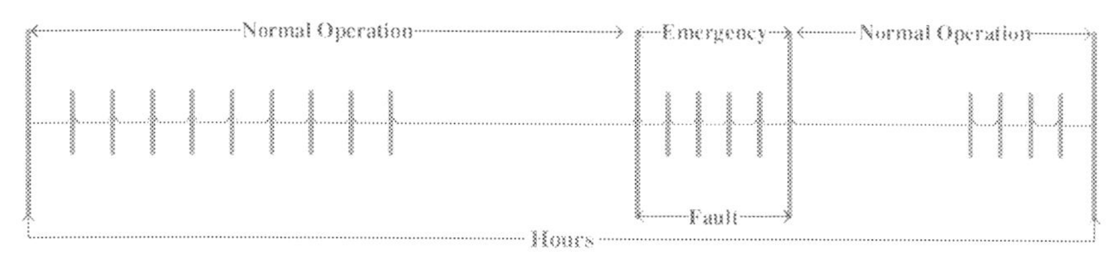

Figure 3.2 Time horizon for MGs operation.

Two aforementioned strategies are utilized to enhance the resilience of MGs after a fault occurs in the network. In this study, the normal operation mode and the emergency mode are used to handle the extreme conditions in the network. Fig. 3.2 illustrates the operation time scheduling for each MG before and after a fault.

According to Fig. 3.2, in the emergency operation mode of MGs the price of electricity in the real time market is at the highest level as compared to the normal operation mode. As a result, the customers have a tendency to shift their potential loads from emergency hours to the hours with lower electricity prices. Due to the discussed potential of the NMGs structure in switching from the interconnected mode to the islanded mode while keeping the connection between MGs and transferring the shiftable loads in emergency hours using DR programs, the proposed schemes could improve the resilience of MGs.

3.3 PROBLEM FORMULATION

3.3.1 Distributed Energy Resources

In this study, WT and PV systems are assumed as nondispatchable sources, while MTs, FCs, and CHPs are dispatchable energy suppliers in the MGs. The uncertainties relating to WTs and PV panels are due to the wind speed and solar radiation variations. To handle the uncertainties of wind speed and solar insolation, two commonly used probability distribution functions (PDFs) termed Weibul [26] and Normal [6] distributions are used in the model. To convert the obtained initial energies to power, formulas in [6] are used for WTs and PVs, respectively.

The generation costs for nondispatchable sources are zero due to the free price of initial energies for WTs and PVs while the mentioned cost for MTs, FCs, and CHPs can be calculated as follows:

$$C^{t}_{g,MT} = \lambda_{g,MT} \times \frac{P^{t}_{g,MT}}{\eta^{t}_{MT}}, \quad \forall t \in T \tag{3.12}$$

$$C_{g,FC}^{t} = \lambda_{g,FC} \times \frac{P_{g,FC}^{t}}{\eta_{FC}^{t}}, \quad \forall t \in T \tag{3.13}$$

$$\lambda_{g,MT} = \lambda_{g,FC} = \frac{C_{nl}}{L} \tag{3.14}$$

$$C_{g,CHP} = C_{g,MT} \times \left(1 - \frac{e_{prec}(\eta_{CHP}^{t} - \eta_{e}^{t})}{\eta_{b}}\right), \quad \forall t \in T \tag{3.15}$$

The maintenance cost of all DGs are defined as follows:

$$C_{OM,WT}^{t} = \lambda_{OM,WT} \times P_{g,WT}^{t}, \quad \forall t \in T \tag{3.16}$$

$$C_{OM,PV}^{t} = \lambda_{OM,PV} \times P_{g,PV}^{t}, \quad \forall t \in T \tag{3.17}$$

$$C_{OM,MT}^{t} = \lambda_{OM,MT} \times P_{g,MT}^{t}, \quad \forall t \in T \tag{3.18}$$

$$C_{OM,FC}^{t} = \lambda_{OM,FC} \times P_{g,FC}^{t}, \quad \forall t \in T \tag{3.19}$$

The generation of dispatchable sources is in the prespecified ranges. Moreover, the ramp rate of generation units is another inequality constraint in each MG.

$$u_{g,MT}^{t} \cdot P_{\min,MT}^{t} \le P_{g,MT}^{t} \le u_{g,MT}^{t} \cdot P_{\max,MT}^{t}, \quad \forall t \in T \tag{3.20}$$

$$u_{g,FC}^{t} \cdot P_{\min,FC}^{t} \le P_{g,FC}^{t} \le u_{g,FC}^{t} \cdot P_{\max,FC}^{t}, \quad \forall t \in T \tag{3.21}$$

$$P_{g,MT}^{t} - P_{g,MT}^{t-1} \le UR_{g,MT}, \text{if } P_{g,MT}^{t} > P_{g,MT}^{t-1}, \quad \forall t \in T \tag{3.22}$$

$$P_{g,FC}^{t} - P_{g,FC}^{t-1} \le UR_{g,FC}, \text{if } P_{g,FC}^{t} > P_{g,FC}^{t-1}, \quad \forall t \in T \tag{3.23}$$

$$P_{g,MT}^{t} - P_{g,MT}^{t-1} \le DR_{g,MT}, \text{if } P_{g,MT}^{t} < P_{g,MT}^{t-1}, \quad \forall t \in T \tag{3.24}$$

$$P_{g,FC}^{t-1} - P_{g,FC}^{t} \le DR_{g,FC}, \text{if } P_{g,FC}^{t} < P_{g,FC}^{t-1}, \quad \forall t \in T \tag{3.25}$$

3.3.2 Load Demand

To describe the load variations, normal PDF is utilized in modeling each MG [4].

3.3.3 Battery Modeling

The batteries are utilized in MGs to store electricity when there is surplus generation. If the output power of DGs in MGs is lower than the total demand, the batteries would begin to discharge. The initial charging batteries are 50% of the total battery capacity [27]. The mathematical model for batteries in each MG can be illustrated using the following relations:

$$0 \leq P^{t}_{BAT,CH,m} \leq P_{BAT,CAP,m} \times (1 - SOC^{t-1}_{BAT,m}) \times \frac{1}{1 - P^{loss}_{BAT,CH,m}},$$
$$\forall t \in T, \quad \forall m \in MG \tag{3.26}$$

$$0 \leq P^{t}_{BAT,DCH,m} \leq P_{BAT,CAP,m} \times SOC^{t-1}_{BAT,m} \times (1 - P^{loss}_{BAT,DCH,m}),$$
$$\forall t \in T, \quad \forall m \in MG \tag{3.27}$$

$$SOC^{t}_{BAT,m} = SOC^{t-1}_{BAT,m} - \frac{1}{P_{BAT,CAP,m}} \times \left(\frac{1}{1 - P^{loss}_{BAT,DCH,m}} \times P^{t}_{BAT,DCH,m} - (1 - P^{loss}_{BAT,CH,m}) \times P^{t}_{BAT,CH,m} \right), \quad \forall t \in T, \forall m \in MG \tag{3.28}$$

$$SOC_{\min,BAT,m} \leq SOC^{t}_{BAT,m} \leq SOC_{\max,BAT,m}, \quad \forall t \in T, \forall m \in MG \tag{3.29}$$

3.3.4 Transaction of Power Among the MGs

One of the important contributions of the proposed EMS is to enable a coordination among the MGs when they cannot fulfill their supply-demand balance. Actually, NMGs provide either technical or economic benefits for MGs to compensate for their deficiencies through sharing their generated energy by local resources with other MGs as well as with the main distribution grid. In this regard, the reliability and sustainability of the power network is enhanced due to the fact that the load interruption decreases considering the environmental issues. Therefore, in this structure, MGs can purchase their required energy from other MGs and the main grid or sell their surplus power to them considering the whole economic and environmental benefits of the network. All the decisions are made by MGCC for each MG. The following constraints are applied to purchased and sold powers:

$$\text{if } P^t_{g,m} - P^t_l > 0 \rightarrow P^t_{pur,m} = 0,\ P^t_{sell,m} > 0,\ \forall t \in T,\ \forall m \in MG \tag{3.30}$$

$$\text{if } P^t_{g,m} - P^t_l < 0 \rightarrow P^t_{pur,m} > 0,\ P^t_{sell,m} > 0,\ \forall t \in T,\ \forall m \in MG \tag{3.31}$$

In sharing power among MGs, the congestion of power in distribution lines should be controlled:

$$0 \leq P^t_{tran,m-n} \leq P^t_{\max,tran,m-n},\ \forall t \in T,\ \forall \{m,n\} \in MG,\ m \neq n \tag{3.32}$$

The costs of purchased and sold powers by each MG can be determined through price coefficients imposed by the real time market.

$$C^t_{pur,m} = \sum_n C^t_{pur,mn} \times P^t_{pur,mn} \quad \forall t \in T,\ \forall \{m,n\} \in \{MG, NW, BAT\},\ m \neq n \tag{3.33}$$

$$C^t_{sell,m} = \sum_n C^t_{sell,mn} \times P^t_{sell,mn} \quad \forall t \in T,\ \forall \{m,n\} \in \{MG, NW, BAT\},\ m \neq n \tag{3.34}$$

The cost difference between the purchased and sold energy provides the cost of power transaction in each MG as follows:

$$C^t_{tran,m} = C^t_{pur,m} - C^t_{sell,m} \quad \forall t \in T,\ \forall m \in MG \tag{3.35}$$

3.4 OBJECTIVE FUNCTION

In this chapter, two different operation modes are defined to handle the unanticipated faults and events in the power network. Before a disturbance occurs in the network, MGs operate in the normal mode, in which MGs can link to each other and the main grid while after a disturbance event, MGs cannot be linked to the main grid and supply their loads from the distribution system; however, MGs can keep their electrical relations with other MGs, which is the most important benefit of the NMGs structure. Therefore, in this section the objective functions of the two modes are discussed.

3.4.1 Normal Operation Mode

In normal performance of MGs, the final objective function contains five terms, namely the costs of generated power by RESs, power transaction through MGs, maintenance, charging and discharging of battery packs, and pollutions. It should be mentioned that the costs related to charging or discharging of batteries are embedded in the power transaction cost.

$$\begin{aligned} &\text{Min } OF_m = \sum\nolimits_t (Cost^t_{op,m} + Cost^t_{em,m}), \quad \forall t \in T, \\ &\forall m \in \{MG1, MG2, MG3\} \end{aligned} \tag{3.36}$$

$$\begin{aligned} &Cost^t_{op,m} = C^t_{g,m} + C^t_{tran,m} + C^t_{OM,m}, \quad \forall t \in T, \\ &\forall m \in \{MG1, MG2, MG3\} \end{aligned} \tag{3.37}$$

$$\begin{aligned} &Cost^t_{em,m} = \sum\nolimits_k \gamma k \times \left(\sum\nolimits_u \rho_{uk} \times P^t_{g,u} \right), \quad \forall t \in T, \\ &\forall m \in \{MG1, MG2, MG3\}, \forall k \in \{CO_2, S\ O_2, NO_x\}, \\ &\forall u \in \{WT, PV, FC, MT, CHP\} \end{aligned} \tag{3.38}$$

The proposed objective function is minimized while respecting some equality and inequality constraints. The inequality constraints were discussed in the previous section. In the case of equality constraints, the total generation power along with the exchanged power and the charge or discharge state of battery packs must meet the predicted load demand of each MG at any given time.

$$\begin{aligned} &P^t_{l,m} = P^t_{g,m} + P^t_{tran,m} + P^t_{CH,m} + P^t_{DCH,m}, \\ &\forall t \in T, \forall m \in \{MG1, MG2, MG3\} \end{aligned} \tag{3.39}$$

The generated powers by MTs, FCs, and CHPs, battery charge or discharge, and power exchange with other MGs and with distribution network are considered as decision variables in each MG in each hour. In order to cope with the uncertainties of WTs, PVs, and loads, Monte Carlo simulation (MCS) method is used. This study will make use of a scenario-based method to cover the uncertainty of the problem [28].

After producing some scenarios for the mentioned uncertain inputs, the system is analyzed under these scenarios as deterministic inputs.

3.4.2 Emergency Operation Mode

In this work in order to cope the natural disturbance in the system, the normal operation mode of MGs will be switched to emergency mode after a fault. In emergency operation mode, MGs have an ability to link other MGs without linking to the distribution network. Therefore, MGs cannot purchase their required energy from the main grid or sell the surplus power to it. In addition, the load curtailment is probable during emergency mode. Nevertheless, due to participating of customers in DR programs from one hand and electrical connection of MGs with each other, the most loads can be survived through shifting to other intervals or being supplied by other MGs. As a result, the cost of load curtailment should be included in the objective function. It should be mentioned that in the emergency period which the MGs operate in islanded mode, MGs cannot have a connection with the main grid. So, the purchased and sold power costs should be updated:

$$C^t_{pur,m} = \sum_n C^t_{pur,mn} \times P^t_{pur,mn} \quad \forall t \in T, \forall \{m,n\} \in \{MG, BAT\}, \quad m \neq n \tag{3.40}$$

$$C^t_{sell,m} = \sum_n C^t_{sell,mn} \times P^t_{sell,mn} \quad \forall t \in T, \forall \{m,n\} \in \{MG, BAT\}, \quad m \neq n \tag{3.41}$$

Then, the objective function can be defined as follows:

$$\text{Min } OF_m = \sum_t (Cost^t_{op,m} + Cost^t_{em,m} + Cost^t_{shed,m}), \quad \forall t \in T, \forall m \in \{MG1, MG2, MG3\} \tag{3.42}$$

$$Cost^t_{op,m} = C^t_{g,m} + C^t_{tran,m} + C^t_{OM,m}, \quad \forall t \in T, \forall m \in \{MG1, MG2, MG3\} \tag{3.43}$$

$$Cost^t_{shed,m} = C^t_{shed} \times P^t_{shed,m}, \quad \forall t \in T, \forall m \in \{MG1, MG2, MG3\} \tag{3.44}$$

3.4.3 Resilience Index

In this research, in order to show the resilience of MGs during the emergency operation mode, a resilience index is deployed which is defined as the amount of load survived during the disturbance event as follows:

$$RI_m^t = 1 - \left(\frac{P_{shed,m}^t}{P_{l,m}^t}\right), \quad \forall t \in T, \quad \forall m \in \{MG1, MG2, MG3\} \tag{3.45}$$

3.4.4 Optimization Procedure

The proposed algorithm solves the problem in two levels: In the first level, each MG separately optimizes the operation costs and schedules its own local generation units by taking the uncertainties of RESs and loads into account. In the second level, after gaining the optimal operation cost of each MG, the MG entities should be coordinated with DNO. In this condition, each MG shares its power with other MGs and the main grid. Then, DNO plays a crucial role in responding to the MGs' requests. Overall, the proposed problem is solved with a bi-level algorithm. In Fig. 3.3, the performance of the proposed EMS is illustrated.

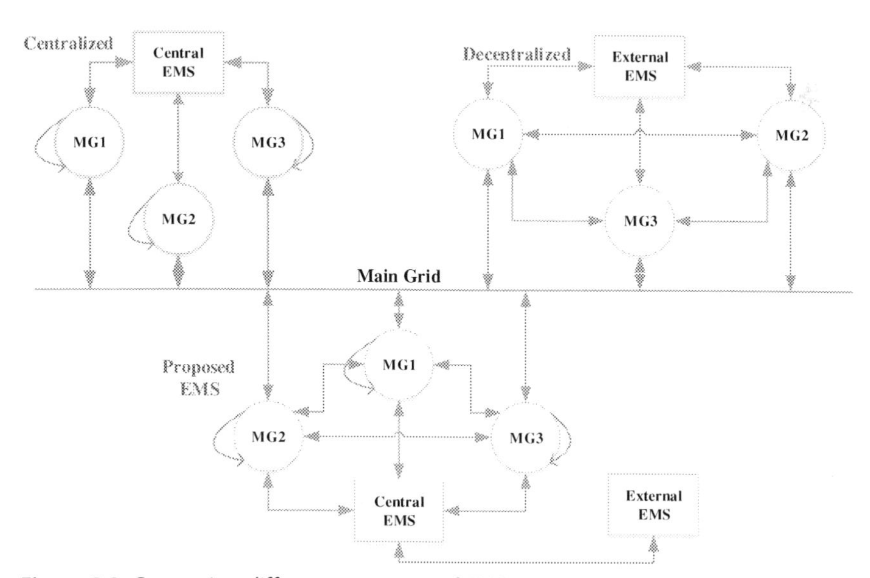

Figure 3.3 Comparing different structures of EMSs.

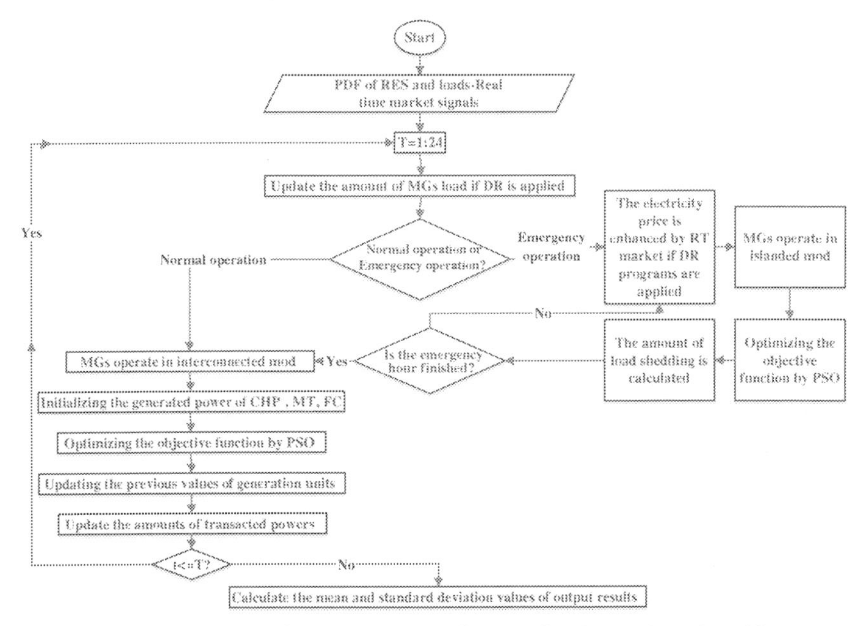

Figure 3.4 The flowchart of implementation for PSO for the proposed problem.

In this work, the particle swarm optimization (PSO) algorithm [29] is used to minimize the objective function regarding various defined constraints. Fig. 3.4 shows the flowchart of the PSO implementation with respect to the proposed problem.

3.5 NUMERICAL RESULTS

In the proposed structure based-on NMGs, there are three MGs with local energy sources and loads. The aim of MG energy management is to minimize the operation cost by taking into account the economic issues using DR programs (DRPs) and environmental constraints. In the case of DR, \$0.034 kWh^{-1} is defined as the flat rate price in electrical energy selling. Moreover, self and cross elasticities are considered as -0.2 and 0.01, respectively. Also, the participants level of loads in any MG can be varied.

The hourly electricity price obtained from the real time market [30] along with the mean value of loads for each MG is shown in Table 3.1. In this work, it is assumed that from 12:00 p.m. to 15:00 p.m. a

Table 3.1 Mean value of MGs' load and consumption prices based on RTP

Hour	Load (kW)			Price	Hour	Load (kW)			Price
	MG1	MG2	MG3	($/kWh)		MG1	MG2	MG3	($/kWh)
1	300.36	277.87	311.08	0.023	13	343.89	344.77	446	0.07
2	287.31	264.17	308.31	0.021	14	374.59	317.35	404.48	0.07
3	283.55	253.42	290.09	0.02	15	346.8	313.31	390.61	0.07
4	305.25	274.16	334.2	0.019	16	372.82	359.38	407.45	0.061
5	273.5	255.15	300.48	0.02	17	458.05	452.04	602.02	0.062
6	272.16	257.67	293.78	0.022	18	518.7	464.33	575.03	0.053
7	337.35	294.58	348.98	0.024	19	567.4	468.67	709.36	0.046
8	371.8	333.41	389.42	0.026	20	572.66	422.92	573.83	0.04
9	431.08	372.55	519.1	0.028	21	494.51	451.99	581.53	0.037
10	466.28	418.85	542.73	0.033	22	438.72	423.91	540.36	0.032
11	348.19	328.19	415.65	0.038	23	414.65	359.97	476.6	0.027
12	364.7	325.97	414.86	0.04	24	335.92	337.86	423.7	0.026

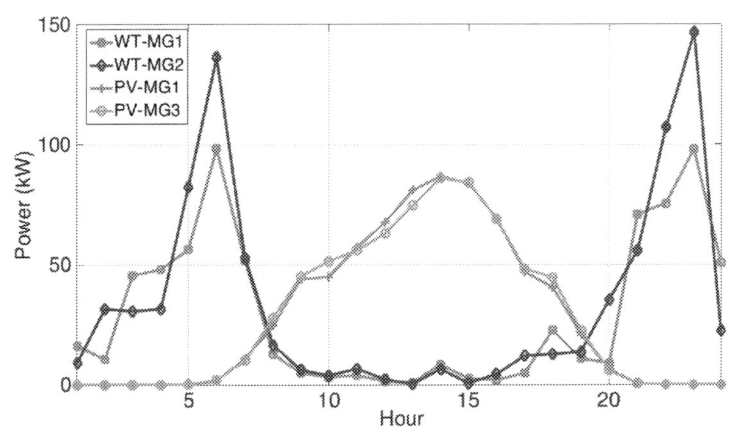

Figure 3.5 Generated power of WTs and PVs based on mean value.

disturbance occurs, so the electricity price in this interval is modified which is considered \$0.07 kWh^{-1}.

In this work, two different scenarios are implemented. In the first scenario, the proposed problem and objective function is optimally solved without considering DR programs while the second scenario considers the effects of RTP program on the resilience of MGs. In both scenarios, MGs operate in normal operation mode before a disturbance event in the network while after a fault the operation of MGs is switched to emergency operation and after finishing the event, the MGs return to normal mode again.

The amount of generated power by nondispatchable energy sources (WTs and PVs) is illustrated in Fig. 3.5 based on the mean value.

The load profile of MGs can be observed in Fig. 3.6 concerning RTP program and without DR. In Fig. 3.6 when the RTP program is applied to the original load curve, the electricity consumption reacts to the price variations in each hour. In valley times, the amount of load increases and in high rate price times, there is less tendency of power consumption. During the emergency condition, due to the high rate price comparing to other intervals, the customers are not eager to consume energy. After gaining the load curve for each MG, the PSO algorithm tries to minimize the cost function.

After optimizing the objective function by PSO algorithm, the outputs of the power scheduling problem for a 24-hour period are shown in Table 3.2 based on the mean value. In the table, the generated power by

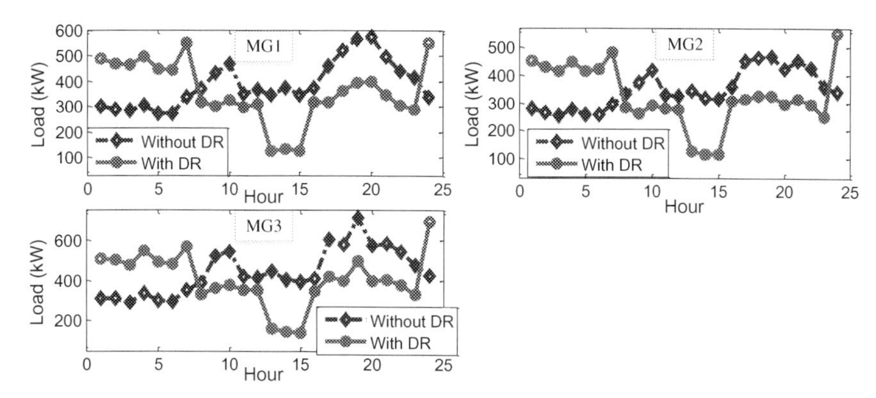

Figure 3.6 Comparing the load profile with and without considering DR in 3 MGs based on mean value.

each MG as well as the traded power with different MGs and main grid can be found, in which the results obtained by RTP are compared to the outputs of typical solution without using DR program.

As it is seen in Table 3.2, the gained results for generated and transacted powers are optimal outputs in 24 hours period scheduling. The negative numbers in transacted power describe the fact that a MG tends to sell its power to other MGs, or main grid, or charge the battery complex to meet the supply-demand balance in any given time.

The commitment of each generation unit belongs to the MGs is shown in Fig. 3.7. Besides, the amount of transaction of power by all MGs is illustrated in Fig. 3.8. A it is observed in Fig. 3.8, the most MGs during emergency hours are able to sell their generated power after participating in DR programs, because the load level of MGs has significantly been decreased.

In Table 3.3, the operation cost is described according to the mean value in a 24 hours period for two scenarios.

According to the Table 3.3, the total reduction for operation cost using RTP is significant in an emergency period. According to this table, participating customers in the DR programs cause a significant reduction in operation costs of MGs during emergency conditions. After occurring a fault in the network and considering the effects of DR, the load profile of all MGs is modified regarding great changes in the real time price. While, without considering DR programs, the customers continue their consumption even after a fault in the network.

Table 3.2 Optimal results of networked MGs' power scheduling in 24 h period based on mean value in various scenarios

Hour	MG1				MG2				MG3			
	Generated power (kW)		Transacted power (kW)		Generated power (kW)		Transacted power (kW)		Generated power (kW)		Transacted power (kW)	
	No DR	With DR	No DR	With DR	No DR	With DR	No DR	With DR	No DR	With DR	No DR	With DR
1	164	195	137	295	283	248	−6	205	276	230	35	277
2	156	156	131	312	258	240	7	190	277	204	32	299
3	233	216	50	246	273	264	−20	149	272	217	18	256
4	195	237	110	261	329	267	−55	180	264	254	70	291
5	230	190	44	256	354	296	−99	120	241	255	59	235
6	262	236	11	208	336	343	−78	78	257	219	37	260
7	256	286	82	264	301	307	−7	173	247	312	102	257
8	213	188	159	128	232	255	101	28	277	227	112	104
9	256	246	175	53	222	216	151	43	272	315	247	45
10	179	221	287	103	312	199	107	92	266	265	277	112
11	185	251	163	45	221	218	108	61	253	303	162	50
12	250	282	114	28	260	231	66	46	265	323	150	29
13	234	265	25	-29	299	240	12	−23	311	281	34	−26
14	268	269	40	−33	277	206	5	−22	370	338	6	−30
15	241	225	37	−24	264	226	20	−25	396	338	−23	−33
16	252	229	121	88	264	226	96	80	359	332	48	15
17	211	205	247	114	233	255	219	59	280	313	322	106
18	214	260	305	101	232	217	232	106	234	290	341	110
19	177	198	390	196	263	226	206	100	299	295	411	199
20	178	191	395	207	315	241	108	53	248	247	326	152
21	223	259	272	85	301	294	151	21	239	269	343	135
22	238	238	201	67	316	327	108	−33	230	233	311	143
23	257	304	157	−16	387	313	−27	−63	282	235	195	96
24	191	215	145	333	272	277	65	274	215	220	209	471

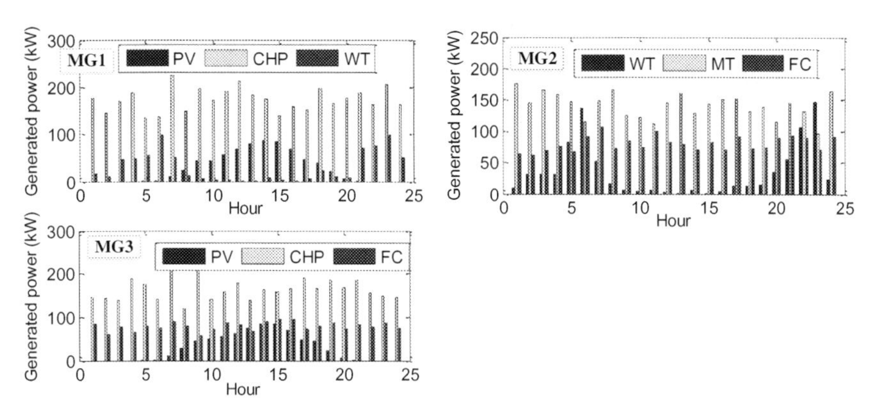

Figure 3.7 Generation units commitment in each MG based on mean value considering DR program.

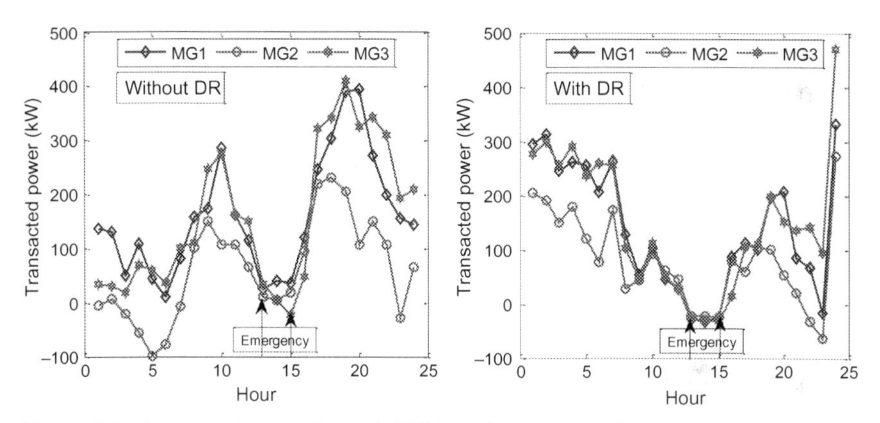

Figure 3.8 Transacted power in each MG based on mean value.

The load consumption cost can be seen in Fig. 3.9. The huge load cost reduction is due to the reaction of the customers to high rate price during the emergency hours, while the customers are indifferent to the emergency conditions when they ignore the DR programs.

To compare the results obtained by the PSO algorithm, stochastic optimization is utilized to show the accuracy of the results. In Fig. 3.10, the summation costs of operation and pollution of all MGs indicate the total cost of the network in any given time.

In Table 3.4 the value of objective function obtained by PSO is compared with stochastic optimization in two scenarios based on mean and standard deviation values.

Table 3.3 Operation cost of MGs with and without considering DR

Hour	MG1			MG2			MG3		
	Without DR($/h)	With DR($/h)	Reduction (%)	Without DR($/h)	With DR($/h)	Reduction (%)	Without DR($/h)	With DR($/h)	Reduction (%)
1	41.48	80.96	0.00	64.47	98.23	0.00	44.87	88.59	0.00
2	41.33	79.88	0.00	58.95	90.89	0.00	41.65	87.67	0.00
3	33.79	73.33	0.00	57.34	88.07	0.00	37.17	80.79	0.00
4	39.31	78.19	0.00	64.76	94.54	0.00	47.33	90.19	0.00
5	31.08	70.92	0.00	56.03	83.82	0.00	39.48	81.18	0.00
6	27.95	67.23	0.00	49.34	77.72	0.00	41.72	82.54	0.00
7	43.32	84.94	0.00	63.46	99.72	0.00	50.44	92.93	0.00
8	55.77	46.44	16.73	73.54	63.62	13.49	57.65	51.50	10.68
9	65.34	36.74	43.78	84.82	58.16	31.43	91.22	46.84	48.65
10	83.97	45.57	45.73	95.43	65.96	30.89	95.31	58.49	38.63
11	53.43	36.03	32.56	73.47	61.43	16.39	66.03	49.46	25.10
12	51.22	35.10	31.47	75.04	62.20	17.12	63.64	45.51	28.49
13	172.42	23.26	86.51	128.64	51.99	59.58	214.23	29.80	86.09
14	148.03	21.74	85.31	125.14	42.75	65.84	98.54	37.64	61.80
15	149.55	19.81	86.75	115.65	47.55	58.88	77.46	37.75	51.27
16	52.69	41.75	20.75	80.32	68.34	14.91	54.08	46.23	14.52
17	78.02	45.19	42.08	100.88	71.26	29.37	109.33	60.27	44.87
18	92.30	49.39	46.49	104.59	72.55	30.64	107.79	59.98	44.35
19	107.36	64.84	39.61	105.34	71.81	31.83	129.65	81.46	37.17
20	107.44	65.24	39.28	93.18	64.26	31.04	103.32	63.83	38.22
21	85.25	45.36	46.79	97.08	66.68	31.31	106.36	64.57	39.29
22	70.18	39.85	43.22	85.07	58.30	31.47	101.14	58.88	41.78
23	60.84	28.92	52.46	68.14	42.40	37.77	79.58	51.08	35.81
24	50.95	90.58	0.00	74.17	118.37	0.00	70.80	125.91	0.00

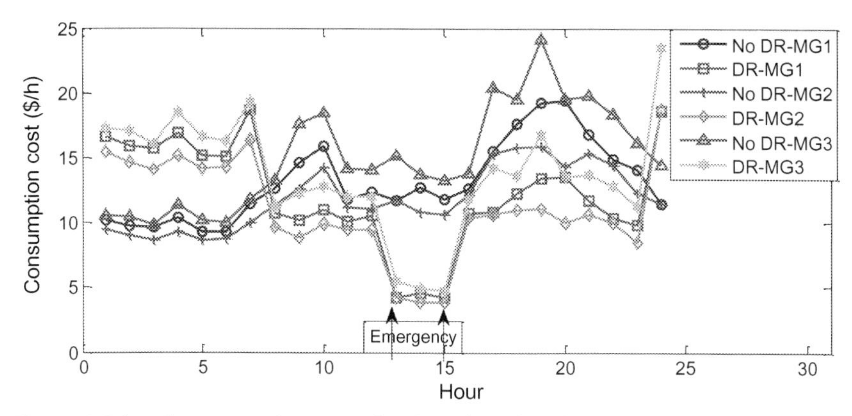

Figure 3.9 Load consumption cost of MGs with and without considering DR.

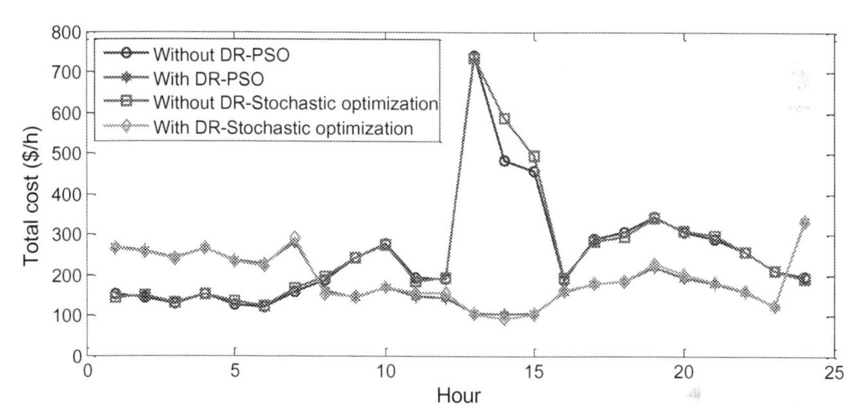

Figure 3.10 The value of objective function for each hour considering different scenarios.

Table 3.4 The value of cost function obtained by two methods in different scenarios

Method	Mean		Standard deviation	
	Without DR	With DR	Without DR	With DR
PSO	6135.5	4583.3	167.49	64.21
Stochastic optimization	6282.9	4601.5	188.19	65.24

In Table 3.5, the resilience index is analyzed during the emergency interval in two scenarios.

According to Table 3.5, the impacts of DR programs are appreciable, and without using DR programs, the total percentages of loads in MG1, MG2, MG3 which cannot be survived are approximately 24, 10, and 22

Table 3.5 Resilience index during emergency condition in different scenarios

Hour	MG1		MG2		MG3	
	Without DR	With DR	Without DR	With DR	Without DR	With DR
13	0.763	1	0.901	1	0.778	1
14	0.831	1	0.889	1	0.920	1
15	0.796	1	0.904	1	0.950	1

for hour 13; while if the customers participate in the DR program, the amount of load shedding is zero. It should be mention that with increasing amounts of loads, the loads that cannot survive will increase.

3.6 CONCLUSION

In this work, two important strategies are developed to enhance the resilience of MGs after a disturbance event in the power network. In the proposed EMS for NMGs, MG entities can operate both in islanded and interconnected modes while taking the uncertainties of RESs and loads into account. In addition, RTP as an efficient DR program plays a significant role in improving the resilience considering the economic aspects of the MGs especially after the occurrence of a natural disaster in the network. In this research, MGs serve in normal operation mode before and after a disturbance event in which MGs can operate in either islanded or interconnected mode based on the economic consideration. During the disturbance, MGs enter into an emergency condition in which they lose their links with the main grid and operate in the islanded mode; however, MGs can connect to other MGs to share energy and maintain their own supply-demand balance. In the proposed procedure for the optimization, PSO algorithm solves the problem considering the uncertainties through generating different scenarios by the MCS method. The objective function minimized by PSO is compared with the stochastic optimization.

REFERENCES

[1] K.P. Schneider, F.K. Tuffner, M.A. Elizondo, C.C. Liu, Y. Xu, D. Ton, Evaluating the feasibility to use microgrids as a resiliency resource, IEEE Trans. Smart Grid 8 (2) (2017) 687–696.

[2] Framework for establishing critical infrastructure resilience goals, Nat. Infrastruct. Advisory Council, USA, 2010.

[3] N. Nikmehr, S.N. Ravadanegh, Reliability evaluation of multi-microgrids considering optimal operation of small scale energy zones under load-generation uncertainties, Int. J. Electr. Power .Energy Systems 78 (2016) 80–87.

[4] N. Nikmehr, S.N. Ravadanegh, A study on optimal power sharing in interconnected microgrids under uncertainty, Int. Trans. Electr. Energy Systems 26 (1) (2016) 208–232.

[5] A. Hussain, V.H. Bui, H.M. Kim, A resilient and privacy-preserving energy management strategy for networked microgrids, IEEE Trans. Smart Grid PP (99) (2016) 1.

[6] N. Nikmehr, S.N. Ravadanegh, Optimal power dispatch of multi-microgrids at future smart distribution grids, IEEE Trans. Smart Grid 6 (4) (2015) 1648–1657.

[7] Z. Wang, B. Chen, J. Wang, M.M. Begovic, C. Chen, Coordinated energy management of networked microgrids in distribution systems, IEEE Trans. Smart Grid 6 (1) (2015) 45–53.

[8] Z. Wang, B. Chen, J. Wang, C. Chen, Networked microgrids for self-healing power systems, IEEE Trans. Smart Grid 7 (1) (2016) 310–319.

[9] C. Chen, J. Wang, F. Qiu, D. Zhao, Resilient distribution system by microgrids formation after natural disasters, IEEE Trans. Smart Grid 7 (2) (2016) 958–966.

[10] S. Maharjan, Y. Zhang, S. Gjessing, O. Ulleberg, F. Eliassen, Providing microgrid resilience during emergencies using distributed energy resources, in: 2015 IEEE Globecom Workshops (GC Wkshps), 2015, pp. 1–6.

[11] A. Gholami, T. Shekari, F. Aminifar, M. Shahidehpour, Microgrid scheduling with uncertainty: the quest for resilience, IEEE Trans. Smart Grid 7 (6) (2016) 2849–2858.

[12] Z. Wang, J. Wang, Service restoration based on ami and networked mgs under extreme weather events, IET Gen. Transm. Distribut. 11 (2) (2017) 401–408.

[13] A. Castillo, Microgrid provision of blackstart in disaster recovery for power system restoration, in: 2013 IEEE International Conference on Smart Grid Communications (SmartGridComm), 2013, pp. 534–539.

[14] P. Siano, Demand response and smart grids—a survey, Renew. Sustain. Energy Rev. 30 (Suppl. C) (2014) 461–478.

[15] N. Nikmehr, S. Najafi-Ravadanegh, A. Khodaei, Probabilistic optimal scheduling of networked microgrids considering time-based demand response programs under uncertainty, Appl. Energy 198 (Suppl. C) (2017) 267–279.

[16] X. Yang, X. He, J. Lin, W. Yu, Q. Yang, A novel microgrid based resilient demand response scheme in smart grid, in: 2016 17th IEEE/ACIS International Conference on Software Engineering, Artificial Intelligence, Networking and Parallel/Distributed Computing (SNPD), 2016, pp. 337–342.

[17] H. Gao, Y. Chen, Y. Xu, C.C. Liu, Resilience-oriented critical load restoration using microgrids in distribution systems, IEEE Trans. Smart Grid 7 (6) (2016) 2837–2848.

[18] S.D. Manshadi, M.E. Khodayar, Resilient operation of multiple energy carrier microgrids, IEEE Trans. Smart Grid 6 (5) (2015) 2283–2292.

[19] E. Pashajavid, F. Shahnia, A. Ghosh, Development of a self-healing strategy to enhance the overloading resilience of islanded microgrids, IEEE Trans. Smart Grid 8 (2) (2017) 868–880.

[20] A. Hussain, V.H. Bui, H.M. Kim, Optimal operation of hybrid microgrids for enhancing resiliency considering feasible islanding and survivability, IET Renew. Power Gen. 11 (6) (2017) 846–857.

[21] D.E. Olivares, C.A. Cañizares, M. Kazerani, A centralized optimal energy management system for microgrids, in: 2011 IEEE Power and Energy Society General Meeting, 2011, pp. 1–6.

[22] B.M. Radhakrishnan, D. Srinivasan, A multi-agent based distributed energy management scheme for smart grid applications, Energy 103 (2016) 192–204.
[23] A.H. Mohsenian-Rad, A. Leon-Garcia, Optimal residential load control with price prediction in real-time electricity pricing environments, IEEE Trans. Smart Grid 1 (2) (2010) 120–133.
[24] P.C. Reiss, M.W. White, Household electricity demand, revisited, Rev. Econ. Stud. 72 (3) (2005) 853–883.
[25] D.S. Kirschen, G. Strbac, Fundamentals of Power System Economics, John Wiley & Sons, Chichester, 2004.
[26] N. Nikmehr, S. Najafi-Ravadanegh, Optimal operation of distributed generations in micro-grids under uncertainties in load and renewable power generation using heuristic algorithm, IET Renew. Power Gen. 9 (8) (2015) 982–990.
[27] S. Wang, Z. Li, L. Wu, M. Shahidehpour, Z. Li, New metrics for assessing the reliability and economics of microgrids in distribution system, IEEE Trans. Power Systems 28 (3) (2013) 2852–2861.
[28] R. Jabbari-Sabet, S.-M. Moghaddas-Tafreshi, S.-S. Mirhoseini, Microgrid operation and management using probabilistic reconfiguration and unit commitment, Int. J. Electr. Power Energy Systems 75 (2016) 328–336.
[29] J. Kennedy, Particle swarm optimization, Encyclopedia of Machine Learning, Springer, New York, 2011, pp. 760–766.
[30] Hourly ameren corporation energy price, https://www2.ameren.com/RetailEnergy/rtpDownload.

CHAPTER 4

The Use of Hybrid Neural Networks, Wavelet Transform and Heuristic Algorithm of WIPSO in Smart Grids to Improve Short-Term Prediction of Load, Solar Power, and Wind Energy

Naser Nourani Esfetanaj[1] and Sayyad Nojavan[2]
[1]Sahand University of Technology, Tabriz, Iran
[2]University of Tabriz, Tabriz, Iran

4.1 INTRODUCTION

With the ever-increasing electricity demand throughout the world, fossil fuel reduction and higher level emissions and the related costs, electricity suppliers have been shifting toward green energy alternatives (i.e., renewable energies). Wind energy and solar power are of great importance among renewable energy resources as they are in abundance and have relatively lower costs. However, one of the main difficulties regarding power production from these energy resources is their uncertainty as well as the corresponding power production from their power plants. Electricity cannot be stored in large-scale and power production/distribution management should plan, operate, and invest in an optimized way based on electricity supply/demand. Thus, in a future planning of the system, prediction of load and solar/wind power production are indispensable and the related prediction error should be reduced [1]. Required data for optimum planning in smart distribution systems is available through predictions of electricity demand, solar as well as wind power production in given intervals. Therefore, the first and most important step in smart distribution system planning is to have sufficient and complete data of electricity demand and its logical growth prediction with regard to different influential factors in electricity consumption as well as to understand the

Operation of Distributed Energy Resources in Smart Distribution Networks
DOI: https://doi.org/10.1016/B978-0-12-814891-4.00004-7

amount of power produced by distributed generations (DGs), i.e., solar and wind power, in smart systems. Obviously, because any types of decision-making in this regard depends highly upon having related information about electricity load in different places/times in the system as well as understanding the amount of wind/solar power. Electricity consumption is a complex and nonlinear function of various parameters such as meteorological conditions (i.e., temperature, humidity, illumination, wind velocity, etc.). Additionally, there exists a specific load curve for each day in a week. Load consumption curves are also different for weekdays and weekends. And, by the time, due to population and economic growth of the community, load demand increases; otherwise, it may reduce via optimization of electric systems and economic measures taken by main electricity consumers. Another barrier for integration of solar/wind powers to the smart distribution system is unavailability of power production and its fast changes. As the output power from solar/wind power plants depend highly on peripheral conditions, prediction of output power of them is along with errors. This type of error affects system performance. Therefore, the less the amount of prediction error, the less the amount of negative impact of power unbalance due to solar/wind power [2,3]. Predictions are often completed in four time intervals: very short-term, short-term, mid-term, and long-term intervals. Physical and statistical methods are considered as the main and basic ones that have been widely reported in the literature. The former uses physical data of the region where solar/wind farms are positioned to construct prediction model. However, the latter usually uses historical data to construct a model, and the prediction accuracy will be varying considering the model parameters. Statistical methods are divided into two groups: time-series and artificial neural networks. Time-series method is applicable for a short-time interval and artificial neural networks are commonly used to predict power, wind velocity, and solar power. Compared to time-series, having no complicated mathematical relationships for model description and higher learning ability are among benefits of neural networks. Considering nonlinear relationship between historical data and power as well as wind velocity, we should use methods with higher capability in modeling nonlinear relationship between inputs and outputs for the purpose of accurate prediction. Neural networks are among the methods that are able to construct a nonlinear model through the learning process. Besides the abovementioned methods, the combination of above methods and meta-heuristics are recently reported in the literature. The aim of combining prediction methods is to improve the accuracy of predictions by taking the advantage of merits of the available techniques. Considering that each method may be sensitive to some conditions, thus the use of other methods may reduce the risk during an unexpected event. A complete literature review can be found in Refs. [4,5]. Authors in [6,7] considered hourly data and proposed WNN, a weighted nearest neighbor approach for simultaneously predicting a 24-hour values for the next day. To classify a large input data set, for the prediction of load demand, a fuzzy logic load forecasting model ANN is generally developed. Kuihe Yang et al., have proposed a method to simplify system structure and enhance forecasting precision [8]. In addition, Huang et al. [9] proposed a moving average procedure with particle swarm optimization (PSO) to cope with a highly nonlinear demand profile by establishing a short history knowledge about the demand trends. Dynamic and hybrid methods were also proposed to deal with highly nonlinear short term forecasting in power generation with solar photovoltaic (Almonacid et al. [10] and load demand Hu et al. [11]). Due to a nonlinear relation between historical data and wind's speed and power, methods which are able to model nonlinear relationship between inputs and outputs should be utilized for precise predictions [12,13]. Neural networks are among the methods which are able to develop nonlinear models through a learning process [12]. In addition to the aforementioned methods, recently, hybrid methods and metaheuristics as well as novel techniques have been reported in literature [14,12]. PV generation forecast methods can be broadly classified into four approaches, i.e., statistical approach, artificial intelligence (AI) approach, physical approach, and hybrid approach. Statistical approaches are based on data-driven formulation using historical measured data to forecast solar time series [15]. AI approaches utilize advanced AI techniques, such as artificial neural networks (ANNs), to construct solar forecasters, which can be further classified into the category of the statistical approach [16]. This paper proposes a novel hybrid method to reduce prediction error. Considering that historical data are used in statistical methods, thus, we are dealing with a data-mining problem. Data-mining comprises of three stages: data preparation, model learning, and model description and interpretation. The main step in data preparation is data extraction from data resources. Here, a complete set of meteorological conditions are used as the fundamental data. The second step is to process the extracted data. So, different methods are employed to prepare data for learning a model. For this purpose, data normalization methods and wavelet transform are utilized. In the learning stage, using various algorithms and considering data nature, it is tried to determine data arrangement and then propose in a given format as a hidden knowledge in data. For data learning, we should use appropriate methods. Accordingly, a proper predictor is employed to improve the result and help the main predictor. Radial networks such as RBF can be used as an elementary predictor because they have special capability in modeling nonlinear relationships and finding local solutions. For the main predictor, one can use combination of several neural networks such as MLP that makes use of various methods for the learning purpose. MLP neural networks are able to find global solutions and model nonlinear behavior. The combination of RF and MLP leads to coverage of all global/local solutions. Metaheuristics have higher speed and capability in finding global solutions.

4.2 HYBRID NEURAL NETWORK

Artificial neural network, operating on the basis of biological neurons characteristics, consists of interconnected processing elements that can be considered as an alternative for conventional computational techniques. Among the most important features of these networks, one can refer to finding available complicated mapping between input-output without difficult programming and extracting linear/nonlinear relationships between available data through the training process. Furthermore, these networks have parallel structures and thus they can complete processing a task very fast. The main capability of neural networks, leading to its widespread employment, is pattern recognition and learning input-output relationship. A simple neural network is comprised of several neurons which are connected to each other. The structure of a neuron is depicted in Fig. 4.1. As seen, each neuron consists of a number of inputs (scalars of

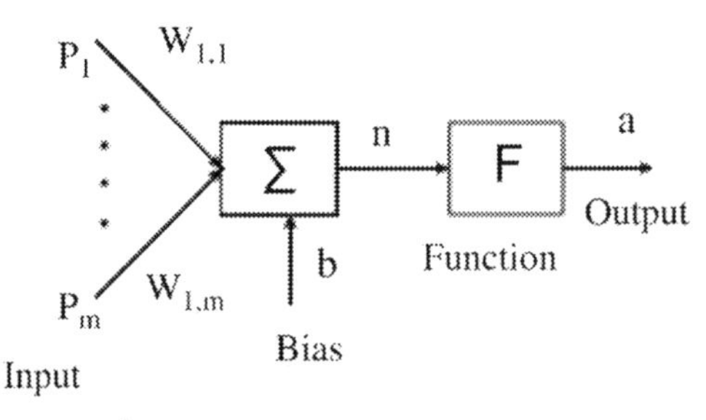

Figure 4.1 The structure of a neuron.

p1...pm), an output (1), a biased sentence, and a stimulation function (F). The amount of impact by each inputs on the output is determined through their $n = \sum_{i=1}^{m} P_1 W_{1i} + b$ corresponding scalar "w" called weight. The neuron output is obtained by $a = F(n)$ where. Parameters w and b are adjustable. The stimulation function "F" is defined by the designer which can be linear/nonlinear and its type is determined based on the specific requirement of the problem. Parameters w and b are adjusted according to the type of stimulation function "F" and the type of learning (training) algorithms. In fact, training in a neural network means that parameters w and b are modified in a way that the relationship between neuron's input and output is accommodated with a specific goal. This process implementation is possible with simple computer programming [2]. After initialization of weights and biases, network training is done. The network may be used for approximation of functions, pattern recognition, and/or pattern classification. Training procedure requires a number of expected examples of the network behavior, including the network and the goal. During training procedure, weights and biases are adjusted in order to minimize performance function (prediction error) of the network. Neural network training is continuously done until stabilizing network weights and reaching to an acceptable minimum error.

Hybrid networks are comprised of a combination of several neural networks to increase performance, accuracy, as well as performance integration of various networks (shown in Fig. 4.1). The provided network in this research consists of three back-propagation neural networks each of which has one hidden layer with ten neurons and internal layer transfer function "tansig." The first network is back-propagation network is trained by Marquardt-Levenberg(ML), the second back-propagation network is trained by Broyden, Fletcher, Goldfarb, Shanno(BFGS), and the third network is back-propagation network trained by (BR) Bayesian regularization. For initial data prediction, a nonlinear predictor of radial basis function (RBF) is used due to the nonlinear nature of data. This RBF network performs an initial data prediction first, and the initial prediction is then employed to the first network as an input. In the next steps, being trained, the first network completes prediction and the prediction is employed to the second neural network along with final weights. The second network also starts training with first network's weights and uses the same method to train the network. Finally, it provides a prediction from a goal of interest too. This prediction is given together with final

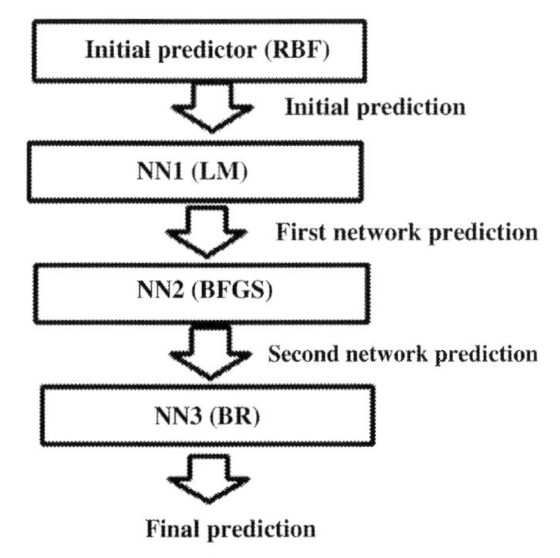

Figure 4.2 Hybrid neural network [1].

weights to the third neural network as an input. After starting training task with second network weights, the third network is trained with the mentioned method and finally provides a prediction of the goal of interest used for final prediction to get network error [1,17] (Fig. 4.2).

4.3 PARTICLE SWARM OPTIMIZATION ALGORITHM

PSO algorithm is inspired by bird's flying pattern. Two aspects have been considered in modeling available arrangement in swarm movement: one aspect is social interaction among swarm members, and the other one is individual privilege. In the first aspect, all members of the swarm are required to modify their direction following the best individual in the group. In the second aspect, each of the individuals are required to store their best individual position and simultaneously move toward their best experienced position. In objective optimization, finding the optimum solution is based on problem variables. An array of problem variables to be optimized is developed, that are named particles. In optimization of N_{var}-dimension problem, a particle is a row array with N_{var} arrays given by:

$$\text{particle} = [p_1, p_2, \ldots, p_{N\text{var}}] \tag{4.1}$$

For algorithm initialization, a number of particles should be created. Thus, total particles matrix is constructed by random.

$$\text{particle} = \begin{bmatrix} \text{particle}_1 \\ \text{particle}_2 \\ \vdots \\ \text{particle}_N \end{bmatrix} = \begin{bmatrix} p_{1,1}, p_{2,1}, \ldots, p_{N_{\text{var}},1} \\ p_{1,2}, p_{2,2}, \ldots, p_{N_{\text{var}},2} \\ \vdots \\ p_{1,N}, p_{2,N}, \ldots, p_{N_{\text{var}},N} \end{bmatrix} \tag{4.2}$$

Cost of each particle is obtained by evaluation of function "f" in terms of $p_1, p_2, \ldots, p_{N_{\text{var}}}$. Thus,

$$\text{cost}_i = f(p_1, p_2, \ldots, p_{N_{\text{VAR}}})i = 1, 2, 3, \ldots N \tag{4.3}$$

A particle with the lowest cost is considered as the best global solution. It should be mentioned that initial velocity for each particle is constructed randomly.

$$V = \begin{bmatrix} v_1 \\ v_2 \\ \vdots \\ v_N \end{bmatrix} = \begin{bmatrix} v_{1,1}, v_{2,1}, \ldots, v_{N_{\text{var}},1} \\ v_{1,2}, v_{2,2}, \ldots, v_{N_{\text{var}},2} \\ \vdots \\ v_{1,N}, v_{2,N}, \ldots, v_{N_{\text{var}},N} \end{bmatrix} \tag{4.4}$$

Once initial population is created and an initial velocity is considered for each particle, each particle should be calculated based on its position. Each particle modifies its velocity based on the best solution obtained in the swarm and the best past position. In time unit variations, this velocity is added to the particle position and thus a new position of the particle is achieved. As seen in Fig. 4.3, particle's velocity in each step is calculated by (4.5) and particle position is updated.

$$v_{k+1}^i = wv_k^i + c_1.r_1.(p_k^i - x_k^i) + c_2.r_2.(p_k^g - x_k^i) \tag{4.5}$$

$$x_{k+1}^i = x_k^i + v_{k+1}^i \tag{4.6}$$

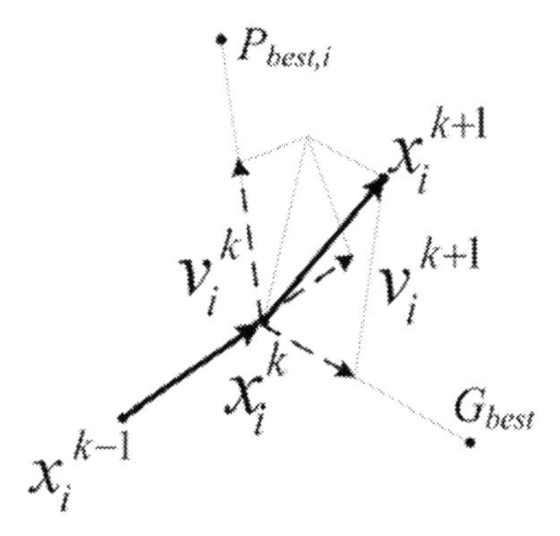

Figure 4.3 Particle's position and velocity updating process.

where r_1 and r_2 are random numbers in [0, 1]. p_k^g, p_k^i are the best position obtained by particle "*p*" and the best solution achieved so far, respectively.

Factors r_1 and r_2 are called learning factors, denoting handling amount of particle from the best individual experience and swarm experiences, respectively. Totally, the value of these factors should be less than four. W is the weighting inertia. Higher values of W lead to broadening of search space regardless of individual and swarm experiences; on the other hand, lower weighting inertia results in contracted search space around current position. After obtaining new velocity, each particle moves toward its new position. If a particle obtains a position better than the best past position, this position will be considered as the best individual position. In addition, if this solution is better than the best position in previous steps, it will be considered as the best global solution. In order to avoid of early divergence or convergence, the maximum and minimum velocities are limited. [18,19].

1. **PSO**

$$w = 1 \tag{4.7}$$

2. **Modified PSO algorithm (MPSO) relationships are as follows** [20]:

$$w = (w_{\max} - (((w_{\max} - w_{\min}) * it)/\text{max}it)) \tag{4.8}$$

it = iteration in this step, and max it = max iteration.

3. **Weight Improved PSO (WIPSO)** [21]:

$$w = (w_{\max} - (((w_{\max} - w_{\min}) * it)/\text{max}it)) \tag{4.9}$$

$$w = w_{\min} + w * rand \tag{4.10}$$

$$c_1 = c_{1\max} - (((c_{1\max} - c_{1\min})/\text{max}it) * it) \tag{4.11}$$

$$c_2 = c_{2\max} - (((c_{2\max} - c_{2\min})/\text{max}it) * it) \tag{4.12}$$

Parameters of PSO, MPSO, and WIPSO algorithms are expressed in "Table 4.1." Conventional PSO algorithms have constant weights during optimization process; while, modified evolutionary algorithms fulfill particles updating process through a method based on improving weighting inertia. This leads to finding appropriate positions and, in turn, optimal solutions compared to the original algorithm. For example, this updating process for WIPSO is done in relationships (4.9)–(4.12) and in each step, inertia weight depends on maximum iteration of the algorithm, current

Table 4.1 Parameters of PSO, MPSO, and WIPSO algorithms

Parameters	PSO	MPSO	WIPSO
Max iteration	100	100	100
Number of population	200	200	200
Initial weight	1	1	1
C_1	2	2	–
$C_{1\min}$	–	–	1.5
$C_{1\max}$	–	–	2.2
C_2	2	2	–
$C_{2\min}$	–	–	1.5
$C_{2\max}$	–	–	2.2
w_{damp}	0.99	–	–
$w_{\min}$	–	0.4	0.3
$w_{\max}$	–	0.9	1.2

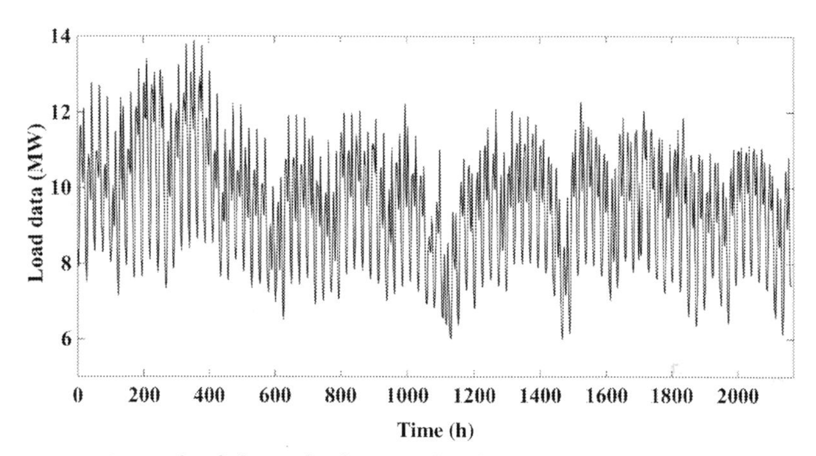

Figure 4.4 A sample of electric load curve related to summer 2013.

iteration, maximum and minimum inertia weights in order to complete population update (Table 4.1).

4.4 DATA SELECTION

The first step in electric load prediction is gathering historical loads of the system under study. Because much of referenced research made use of national-wide (Iran) data and there existed no standard system for simulation purposes, thus, in this paper, we used load data of a smart distribution system of Southern Alberta, Canada for recent years. A sample curve of electric load, solar power, and wind energy of this system for a 3-month period in 2013 are depicted in Figs. 4.4–4.6, respectively [29,31].

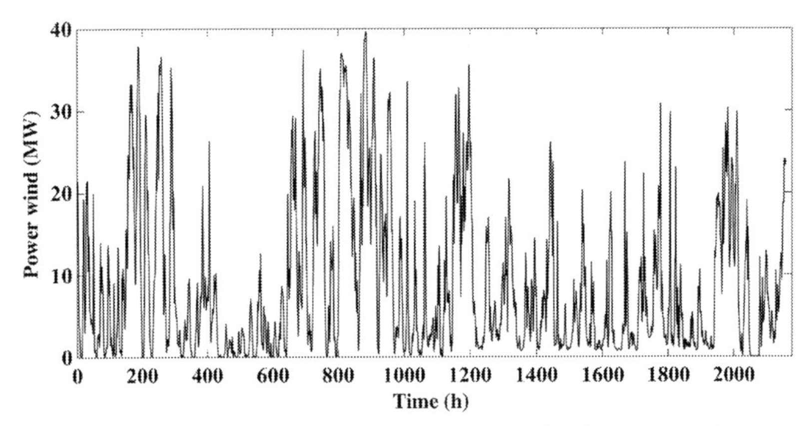

Figure 4.5 A sample of wind power generation curve related to summer 2013.

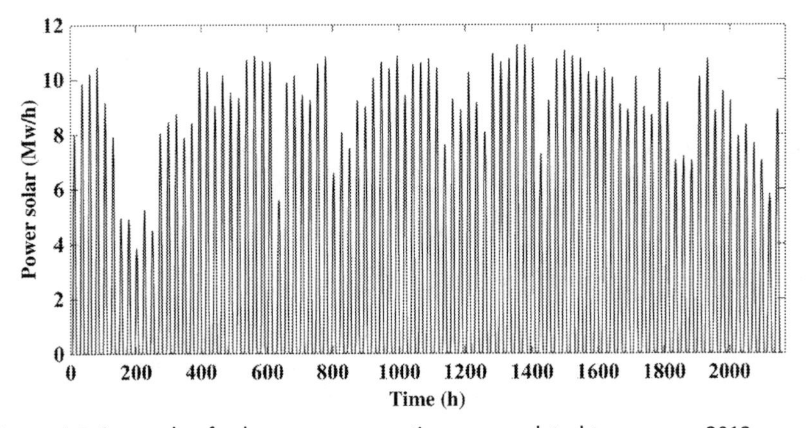

Figure 4.6 A sample of solar power generation curve related to summer 2013.

Data should be prepared in different stages for load prediction with various networks in the input of neural network. These stages are:

1. irrelevant data recognition,
2. abundant data recognition.

For the first stage, irrelevant data recognition is of great importance because training neural network is along with higher error. Thus, the following statistical method is employed:

$$M(x) - 3 * \delta < x < M(x) + 3 * \delta \tag{4.13}$$

where:

M: data mean,

δ: data standard deviation,

x: data.

Therefore, data with three time's higher standard deviation with mean distance are defined and replaced with the mean itself.

For the second stage, correlation coefficient is a statistical tool to determine the type and degree of a quantitative variable with another quantitative variable. Correlation coefficient is one of the criteria used to define correlation between two variables. The correlation coefficient shows the extremity of a relationship as well as the type of relation (direct or reversed). This coefficient value is between −1 and 1; it is equal to zero if no correlation is found between two variables.

$$corr(X, Y) = \frac{\mathrm{cov}(X, Y)}{\delta_X \delta_Y} = \frac{E[(X - M_X)(Y - M_Y)]}{\delta_X \delta_Y} \tag{4.14}$$

where E is mathematical expectation operator, cov is covariance, corr is correlation, and sigma is standard deviation symbol. To predict hourly load during round-the-clock, one should use historical data. A question raised here is that which combination of previous data should be used for short-term load prediction. There is certainly a limited number of historical data that are significantly important for hourly load prediction. In addition, higher number of inputs in the system under study for prediction purpose, including fuzzy or neural, leads to complication and thus difficulty in problem solving. Then, in this paper, sensitive analysis is used along with correlation coefficients of vectors to select the most effective data for a set of system inputs. The highest correlation between load consumption and power generated from solar/wind resources and related indices in past hours are expressed in Table 4.2. Power is correlated with

Table 4.2 Correlation analysis to determine searching motor inputs

Type of data	WP	*T*	WD	WS	AP
WP in time (*t*)	*t*-1 to *t*-8	*t*-1 to *t*-4	*t*-1 to *t*-6	*t*-1 to *t*-6	*t*-1 to *t*-6
Type of data	PL	*T*	—	—	—
PL	*t*-1, *t*-2, *t*-23, *t*-24, *t*-25, *t*-168	*t*-4, *t*-5, *t*-6, *t*-7	—	—	—
Type of data	SP	RP	—	—	—
SP	*t*-1, *t*-2, *t*-3	*t*-1, *t*-2, *t*-3	—	—	—

powers in previous hours; while wind power is correlated with powers and velocities in previous hours. And, solar power is correlated with powers and temperatures in previous hours.

4.5 PREPARATION OF DATA

4.5.1 Wavelet Transform

Wavelet transform is a mathematical approach widely used for signal processing applications. It can decompose special patterns hidden in mass of data. Regarding the prediction issue through time series and neural networks, we need modeling task. Neural networks as a general estimator in estimation of extremely nonlinear systems have limited capability [22]. Wavelet transform has the ability to simultaneously display functions and manifest their local characteristics in time-frequency domain. The use of these characteristics facilitates training of neural networks with accuracy to model extremely nonlinear signals [23]. Wavelet transforms are mainly divided into two groups [24]: continuous wavelet transform (CWT) and discrete wavelet transform (DWT). In CWT, if scale and displacement parameters are continuous, CWT will be a very slow transform with extra and useless data due to overlapping feature and duplicity of neighboring data [25]. Thus, DWT is used in this paper. Eqs. (4.15) and (4.16) express CWT and DWT, respectively [26].

CWT:

$$W(a,b) = \frac{1}{\sqrt{a}} \int_{-\infty}^{+\infty} f(x)\varphi\left(\frac{x-b}{a}\right) dx \tag{4.15}$$

where a is scale parameter, b is transform parameter, and φ is mother wavelet.

$$W(m,n) = 2^{-\left(\frac{m}{2}\right)} \sum_{t=0}^{T-1} f(t)\varphi\left(\frac{t-n2^m}{2^m}\right) \tag{4.16}$$

where T is signal length, transform and scale parameters are a function of integer values ($a = 2^m$, $b = n2^m$). Stephane Mallat multidecomposition theory has been often used in the literature in order to employ DWT [27]. This method comprises of two basic steps: decomposition and composition. Fig. 4.7 depicts decomposition and composition steps. In the first step, a signal is decomposed into two high and low frequency components. Then, high frequencies are retained; while, low frequencies are decomposed again into two high and low frequencies. High frequencies are called

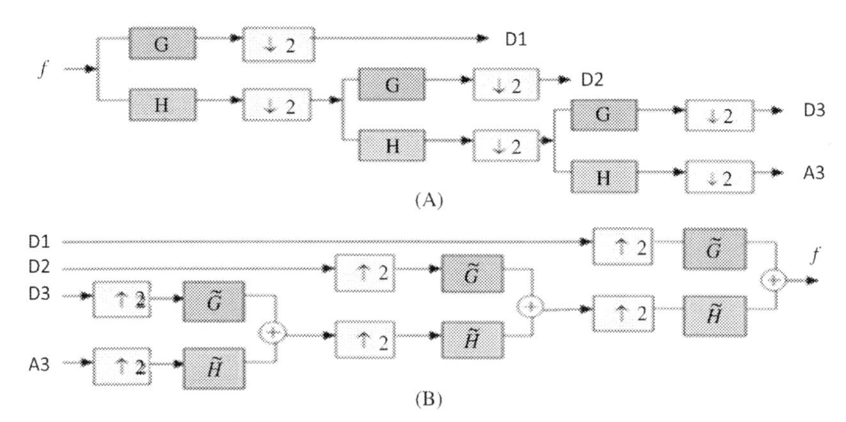

Figure 4.7 Depicts (A) decomposition step and (B) composition step in Stephane Mallat multidecomposition theory.

signal details and low frequencies are an approximation of the signal. In the composition step, the decomposition process is done in reversed fashion. The set of wavelet basic functions (such as marlette, har, and mexican hat) is also named family [28]. Within this family. Daubechies has had better results. In this paper, second order Daubechies (db2) are used as mother wavelet. Fig. 4.7 depicts (A) decomposition step and (B) composition step in Stephane Mallat multidecomposition theory [27].

4.5.2 Normalization

Because most of performance functions used in neural networks, including logistic sigmoid, hyperbolic tangent, two-pole functions, etc., are within [0–1] or [−1 and 1], it is, therefore, essentially required to normalize network inputs in a way that their values are in these intervals. By this, saturation of neurons is avoided because even large changes in the input leads to small variations in the input.

Thus, following load data are used for normalization:

$$x_n = \frac{x - x_{\min}}{x_{\max} - x_{\min}} \tag{4.17}$$

where x_n is normalized load, $x_{\min}$ is the lowest value of input data, $x_{\max}$ is the highest value of input data and x is real value load.

Obviously, output of neural network will be a normalized value in a similar way. The following relationship is used to convert it into a real value of the load:

$$x = x_{\min} + x_n(x_{\max} - x_{\min}) \tag{4.18}$$

4.6 EVALUATION CRITERION FOR THE OBTAINED RESULTS

Considering that all predictions have essentially prediction errors. Thus, implementation procedure of error criteria is of great importance for the prediction purposes. Various criteria are reported in the literature that can be divided into two groups: first-order and second-order criteria. Using one criterion is not sufficient to compare various prediction methods. Rather, several criteria are required to obtain the appropriate method. First-order criteria provide similar results. Also, second-order criteria provide similar outcomes. The latter shows higher sensitivity compared to the former in terms of the values of errors. These criteria have no equal domains for comparison. Thus, for the comparison purpose, at least one criterion should be selected from each one as a representative. These criteria are defined as vertical difference from real values of predictions. Based on the above definitions, two different criteria are used in this paper for predictions:

Mean absolute percentage error (MAPE) criterion is related to first-order error. That is, it provides the mean value of absolute error of prediction for the models. For example, if wind power is forecasted, this criterion is equal to a part of energy that is deviated from real value in the prediction interval. In addition, normalized root mean square error (NRMSE) criterion is related to second-order error with higher variations when prediction error is high.

In order to assess performance of designed system load prediction, three criteria are used: MAPE, normalized mean absolute error (NMAE), mean absolute error (MAE) [25,26].

$$MAPE(\%) = \frac{1}{NH}\sum_{t=1}^{NH}\frac{|WP(t) - WPF(t)|}{\overline{WP(t)}} \times 100 \tag{4.19}$$

$$NRMSE(\%) = \sqrt{\frac{1}{NH}\sum_{t=1}^{NH}\left(\frac{WP(t) - WPF(t)}{WP_N}\right)^2} \times 100 \tag{4.20}$$

$$NMAE(\%) = \frac{1}{NH}\sum_{t=1}^{NH}\frac{|WP(t) - WPF(t)|}{WP(t)} \times 100 \tag{4.21}$$

$$MAE(\%) = \frac{1}{NH}\sum_{t=1}^{NH}|WP(t) - WPF(t)| \times 100 \tag{4.22}$$

NH denotes predicted hours, *WPF*(t) denotes predicted value, *WP*(t) denotes real value, PN denotes maximum power of loads, load power is 105 MW, wind power is 40 MW, and solar power is 35 MW [29,31].

4.7 PREDICTION MOTOR

Prediction motor comprises of two parts: elementary predictor and main predictor.

Elementary predictor is formed by a RBF trying to find local solutions. Main predictor consists of three in-series MLPs using learning algorithms of BR, BFGS, and LM that are connected through PSO algorithm for optimization of weights. MLP neural network has a hidden layer with ten neurons selected by trial-and-error technique.

Fig. 4.8 shows main predictor motor. Predictors are within time intervals of 1-hour and for time periods of 24-hour. The effect of combining various inputs on simulation results will be further assessed. In the following, this method will be evaluated.

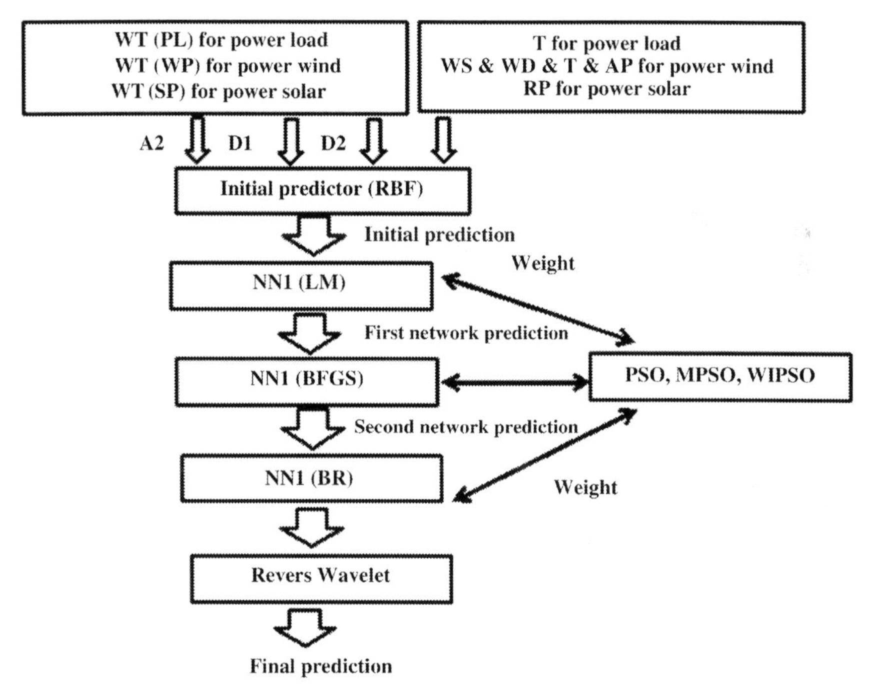

Figure 4.8 Main predictor motor for solar/wind power prediction and power consumption in smart distribution system.

1. First step is related to data preparation. The type of input data for systems consumed power prediction is in the form of two types of data (load power and temperature) with previous data inputs in total of ten and for wind prediction in the form of five types of data (wind power, wind speed, wind direction, temperature, air pressure). In terms of the number, in total, 30 and for solar prediction two types of data (solar power and irradiation power) which are six in terms of previous data related to past hours that has the highest effect on solar power.
2. In the second step, elementary predictor is utilized through RBF neural network.
3. In the third step, the output of RBF neural network that is an elementary prediction of wind power and solar power plants is added to total wind power data.
4. In the fourth step, due to positive impact of data normalization of the performance of neural networks, the output data in third step are normalized along with other meteorological data.
5. The fifth step is related to main predictor. Main predictor is comprised of three MLP neural networks by learning algorithms of BR, BFGS, and LM. In each of these networks, after finding weights in training phase, weights of neural network are considered as the input of metaheuristic algorithm. After getting optimized solution by the algorithms, its final values are regarded as the final weights of neural network. These weights are the initial weights of the next neural network. This process is iterated for all of these three neural networks. The input data in this step are the output of the fourth step. In each of these three neural network, we will have better prediction of wind power, solar power ,and power consumption. In prediction phase, the output of each of these networks is considered as the input along with other inputs for the next neural network.
6. In the sixth step, by employing reversed normalization of the signal of wind power, solar power and power consumption, the real predicted values are obtained.
7. The above method is iterated ten times, and validation and training is performed each time with the rate of (10, 10, 80). And finally, mean predicted value is considered as the final result.

4.8 SIMULATION

Predictions are performed in 1-hour intervals for 24-hour time periods. Tables 4.3–4.5 provide simulation results in a 24-hourr time period for

Table 4.3 Mean prediction for the proposed method in power consumption

Seasons	Average of 5 time running	Methods					
		RBF + LM	RBF + LM + WT	RBF + HNN + WT	RBF + HNN + WT + PSO	RBF + HNN + WT + MPSO	RBF + HNN + WT + WIPSO
December	MAPE	4.225	4.225	3.752	3.2203	2.820	1.974
	NRMSE	4.630	4.630	4.262	3.984	3.002	2.227
	NMAE	3.807	3.807	3.381	3.01	2.750	1.779
	MAE	42.74	42.74	65.23	40.02	28.25	19.97
July	MAPE	6.504	6.504	6.369	4.704	3.802	1.720
	NRMSE	4.958	4.958	5.269	4.02	4.020	1.795
	NMAE	4.958	4.958	6.304	3.89	3.701	1.381
	MAE	55.67	55.67	61.9	43.01	35.50	15.50
May	MAPE	6.369	6.369	5.025	4.23	2.50	1.293
	NRMSE	5.269	5.269	8.265	4.12	2.750	1.265
	NMAE	6.304	6.304	8.501	4.25	2.4	1.031
	MAE	61.9	61.9	62.25	42.8	24.2	11.57
October	MAPE	3.086	3.086	3.752	3.65	4.002	2.086
	NRMSE	4.262	4.262	4.262	3.55	4.101	3.060
	NMAE	3.381	3.381	3.381	3.25	3.902	2.336
	MAE	65.25	65.25	65.25	44	22	21.22

Table 4.4 Mean prediction for the proposed method in solar power

Seasons	Average of 5 time running	Methods					
		RBF + LM	RBF + LM + WT	RBF + HNN + WT	RBF + HNN + WT + PSO	RBF + HNN + WT + MPSO	RBF + HNN + WT + WIPSO
December	MAPE	10.9171	7.9127	6.4105	5.4360	5.0586	3.7567
	NRMSE	3.4135	1.9775	1.5283	1.2397	1.2190	0.8740
	NMAE	1.2977	0.9406	0.7620	0.6462	0.6013	0.4466
	MAE	0.4542	0.3292	0.2667	0.2262	0.2105	0.1563
July	MAPE	9.4333	7.4303	6.4289	5.4274	4.0253	3.6248
	NRMSE	3.3255	2.2321	1.7216	1.2752	0.8506	0.7887
	NMAE	1.1214	0.8833	0.7642	0.6452	0.4785	0.4309
	MAE	0.3925	0.3091	0.2675	0.2258	0.1675	0.1508
May	MAPE	9.7089	7.5057	6.5042	5.5028	4.5013	4.0006
	NRMSE	3.4621	2.2476	1.7286	1.2668	0.9497	0.8975
	NMAE	1.1541	0.8922	0.7732	0.6541	0.5351	0.4756
	MAE	0.4039	0.3123	0.2706	0.2289	0.1873	0.1664
October	MAPE	9.0995	6.9964	6.5958	5.5943	4.5928	3.3911
	NRMSE	3.4252	2.2538	2.0379	1.5204	1.0727	0.8095
	NMAE	1.0817	0.8317	0.7841	0.6650	0.5460	0.4031
	MAE	0.3786	0.2911	0.2744	0.2328	0.1911	0.1411

Table 4.5 Mean prediction for the proposed method in wind power

Seasons	Average of 5 time running	Methods					
		RBF + LM	RBF + LM + WT	RBF + HNN + WT	RBF + HNN + WT + PSO	RBF + HNN + WT + MPSO	RBF + HNN + WT + WIPSO
December	MAPE	8.9637	5.5133	5.707	3.0015	2.2146	1.9842
	NRMSE	7.4053	4.7613	4.961	3.829	2.4262	1.8773
	NMAE	5.9185	3.6389	3.7682	2.8744	1.655	1.3101
	MAE	15.6248	9.6104	9.948	7.5885	4.3692	3.4587
July	MAPE	12.3854	9.1248	7.1178	4.0178	3.8053	2.8145
	NRMSE	6.2205	4.5872	3.3225	2.8236	2.3078	1.3984
	NMAE	5.1229	3.7742	2.6611	2.0129	1.8522	1.1642
	MAE	13.5245	9.9639	7.0252	5.3141	4.8897	3.0734
May	MAPE	13.3235	9.1156	5.3808	4.6431	3.9604	2.6616
	NRMSE	8.8973	6.144	5.7047	4.2272	3.251	2.2167
	NMAE	7.2463	4.9577	4.0975	2.9043	2.1539	1.4382
	MAE	19.1301	13.0882	10.65	7.6674	5.5664	3.7968
October	MAPE	7.9645	4.3929	4.9123	3.7871	2.3425	1.773
	NRMSE	5.5652	3.0085	3.3407	2.7825	1.7453	1.3712
	NMAE	4.3035	2.3737	2.6543	2.0463	0.958	0.8513
	MAE	11.3613	6.2664	7.1911	5.4022	2.5291	2.2479

power consumption, solar power generation and wind power generation, respectively. These simulations are done for 4 days in four seasons. Simulations are performed five times and the mean of results for various error criteria are assessed. The following methods were compared with the proposed method (RBF + HNN + WT + WIPSO): RBF + LM, RBF + LM + WT, RBF + HNN + WT, RBF + HNN + WT + PSO, RBF + HNN + WT + MPSO. Related days of simulations were December 20th, July 10th, May 10th, and October 10th. Evaluating Tables 4.3–4.5 shows that the proposed method is absolutely dominant over the other methods and the resulting prediction error in all seasons is relatively lower. The analysis of prediction results reveal that WIPSO is better than other simulation methods in terms of optimization of neural network weights and finding global optimum solution. For instance, mean MAPE value for the proposed method is lower than RBF + HNN + WT + MPSO method for power consumption prediction (46%), wind power (30%), and solar power (18.73%). Comparing second and third methods reveal that the impact of in-series combination of neural networks on prediction accuracy is high (19.06% reduction in power consumption, and 17.8% in wind power, and 13.08% in solar power in mean MAPE criterion).

Fig. 4.9 illustrates prediction results of the power consumption on May 20, 2013 along with prediction error [29,30]. Fig. 4.10 shows prediction error of power consumption by the proposed algorithm on December 20, 2013. Fig. 4.11 shows prediction results for solar power on

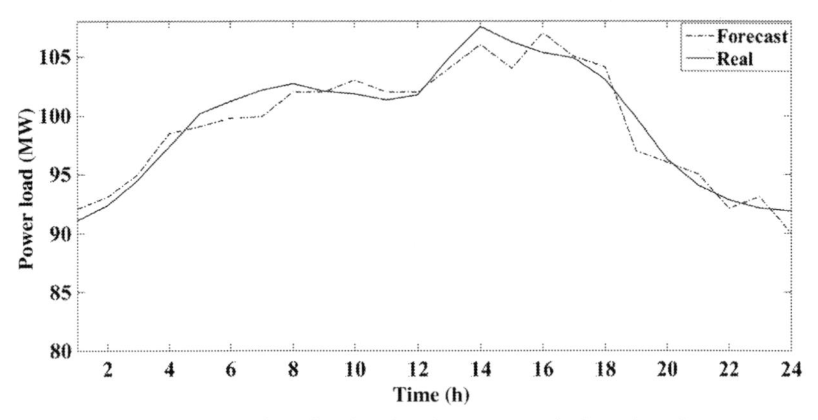

Figure 4.9 Real value, predicted value by the proposed algorithm for power consumption on December 20, 2013.

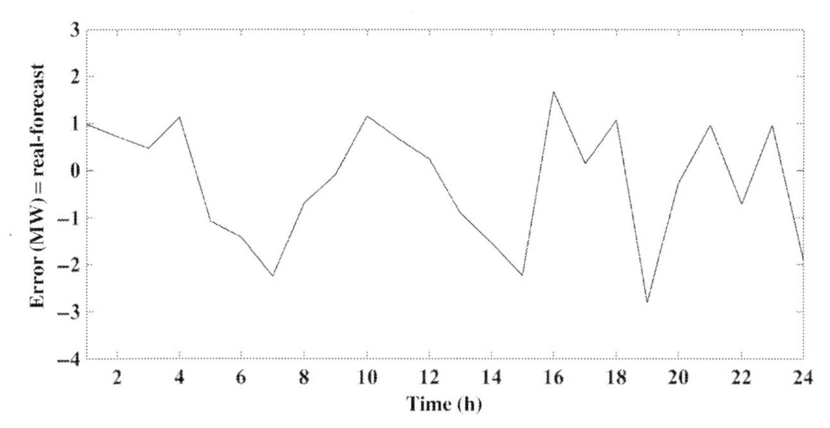

Figure 4.10 Prediction error of power consumption by the proposed algorithm in December 20, 2013.

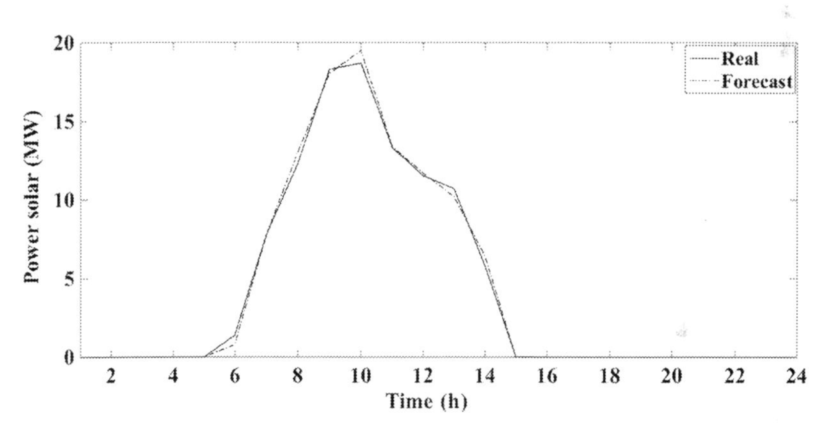

Figure 4.11 Real value, predicted value by the proposed algorithm for solar power on December 20, 2013.

May 20, 2013 along with prediction error [30,31]. Fig. 4.12 illustrates prediction error of solar power by the proposed algorithm on December 20, 2013. Fig. 4.13 depicts prediction results for wind power on May 20, 2013 along with prediction error. Fig. 4.14 illustrates prediction error of wind power by the proposed algorithm on December 20, 2013 [29,30].

Fig. 4.15, compares mean MAPE for various prediction methods for wind power, solar power, and power consumption. Simulations are carried out in MATLAB using a PC with the following features: 6 GB Ram, 4-core processors, 2.3 GHz. The average simulation time for a 24-hour time period is approximately three mines.

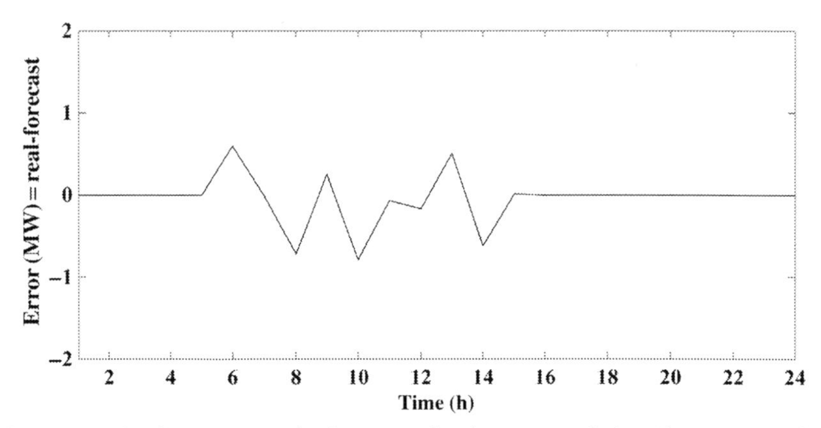

Figure 4.12 Prediction error of solar power by the proposed algorithm in December 20, 2013.

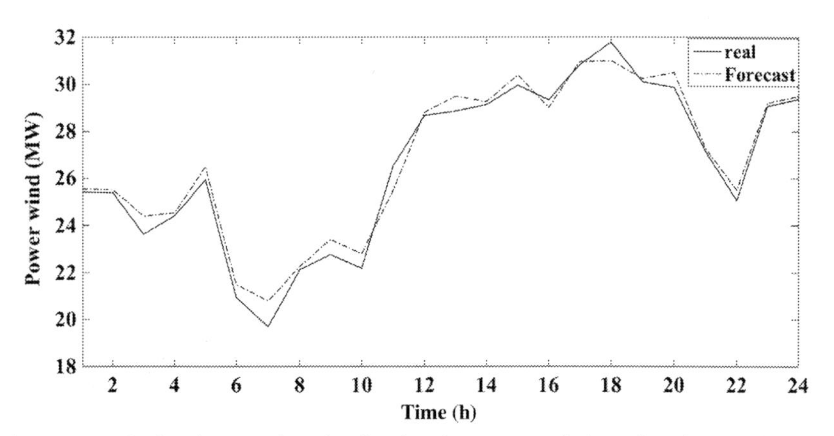

Figure 4.13 Real value, predicted value by the proposed algorithm for wind power on December 20, 2013.

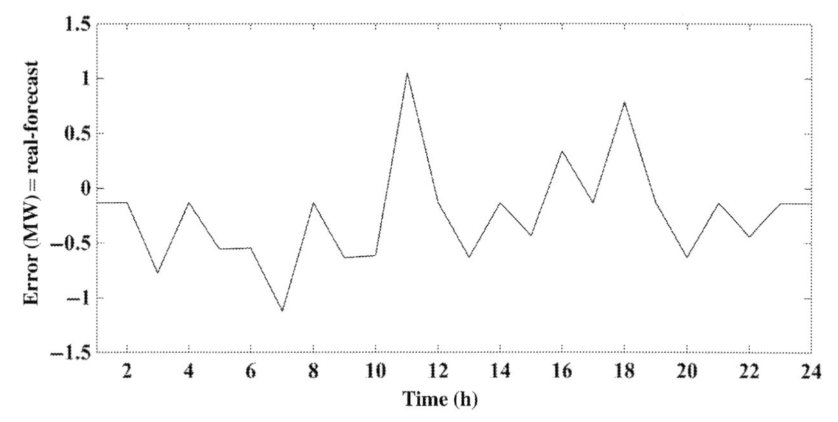

Figure 4.14 Prediction error of wind power by the proposed algorithm on December 20, 2013.

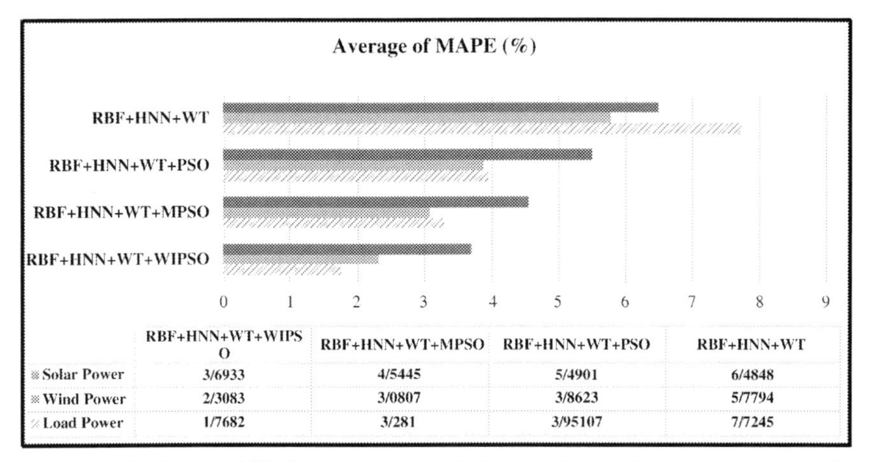

Figure 4.15 Mean MAPE for various prediction methods for wind power, solar power, and power consumption.

4.9 CONCLUSION

The proposed method in this paper is a hybrid technique comprising of radial basic function (RBF) neural networks as the elementary predictor to find local solutions. Combination of three MLP neural networks via various learning methods by WIPSO is used for final prediction and modeling nonlinear behavior of wind power, solar power, and load curves. The input data of the algorithm are: wind power, wind velocity, load amount, solar power, and irradiation intensity. Training network via evolutionary algorithms leads to improved prediction performance and reduced prediction error. In the next steps, the use of hybrid method improved prediction performance for the networks. It also makes use of the advantages of all networks and thus decreases prediction error. The advantage of using the proposed methods is obvious in error reduction, better load pattern recognition in prediction task, integration of various operations of a network, and appropriate approximation. Comparing simulation results of the proposed method with those of the other three techniques reported in the literature reveals the proper selection of the structure and its superior performance. For better comparison, these simulations are performed in 1-hour time intervals for 24-hour time periods. These simulations are done for 4 different days in four seasons in 2013 and for various meteorological conditions in a Southern Alberta, Canada smart distribution network. It is promising that providing an appropriate planning method in smart systems is done for proper supply/demand adaptation in order to have systems with less dependability to upstream networks.

REFERENCES

[1] N. Amjady, Senior Member, IEEE, Farshid Keynia, Member, IEEE" Hamidreza Zareipour, Senior Member, IEEE, IEEE Transactions Wind Power Prediction by a New Forecast Engine Composed of Modified Hybrid Neural Network IEEE Transactions on Sustainable Energy, vol. 2, no. 3, July 2011.
[2] A.K. Sinha, Short Term Load Forecasting Using Artificial Neural Networks, in Proc. Of IEEE Int. Conf. on Industrial Technology, Goa, India, 2000, vol. 1, pp. 548–553.
[3] M. Lei, L. Shiyan, J. Chuanwen, L. Hongling, Z. Yan, A review on the forecasting of wind speed and generated power, Renew. Sustain. Energy Rev. 13 (2009) 915–920.
[4] S.S. Soman, H. Zareipour, O. Malik, P. Mandal, A review of wind power and wind speed forecasting methods with different time horizons. In North American Power Symposium (NAPS). IEEE; 2010. p. 1–8.
[5] J. Zack, Overview of Wind Energy Generation Forecasting Draft Report for NY State Energy Research and Development Authority and for NY ISO, True Wind Solutions LLC, Albany (NY, USA), 2003, p. 17.
[6] A. Troncoso, J.M. Riquelme, J.C. Riquelme, J.L. Martinez, A. Gomez, Electricity market price forecasting based on weighted nearest neighbor techniques, IEEE Trans. PowerSyst. 22 (2007) 1294–1301.
[7] F. Martínez-Álvarez, A. Troncoso, J.C. Riquelme, J.S. Aguilar-Ruiz, Energy time series forecasting based on pattern sequence similarity, IEEE Trans. Knowl. Data Eng. 23 (2011) 1230–1243.
[8] K. Yang, L. Zhao, Application of Mamdani Fuzzy System Amendment on Load Forecasting Model, Symposium on Photonics and Optoelectronics; 14–16 Aug. 2009. vol.4, no., p. 1–4.
[9] C.M. Huang, C.J. Huang, M.L. Wang, A particle swarm optimization to identifying the ARMAX model for short-term load forecasting, IEEE Trans. Power Syst. 20 (2) (2005) 1126–1133. Available from: https://doi.org/10.1109/TPWRS.2005.846106.
[10] F. Almonacid, P. Pérez-Higueras, E.F. Fernández, L. Hontoria, A methodology based on dynamic artificial neural network for short-term forecasting of the power output of a {PV} generator, Energy Convers. Manage 85 (0) (2014) 389–398. Available from: https://doi.org/10.1016/j.enconman.2014.05.090.
[11] Z. Hu, Y. Bao, T. Xiong, Comprehensive learning particle swarm optimization based memetic algorithm for model selection in short-term load forecasting using support vector regression, Appl. Soft Comput. 25 (0) (2014) 15–25. Available from: https://doi.org/10.1016/j.asoc.2014.09.007.
[12] H. Chitsaz, N. Amjady, H. Zareipour, Wind power forecast using wavelet neural network trained by improved Clonal selection algorithm, Energy Convers. Manage. 89 (2015) 588–598.
[13] F. Thordarson, H. Madsen, H.A. Nielsen, P. Pinson, Conditional weighted combination of wind power forecasts, Wind Energy 13 (2010) 751–763.
[14] J. Jung, R.P. Broadwater, Current status and future advances for wind speed and power forecasting, Renew. Sustain. Energy Rev. 31 (2014) 762–777.
[15] P. Bacher, H. Madsen, H.A. Nielsen, Online short-term solar power forecasting, Solar Energy 83 (10) (Oct. 2009) 1772–1783.
[16] A. Sfetsos, A.H. Coonick, Univariate and multivariate forecasting of hourly solar radiation with artificial intelligence techniques, Solar Energy 68 (2) (Feb. 2000) 169–178.

[17] N. Amjady, A. Daraeepour, F. Keynia, Day-ahead electricity price forecasting by modified relief algorithm and hybrid neural network, IET Gener. Transm. Distrib. 4 (2010) 432–444.
[18] D. Pandey, J.S. Bhadoriya, Optimal placement & sizing of distributed generation (DG) to minimize active power loss using particle swarm optimization (PSO), Int. J. Sci. Technol. Res. 3 (2014).
[19] Z. Bashir, M. El-Hawary, Applying wavelets to short-term load forecasting using PSO-based neural networks, IEEE Trans. Power Systems. 24 (2009) 20–27.
[20] P. Somasundaram, N. Muthuselvan, A modified particle swarm optimization technique for solving transient stability constrained optimal power flow, J. Theor. Appl. Inf. Technol. 13 (2010).
[21] P. Vu, D. Le, N. VO, J. Tlusty, A novel weight-improved particle swarm optimization algorithm for optimal power flow and economic load dispatch problems. IEEE PES T&D2010. IEEE 2010. pp. 1–7.
[22] N. Amjady, F. Keynia, H. Zareipour, Wind power prediction by a new forecast engine composed of modified hybrid neural network and enhanced particle swarm optimization, Sustain. Energy IEEE Trans. 2 (2011) 265–276.
[23] N. Amjady, F. Keynia, Day-ahead price forecasting of electricity markets by mutual information technique and cascaded neuro-evolutionary algorithm, Power Systems IEEE Trans. 24 (2009) 18–306.
[24] A.J. Conejo, M. Plazas, R. Espinola, A.B. Molina, Day-ahead electricity price forecasting using the wavelet transform and ARIMA models, Power Systems, IEEE Trans. 20 (2005) 1035–1042.
[25] P. Mandal, H. Zareipour, W.D. Rosehart, Forecasting aggregated wind power production of multiple wind farms using hybrid wavelet-PSO-NNs, Int. J. Energy Res. 38 (2014) 1654–1666.
[26] N. Amjady, F. Keynia, Short-term load forecasting of power systems by combination of wavelet transform and neuro-evolutionary algorithm, Energy 34 (2009) 46–57. t-1.
[27] S.G. Mallat, A theory for multiresolution signal decomposition: the wavelet representation, Pattern Anal. Mach. Intellig. IEEE Trans. 11 (1989) 674–693.
[28] C. Lei, L. Ran, Short-term wind speed forecasting model for wind farm based on wavelet decomposition. Electric Utility Deregulation and Restructuring and Power Technologies, 2008 DRPT 2008 Third International Conference on. IEEE2008. pp. 2525–2529.
[29] A. E. S. O. (AESO), Wind Power Integration, [Online]. Available: http://www.aeso.ca/gridoperations/20544.html.
[30] Canada's National Climate Archive, Available: http://climate.weatheroffce.gc.ca/climateData.
[31] Solar Photovoltaic Energy: http://www.nrcan.gc.ca/energy/renewables/solar-photovoltaic.

APPENDIX: TERMS AND DEFINITIONS

Here, all terms mentioned in this paper and their definitions are listed in alphabetical order:

BFGS Broyden–Fletcher–Goldfarb–Shanno
BR Bayesian regularization
CWT continuous wavelet transform
DWT discrete wavelet transform

HNN	hybrid neural network
LM	Levenberg–Marquardt
MAPE	mean absolute percentage error
MLP	multilayer perceptron
NN	neural network
NRMSE	normalized root mean square error
PSO	particle swarm optimization
RBF	radial basis function
RMSE	root mean square error
WPF	wind power forecasting
WT	wavelet transform
WS	wind speed
WD	wind direction
T	temperature
AP	air pressure
RP	radiation power
PL	power load
WP	wind power
SP	solar power
NMAE	normalized mean absolute error
MAE	mean absolute error

CHAPTER 5

Impact of Distributed Energy Resource Penetrations on Smart Grid Adaptive Energy Conservation and Optimization Solutions

Moein Manbachi
The University of British Columbia, Vancouver, BC, Canada

5.1 INTRODUCTION

In recent years, penetration of distributed energy resources (DER) on the customer's side has changed the structure, operation, and performance of distribution grids. DERs such as microcombined heat and power units (μ-CHP), photovoltaics (PV), wind turbines and energy storage systems (ESS) have been employed for diverse operational needs at termination points. Recent energy conservation and optimization solutions are designed to perform loss minimization and energy conservation (e.g., conservation voltage reduction (CVR)) based on collected data from advanced metering infrastructure (AMI) and/or distribution management system (DMS). The penetrations of DERs throughout smart micrograms potentially increase grid complexity and uncertainty levels, which may adversely affect the performance and the accuracy of energy conservation and optimization solutions.

Hence, this book chapter aims to investigate the impact of the penetration of DER penetrations on advanced energy conservation and optimization solutions of smart microgrids. This chapter primarily introduces advanced smart grid adaptive energy conservation and optimization engine (ECOE) utilizing smart grid functionalities such as AMI and CVR. It then reviews advanced command and control topologies for such smart grid adaptive solutions. In the third section, the impact of the penetration of DERs is investigated. Next, a novel smart grid adaptive ECOE, capable of minimizing grid losses, and asset operating costs while

Operation of Distributed Energy Resources in Smart Distribution Networks
DOI: https://doi.org/10.1016/B978-0-12-814891-4.00005-9

maximizing energy conservation benefits using AMI data is proposed. The proposed solution also uses fuzzification technique to provide more flexibility for grid operators to define weighting factors of the objective function subparts based on their grid technical/economic requirements.

To test the accuracy and performance of proposed solution in the presence of different DER penetration scenarios, i.e., normal, high, and very high, a 33-node smart microgrid is employed. The results of this study prove how taking the impact of different DER penetration levels into account could result in more effective and precise energy conservation and optimization solutions for present and/or future smart microgrids.

5.2 ADVANCED SMART GRID ADAPTIVE ENERGY CONSERVATION AND OPTIMIZATION SOLUTIONS

5.2.1 The Roles of Smart Microgrid Functionalities on Energy Conservation and Optimization Techniques

Nowadays, the advent and the expansion of smart grid technologies have facilitated energy efficiency improvement in electric distribution networks. The growth of well-designed intelligence layer over utility assets eases the emergence of the smart grid's fundamental applications [1]. As electric power utilities are faced with diverse technological and economic issues, it is necessary for them to contemplate instant changes and/or upgrades of their technologies, operating plans, and business models. Plenty of electric power utilities have already upgraded and improved the operation of their distribution grids using smart grid technologies such as energy management system (EMS), DMS, substation automation (SA), and AMI in recent years. Others intend to renovate and/or upgrade their grids according to the mentioned technologies. However, there is no doubt that electric power utilities are eager to continue integrating novel smart grid functionalities based on their priorities and road maps, but applying smart grid components and technologies necessitate electric power utilities to seek new optimization and energy saving techniques in-line with employed smart grid technologies. The primary concern of electric power utilities is to find a cost-effective solution for optimal operation of their existing grids and then, to specify the best short-term and long-term plans to update and expand distribution networks according to their smart grid technology development plans.

5.2.1.1 Conservation Voltage Reduction

One of the famed energy saving techniques that has been taken into consideration by many utilities in last two decades is conservation voltage regulation, conservation voltage reduction or CVR. As ANSI C 84.1 [2] standard have defined the acceptable ranges of voltage at termination points (e.g., 114–126V in North America), CVR tries to decrease consumer's voltage level into lower limits of ANSI range, i.e., 114–120 volts, to conserve the energy consumption. CVR is not a new concept as it had been used for many years by electric power utilities for load reduction purposes especially in peak times. It had been common to lower voltage of high consumption loads, such as industrial machines, to decrease aggregated load of the system during peak time. Despite this fact, CVR was revived in another manner when it had been realized that many of power distribution network feeders in the United States are giving a higher voltage level to termination points, i.e., on average, consumers receive 120.5V. The main reasons of this issue lied in electric power utility concerns on supplying flawless power to their customers and significant limitations in their applied technologies. Moreover, it has been revealed that many other distribution utility feeders are working at upper level of ANSI range, which is between 120V and 126V. Hence, electric power utilities contemplated again that by lowering voltage level of consumption, it might be possible to perform worthwhile energy conservation as well as demand reduction plans. Fig. 5.1 depicts CVR voltage reduction compared with conventional voltage regulation.

In 1978, R.F. Preiss and V.J. Warnock studied fifteen different feeders in the presence of 5% voltage reduction for a 4-hour period in different weekdays and achieved almost 3% load drop and less than one-half of one percent energy savings [3]. The second phase of this study published in 1986 showed that the average energy saving resulted from fixed 5% voltage reduction equals to 0.71 per units during 24-hour CVR period in a day [4].

In 1987, D.M. Lauria investigated CVR at northeast electric power utilities that resulted in 1% energy consumption saving through 1% voltage reduction [5]. In brief, due to technical limitations and high costs, most CVR plans in 70th and 80th have not been pursued regularly by the electric power utilities. In 1987, Bonneville Power Administration signed a contract with Pacific Northwest Laboratories to test conservation potential [6]. Hence, in the 1990s, American electric power utilities in the Pacific Northwest tested various CVR scenarios to check its

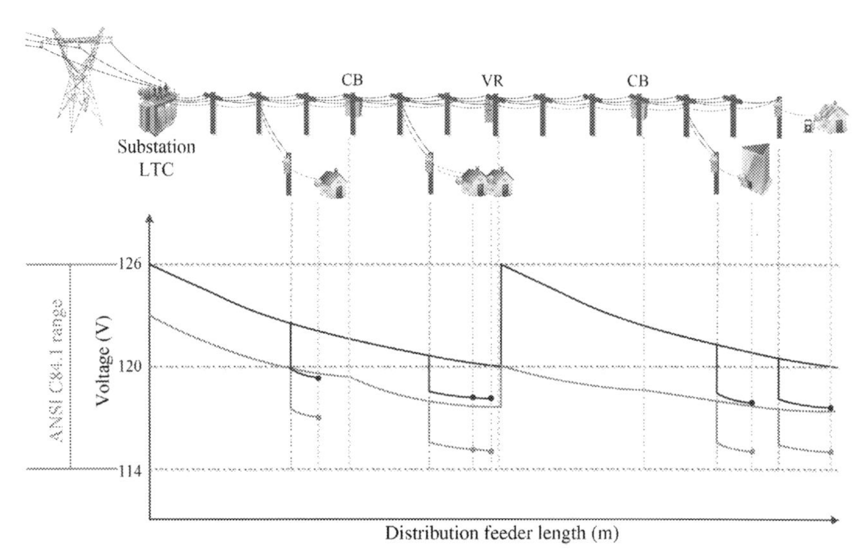

Figure 5.1 Conservation voltage regulation (*bottom curve*) versus conventional voltage regulation technique (*top curve*).

capability on energy conservation. In 1990, CVR implementation on Edison Commonwealth demonstrated that this approach could provide about 1% energy saving by 1.6 voltage reduction [7]. In-line with disparate CVR research projects, new distribution network features such as distribution automation and DMSs emerged in the 1990s [8–10]. These new technologies brought great control and monitoring opportunities for distribution grids. Accordingly, electric power utilities tended to focus more on DA, DMS, and EMS development rather than inventing new CVR approaches in the 1990s.

In the 2000s, new CVR technologies raised up due to developments in voltage regulation of control components and DMS/SCADA infrastructural technologies. T.L. Wilson tried to investigate some of the main concerns caused by CVR implementations such as the under voltage problem and the amount of reduced energy consumption in two electric power utilities in the Pacific Northwest that are being applied to new communication and control technologies [11]. In 2008, a research survey conducted by the Northwest Energy Efficiency Alliance (NEEA) showed that by 1% voltage reduction on average, a 0.7%–0.8% drop in power could be obtained according to several utility experiences [12]. Moreover, this study derived that CVR could save 1%–3% of total energy, 2%–4% of kW demand, and 4%–10% of kVAr demand [13]. According to a comprehensive research held by the Department of Energy's Pacific Northwest National Laboratory (PNNL) in August 2010, CVR could provide peak demand reduction as well as the annual energy reduction of approximately 0.5%–4% depending on feeder specifications [14]. Complete deployment of CVR on all distribution feeders would provide 3.04% reduction in annual energy consumption and if CVR applied only on high value distribution feeders, the annual energy consumption can be reduced up to 2.4% [14]. This study also proves the necessity of accurate load modeling for new CVR techniques. In addition, other reputed research institutes, such as Electric Power Research Institute (EPRI) have worked on CVR strategies for many years. A study on the potential benefits of CVR by EPRI shows that by 3% voltage reduction, it would be possible to typically achieve a 2.1% demand and energy reduction [15]. The abovementioned studies ascertain the fact that CVR could benefit distribution grid in a different manner conforming to grid specification. Today, CVR could be employed through conventional voltage control assets such as load tap changer of transformer (LTC) in distribution substation, voltage regulators (VR) along distribution feeders, conventional reactive power control components such as switched shunt capacitor banks (CB) and new volt-var control components (VVCC) such as smart inverters, etc. Principal advantages of CVR include but are not limited to demand reduction, energy consumption saving, reducing downstream substation and/or feeder overloading, increasing home appliance lifetimes, decreasing complaints on low-voltage and/or high-voltage events and lowering customer's electricity bill. Fig. 5.2 presents the history of key CVR projects.

From the abovementioned studies, it can be concluded that most CVR projects performed pilot and few distribution utilities implemented this approach in real field. Many of electric power utilities preferred to primarily perform CVR pilot projects as they could further perform cost-benefit analysis and get familiar with CVR benefits more. Another reason is due to a term called "lostrevenue" that electric power utilities may face with while performing CVR. As reducing the medium voltage (MV) side of distribution grid leads to lost revenue for electric power utilities (i.e., utilities may sell less power to their consumers) some may consider this as an obstacle for CVR implementation and development.

On the other hand, the electric service provider, i.e., a retail company, and distribution network owner are commonly different in electricity markets. Hence, regulations linked to CVR implementations in

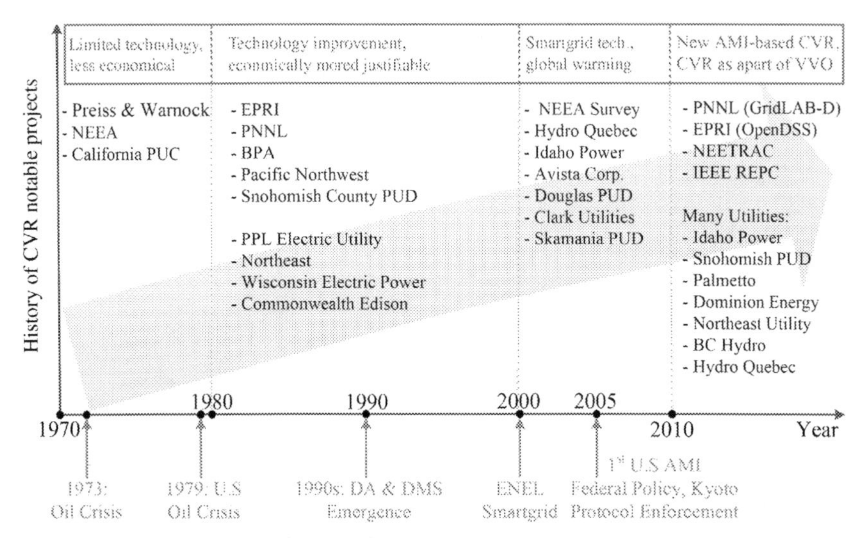

Figure 5.2 Conservation voltage reduction history.

deregulated environments would be much more complicated. Even in several electricity retail markets such as New Zealand [16], generation companies are the owner of retail companies as well. This could raise CVR regulatory issues, as typically generation companies try to increase their revenue through selling more power to customers while duration of usage of distribution network is important for a retailer. Thus, the lack of comprehensive regulations for CVR is one of the barriers of its development especially in deregulated environments.

It is known that the structure and characteristics of each distribution feeder is unique. Hence, benefits resulted by CVR plans could be different from one feeder to another. CVR approach could also be determined according to the availability of control components and load characteristics of a feeder. In general, CVR approaches are getting more sophisticated by growing new technologies. One of the simplest CVR approaches in blind feeders can be done in the presence of substation load tap changer and feeder voltage regulator that are set by line drop compensator (LDC) for end-of-line voltages. In another basic CVR technique that is on-site, voltage regulation is performed only in one specific location of a feeder. In more sophisticated techniques, distribution automation control is used to determine the operating strategies of tap changers, VR, or in some cases CB. This technique can operate without consumer feedback or it can join feedback received from termination points depending on the availability of measurement units at the end of line. Another applicable technique is rule-based CVR that includes independent operation of VVCCs based on preconfigured settings that could be updated frequently by control center through SCADA.

5.2.1.2 Volt-VAr Optimization

One recent tremendous optimization technique for distribution networks is volt-VAr optimization or called VVO. VVO attempts to minimize distribution network losses, improve voltage profile (in some cases) and minimize distribution network operating costs by applying volt-VAR control assets such as transformer load tap changer (OLTC), VR, and switched shunt CB [17]. As CVR control assets could be categorized as control actuators and as CVR and VVO objectives are well-matched, many researchers and/or utilities suggest considering CVR as a part of VVO. Therefore, it is conceivable to deem CVR as a substantial subpart of novel VVO solutions. Consequently, new VVO techniques would be able to perform energy saving and demand reduction besides performing loss and VVCC operating cost minimizations.

Proposing novel smart grid-based VVO solutions became essential as electric power utilities and voltage regulation service companies are seeking more efficient techniques for VVO engine design. As the trend of presenting new and/or future CVR techniques acknowledges CVR as an inseparable part of VVO engine, it would be necessary to investigate requirements and upcoming challenges as well as evaluate pros and cons of these novel energy conservation and optimization approaches for future distribution networks [18].

5.2.1.3 Advanced Metering Infrastructure

AMI provides electric power utilities with a two-way communication system from control center to the meter, as well as the ability to modify customers' different service-level parameters. The expansion of AMI technologies and developments of smart meter installations through smart metering programs provide distribution grids with a great opportunity to capture voltage feedback of termination points. Here, one important question is how many measurement nodes does energy conservation and optimization solution requires. Depending on grid operating condition, smart meter data availability & accessibility and the type of optimization algorithm, required measurement points could differ from one energy conservation and optimization approach to another. As mentioned before, it is possible to use data capturing/filtering techniques to lower AMI costs related to energy conservation and optimization solution especially for centralized approach. In decentralized approach, intelligent agents (IAs) could pick required data of the optimization engine according to their predefined tasks. Another advantage of AMI-based energy conservation and optimization approach is the possibility of monitoring termination points, i.e., the targeted location of voltage reduction, online or in quasi real-time. As most recent smart meters are preconfigured to send consumer data every 15 minutes (or 5 minutes in some cases), it would be conceivable to optimize distribution grid through AMI-based energy conservation and optimization solution for each quasi real-time stage. Another fact on using AMI data that deserves attention is data fluctuations that may occur due to load variations in quasi real-time. It is clear that there is a time gap between data capturing and the operation time of control components commanded by the ECOE. Load variations during and after this time gap could cause inaccurate optimization. Applying speed-up operation techniques through predictive algorithms for load forecasting could be a future solution. In some recent studies, predictive energy conservation and optimization solutions were proposed. In addition, load variations impose a safety margin at the lowest limit of ANSI for AMI-based CVRs that has to be considered as a constraint in the CVR problem. For instance, a CVR approach cannot reduce consumer voltages into 115V or less due to the safety reasons and to support load variations.

5.2.2 Smart Microgrid Command and Control Topologies

5.2.2.1 Real-time Command and Control Topologies

As a part of ECOE, new CVR techniques need to follow VVO topology. Regarding to topology point of view, it is possible to classify new VVO approaches into centralized VVO or decentralized VVO as illustrated in Fig. 5.3.

In centralized VVO, the processing system is placed in a central controller unit such as DMS in the so called utility back office [17]. The DMS uses relevant measurements taken from termination points (i.e., utility subscribers) supplied to it from either field collectors or directly from measurement data management System, to determine the best possible settings for field-bound VVO assets to achieve the desired optimization and conservation targets [17]. These settings are then off-loaded to such assets through existing downstream pipes, such as SCADA network [17]. It can be interpreted from some works that centralized VVO

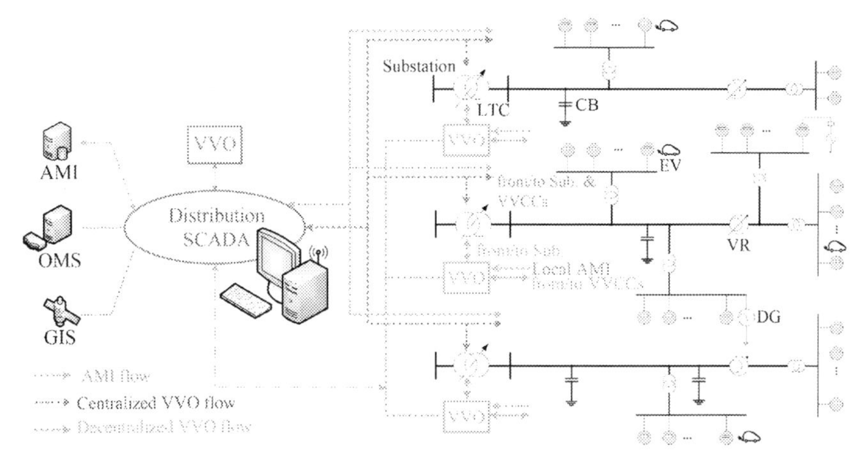

Figure 5.3 Decentralized and centralized volt-VAR control approaches.

approach is in accordance with coordinated control and monitoring of energy management system of distribution networks that is mostly centralized likewise. Hence, it seems that the infrastructure of utilizing centralized VVO is available in many distribution networks.

In contrast, decentralized VVO utilizes optimization engines which are in the field and in close-proximity to the relevant assets to perform optimization according to local attributes of the distribution network [17]. In this case, local measurements do not need to travel from the field to the back-office, and the new settings for VVO/CVR assets are determined locally, rather than from a centralized controller. Moreover, utilizing intelligent systems such as multiagent system (MAS) enable the system to avoid sending excessive data to the VVO engine.

As decentralized VVO approach follows microgrid distributed command and control topology, some may believe that this approach could be developed in-line with microgrid expansion plans within distribution networks. Here, there is no need to transfer a huge amount of data from smart meters located at load premises to the VVO engine. Thus, this approach may lead to less VVO-related AMI cost compared with centralized approach. In both explained VVO topologies, CVR tries to cover a missing regulation zone that is low voltage (LV) secondary side of transformer through existing control components in MV primary side of transformer and/or probable VVCCs in LV secondary employed in some new studies. Algorithm-based CVR techniques such as model-based or heuristics could be applied for both centralized and decentralized VVO/

CVR. In centralized VVO, a huge amount of consumer's data is sent to VVO engine located in control center. Hence, the necessity of using data filtering approaches to avoid SCADA blockage or data tsunami is felt more in centralized VVO rather than decentralized VVO. Contrary, CVR could be accomplished through both mentioned topologies in two different parts of distribution network. In the first method, CVR uses VVCCs located at MV side. In the second method, in-line power regulators (IRPs) are used at LV side of transformer to reduce voltage level of consumption. As this method requires IRP installations, it would be necessary to perform detailed cost-benefit analysis and compare the benefits of this method with the first common technique in the future. As a result, a comprehensive cost-benefit analysis is needed to select centralized or decentralized approach. However, decisions need to be taken according to control command topology and future roadmap regarding smart grid and/or DERs development.

5.2.2.2 Communication Platform Protocols

It is essential to use reliable and secure communication protocols to perform coordinated control, and build a reliable two-way communication system from termination points to energy conservation and optimization solution and from the engine to VVCCs. DNP3 is one of the well-known applicable standards for this aim. Moreover, IEC 61850 [19,20], used in SA and smart grid applications, can also be applied for CVR implementation. IEC 61850-7-420 defines standards for DERs in the system. IEC 61850 90-1 helps substations to communicate with each other. IEC 61850 90-2 supports communications between control center (i.e., ECOE in centralized VVO) and substations. IEC 61850 90-6 provides the required standards for distribution feeder automation systems and IEC 61850 90-7 supports smart grid technologies such as storage, PV and smart inverters. As ECOE sits in the MV substation in decentralized VVO, IEC 61850's main standard could also be used.

The optimization engine can run on a PC representing the automation unit at a high to MV substation. IEC 61850 MMS protocol can be used to gather the status of different switching devices, e.g., circuit breakers, in order to be able to impose the necessary changes in grid configuration in real-time according to the topology of the system. The energy conservation and optimization solution engine can receive measurements and send control commands to remote VVCCs using DNP3 protocol.

DNP3 can be used to show a case that the platform supports multiple kinds of protocols along with IEC 61850 protocols. Moreover, DNP3 is a very popular communication protocol, used by major utilities for wide area monitoring and control in the United States and Canada. A complete revamp of the communication infrastructure would always be carried out in stages and therefore in the transition phases the power systems would be operated partially with the old communication infrastructure (including communication protocols) and partially with the new. In this work a possible method of introduction of IEC 61850 communication protocols is investigated. IEC 61850 communication services are used for automatic service restoration after a fault and updates the optimization engine with real time changes in network topology. Whereas the measurement acquisition from remote measurement units and sending control commands to VVCCs is done through the old communication infrastructure through DNP3 protocol.

5.2.3 Main Objectives and Constraints of Advanced Energy Conservation and Optimization Techniques

5.2.3.1 Main Objective Function

The main objective function of the proposed ECOE can be shown as Eq. (5.1). Here, the objective function seeks to minimize microgrid losses as well as VVCC operating costs and energy conservation (CVR) costs at each operating time interval.

$$Min O.F_{VVO,t} = Min(C_{loss,l,t} + C_{VR,t} + C_{OLTC,t} + C_{SWCB,t} + C_{CVR,t}) \quad (5.1)$$

$$C_{loss,t} = \sum_{l=1}^{l=L} (\alpha_t \times P_{loss,l,t} \times \pi_t \times t) \quad (5.2)$$

The current injection model can be shown as (3). In backward-forward sweep (BFS) load flow technique, the relationship [B] and the bus-injection to branch-current [BIBC] matrices can be gained using Kirchhoff's current law (KCL). Furthermore, the branch-current to bus-voltage [BCBV] matrix can be obtained using Kirchhoff's voltage law (KVL). BFS begins to find load and line currents by initial estimates of node voltages. By calculating current, the voltages of system nodes, i.e., the solution of BFS load flow, can be found via Eqs. (5.3–5.6) in iterative steps till convergence.

$$I_{n,t}^{k} = (\frac{S_{n,t}}{V_{n,t}^{k}})^{*} = (\frac{P_{n,t} + jQ_{n,t}}{V_{n,t}^{k}})^{*} \tag{5.3}$$

$$[B] = [BIBC][I] \tag{5.4}$$

$$[\Delta V^{k+1}] = [BCBV][B] = [BCBV][BIBC][I^{k}] = [DLF][I^{k}] \tag{5.5}$$

$$[V^{k+1}] = [V^{0}] + [\Delta V^{k+1}] \tag{5.6}$$

BFS converges when the difference of voltages between two consecutive iterations is less than epsilon (typically less than 0.0001). By knowing the final voltage and currents of nodes, the active and reactive power losses can be obtained. To find VVCC operating costs and energy conservation (CVR) costs, Eqs. (5.7–5.10) can be applied.

$\forall t \in T$:

$$C_{VR,t} = \sum_{r \in R} \beta_t \times C_{r,t} \times X_{VR,r,t} \tag{5.7}$$

$$C_{OLTC,t} = \sum_{w \in W} \gamma_t \times C_{w,t} \times X_{OLTC,w,t} \tag{5.8}$$

$$C_{swCB,t} = \sum_{c \in C} \delta_t \times C_{c,t} \times Q_{inj,c,t} = \sum_{c \in C} \delta_t \times C_{c,t} \times \rho_{c,t} \times q_{inj,c,t} \tag{5.9}$$

$$C_{CVR,t} = \sum_{n=1}^{n=N} \mu_t \times C_{S.E,t} \times \frac{1}{\Delta E_{n,t}} \times t = \sum_{n=1}^{n=N} \mu_t \times C_{S.E,t} \times \frac{1}{CVR_{f,t} \times \Delta V_{n,t}} \times t \tag{5.10}$$

As any energy conservation study requires an efficient technique to model AMI loads with high precision, ZIP load modeling (a combination of constant impedance, constant current, and constant power), presented in Eqs. (5.11) and (5.12), is used in this study to determine the active and reactive power of loads according to their real characteristics at each quasi real-time operating interval.

$$P_{n,t} = P_{0n,t}(Z_p(\frac{V_{n,t}}{V_{0n,t}})^2 + I_p(\frac{V_{n,t}}{V_{0n,t}}) + P_p) \tag{5.11}$$

$$Q_{n,t} = Q_{0n,t}(Z_q(\frac{V_{n,t}}{V_{0n,t}})^2 + I_q(\frac{V_{n,t}}{V_{0n,t}}) + P_q) \tag{5.12}$$

To evaluate the effectiveness of energy conservation, the percentage of saved consumed energy and the percentage of voltage reduction shown in Eqs. (5.13–5.15) are calculated.

$$\Delta E\% = (\frac{E_{base} - E_{CVR}}{E_{base}}) \times 100 \tag{5.13}$$

$$\Delta V\% = (\frac{V_{base} - V_{CVR}}{V_{base}}) \times 100 \tag{5.14}$$

$$CVR_f = \frac{\Delta E}{\Delta V} \tag{5.15}$$

5.2.3.2 Constraints

Required operational constraints of proposed AMI-based energy optimization and conservation solution are as follows:

1. Voltage magnitude constraints of nodes:

$$V_{n,t}^{min} \leq V_{n,t} \leq V_{n,t}^{max} \tag{5.16}$$

2. Active/reactive power output:

$$P_{n,t}^{min} \leq P_{n,t} \leq P_{n,t}^{max} \tag{5.17}$$

$$Q_{n,t}^{min} \leq Q_{n,t} \leq Q_{n,t}^{max} \tag{5.18}$$

3. Active and reactive power flow balances:

$$P_{n,t} = P_{Gn,t} - P_{Ln,t} \tag{5.19}$$

$$Q_{n,t} = Q_{Gn,t} - Q_{dn,t} + Q_{CB,c,t} \tag{5.20}$$

4. Thermal limits of lines:

$$S_{l,t} \leq S_{l,t}^{max} \tag{5.21}$$

5. OLTC/VR tap limits:

$$TR_{OLTC-VR} = 1 + tap_{oltc-VR,t}\frac{\Delta V_{OLTC-VR,t}}{100} \tag{5.22}$$

$$tap_{LTC-VR,t} \in \left\{ -tap_{LTC-VR,t}{}^{max}, \ldots, -1, 0, 1, \ldots, tap_{LTC-VR,t}{}^{max} \right\} \tag{5.23}$$

6. Injected reactive power limit:

$$\sum_{c=1}^{c=C} Q_{CB,t}^{c} \leq \sum_{i=1}^{I} Q_{req,t}^{i} = Q_{req}^{max} \tag{5.24}$$

7. μ-CHP/PV constraints:

$$V_{DG,n,t}^{min} \leq V_{DG,n,t} \leq V_{DG,n,t}^{max} \tag{5.25}$$

$$P_{DG,n,t}^{min} \leq P_{DG,n,t} \leq P_{DG,n,t}^{max} \tag{5.26}$$

$$Q_{DG,n,t}^{min} \leq Q_{DG,n,t} \leq Q_{DG,n,t}^{max} \tag{5.27}$$

As mentioned above, this chapter employs fuzzification technique within its optimization algorithm to specify ECOE objective function subpart weights. Respectively,

$\alpha, \beta, \gamma, \delta, \mu,$ are weighting factors of loss, VR, OLTC, CB operation and CVR costs that have to be found using fuzzification. This technique enables the system to find accurate weighting factors for different operating time intervals in accordance with system techno-economic values defined in Eq. (5.28). For instance, Eq. (5.28) explains the fuzzification algorithm for loss cost minimization of ECOE.

$$if: f_{Loss} \leq f_{Loss,dmin,t} \rightarrow \alpha_t = 10$$

$$elseif: f_{Loss} > f_{Loss,dmin,t} \& f_{Loss} < f_{Loss,dmax,t}$$

$$\rightarrow \alpha_t = \frac{\left| \alpha_t - f_{Loss,dmin,t} \right|}{f_{Loss,dmax,t} - f_{Loss,dmin,t}} \times 10$$

$$else: f_{Loss} \geq f_{Loss,dmax,t} \rightarrow \alpha_t = 0 \tag{5.28}$$

If the weighting factor found by ECOE is less than or equal to the minimum desired value of loss ($f_{Loss,dmin}$) the weighting factor would be equal to ten but, if the weighting factor is greater than or equal to the maximum desired value of loss ($f_{Loss,dmax}$), the weighting factor would be equal to zero. If the weighting factor is between the desired minimum and maximum range, the system will find a weighting factor between zero and ten according to Eq. (5.28). Other fuzzification values for other ECOE subparts, such as f_{CVR}, f_{CB}, f_{VR} and f_{OLTC}, have to be defined in smart microgrid-based ECOE by the operator to let the engine find the optimal weighting factors of each objective function subparts based on system operational needs.

By receiving required inputs, ECOE runs its main algorithm and performs load flow within each optimization cycle to avoid constraint violations. The algorithm looks for the optimal solution of microgrid control assets and weighting factors at each iteration step through its optimization engine and fuzzification used within the optimization algorithm. More information regarding the optimization algorithm technique can be found in [21]. When the algorithm converges to the optimal solution, the results will be sent to VVCCs to optimally reconfigure the grid. The above iterative steps are performed for each quasi real-time interval, i.e., every 15 minutes.

5.3 DISTRIBUTED ENERGY RESOURCE PENETRATIONS IN SMART GRIDS

In recent years, environmental and economic concerns as well as power quality, and reliability issues in conventional distribution networks, have led evolutionary changes in generation/load interactions of distribution networks. Recent distribution grids tend to rely on DER generations to meet growing needs of demand with more emphasis on efficiency, reliability, and quality of service. In general, DERs are small electricity and/or heat generating units that can be installed close to loads in order to supply a part of load electricity/heat needs. Most types of DER technologies include generator, energy storage, and load control. In microgrids, loads and DERs are aggregated as a system to provide electricity and heat through a distributed command and control infrastructure.

Typically, microgrids are connected to the traditional grid (macrogrid) but they are able to be disconnected from the traditional grid autonomously and operate in island mode to supply a part of grid without interruption to increase system reliability and resiliency. Microgrids can be comprised of various renewable resources such as solar, wind, distributed

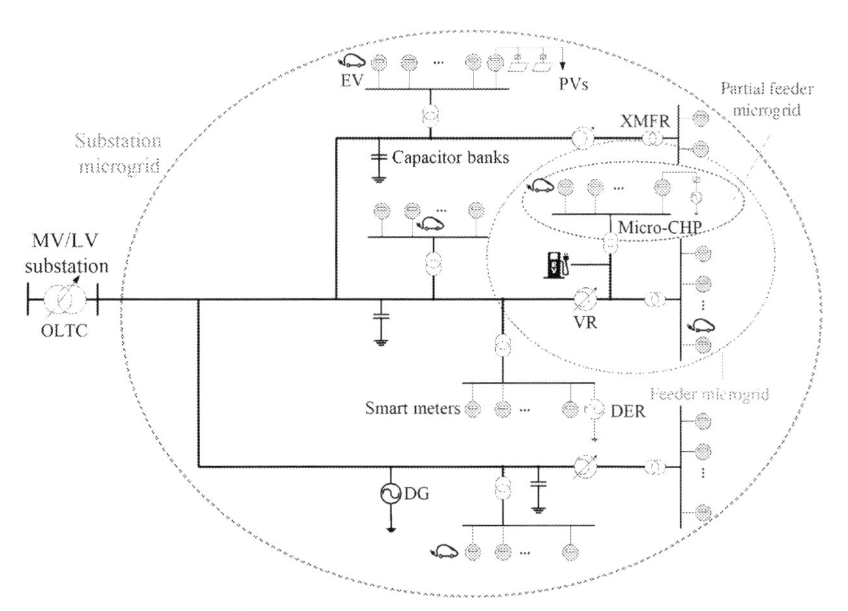

Figure 5.4 Microgrid different operation levels in a typical distribution grid.

generations such as microturbine, natural gas-engines, and DERs such as μ-CHP and ESS. Fig. 5.4 shows different microgrid levels in a typical distribution network. Moreover, employing local energy sources to supply local loads leads to loss reduction and efficiency improvement in distribution grids. In brief, the availability of reliable distributed command and control, communication infrastructure, local generation, power electronics, energy manager, and protection are the key factors that let microgrids to operate semiautonomously. Although there is still a long way to reach to a preemptive self-healing grid, as a long-term target of microgrids, advancements in microgrid component technologies have made new energy optimization and conservation possible.

An energy conservation and optimization solution, as a part of smart distribution network optimization solution has to operate in-line with microgrid key features. As such, new energy conservation and optimization solution techniques require to be integrated with microgrid sources and loads.

Moreover, they need to function in compliance with microgrid control and communication topologies using recent protocols and/or standards. The main aim of microgrid control is to provide control over network and ease the grid with autonomous components. Microgrid control enables interface switch of the system to island microgrid during fault, power quality, and other network unintentional events. By clearing fault and/or event, microgrid control has to reconnect microgrid to the macrogrid autonomously. Moreover, each DER is able to control power in different microgrid operating modes. For instance, DERs employ power vs. frequency droop controls to follow required demand during microgrid island mode of operation. This control provides frequency load-shedding when there is not adequate generation in the system. In addition, each DER has voltage control to ensure system stability. Microgrid voltage control aims to ascertain that there is no large circulating reactive power flowing between DERs, especially in microgrid with high penetration of DERs. For this reason, voltages vs. reactive power droop controllers are employed. Voltage set-point can be defined based on the reactive powers generated by DERs. As kVArs become capacitive, local voltage set-point decreased but inductive kVArs provide voltage rise in set-point.

It has to be stated that microgrid components are following peer-to-peer, plug-and-play architectures to ease system usage, increase the efficiency of the grid as well as increase grid power quality. As such, DERs can be connected through a peer-to-peer structure to their local control systems. This distribution control topology lead the system to a higher level of reliability compared with centralized control or master-slave topologies. On the other hand, microgrid component's plug and play structures let microgrid expansion without restructuring the network. Moreover, reliability and maintenance of a grid with plug and play components are much higher than other grid components with rigid structures. To smarten-up distribution networks, it is possible to control different microgrids that are connected to each other through real-time data of DERs and loads. AMI infrastructure provides the opportunity to capture required data of DERs and loads for control and optimization purposes. Nowadays, microgrid dispatch is no longer based on the number and type of DER units. In conventional control, a control center or dispatching center, dispatches all generating unit active and reactive power outputs to follow system demand. This approach may not be doable in distribution networks with high penetration of generating units. In a distribution system with microgrids, each microgrid can be seen as a unique source with known active/reactive power generation or consumption.

In microgrids with high penetration of sources such as PVs, intermittent generation may cause stability issues. Using DERs can decrease power fluctuations that might be created by these types of sources. Moreover, storage systems are needed for reserve energy when there is no solar radiation. As such, a microgrid including high penetration of PVs is mostly integrated with local generations and storage. The DER in microgrid with high penetration of PVs can control active and reactive power flows between microgrid and distribution grid. Moreover, it can supply power when PVs are not available. On the contrary, the inverter-based energy storage enables generation to be connected as quickly as possible to the microgrid without inverter interference as generation needs fast response of inverter. In microgrid island mode, high penetration of PVs may cause different issues. For instance, microgrid may face with power generation more than the amount that system requires. Energy storage integrated with PVs can help microgrid to store extra power generated by PVs. Another method is using power vs. frequency control to reduce the power output of PVs in light-load islanding operating scenarios. A proper power vs. frequency controller, enables automatic generation reduction of PVs when frequency of the system increases in microgrid light-load island modes. Moreover, it sends commands to the storage to be charged to control PV outputs effectively. In brief, distributed command and control is widely used in microgrids to perform local control over system components and dispatch distribution grids according to microgrid operational requirements.

Reliable and secure bidirectional communication infrastructure is another major feature of successful implementation of smart microgrids. Typically, three communication layers are used in microgrids. Wide area network (WAN) that can be based on 3G/4G, Ethernet, fiber optics, etc. supports transmission grids. For distribution networks, neighbor area network (NAN) and/or local area networks (LAN) are typically used. For customer side, i.e., termination points, home area network (HAN) is used. IEC 61970 focuses on common information model and energy management, IEC 61968 focuses on distribution management, and IEC 61850 is for power automation; those are some of the standards of WAN networks. ANSI C12.22 and IEEE 802 are some of the well-known standards of LAN networks and ZigBee is the most applicable protocol of HANs. The main standard of microgrids is IEEE 1547 that is standard for interconnecting distributed resources with electric power systems [22]. IEEE 1547 covers standards of different microgrid aspects such as design, interconnection, monitoring, and operation of microgrids in normal and island modes. Moreover, IEEE 2030 provides a guide for smart grid interoperability of energy technology and information technology operation with the power system, and end-use applications and loads [23]. The IEC 61850 standard [19] is one of the famed standards for automation of substations through Ethernet LANs in smart grids. IEEE 802.11 [22] shows that with the recent advancement in wireless communication technologies through Wireless LAN (WLAN) or Ethernet LAN, it is now conceivable to employ IEC 61850 standard in distribution level. IEC 61850-7-420 is a part of IEC 61850 standards that intends to give complete object models that are required for DERs [19]. It applies communication services mapped to MMS under IEC 61850-8-1 standard [19].

Therefore, the compatibility of the communication standard/protocol of smart grid-based energy conservation and optimization solution with communication standards/protocols of microgrid technologies and components is necessary. As explained, proposed topology is compatible with smart microgrid distributed command and control topology. Moreover, the control of voltage and reactive powers of smart distribution grids through conventional and new VVCCs can make the optimization engine more efficient. In brief, this research tries to study the impacts of new smart grid-based energy conservation and optimization technology on smart distribution grid as well as the impacts of smart microgrid components on smart grid-based energy conservation and optimization solution. For this reason, different microgrid sources such as DG, energy storage, μ-CHP/PV, and EV are considered as grid components and the impact of them on proposed ECOE is investigated in different studies.

5.3.1 Renewable Resource Penetrations

5.3.1.1 Photovoltaics/μ-CHPs

Penetration of DERs throughout smart grids and/or distribution feeders has raised grid complexities as well as uncertainties that affect conventional monitoring, optimization and control systems. Hence, electric power utilities are trying to verify the impact of DER penetrations on their monitoring, control and optimization systems. As mentioned before, DERs such as μ-CHPs or PVs are used for different operational aims. PV modules supply residential buildings with clean active power while μ-CHP units can provide both heat and electricity. Prior to investigating the impact of different μ-CHP/PV penetration levels on different energy conservation and optimization subparts, it is necessary to initialize μ-CHP and PV types and penetration levels.

In this study, two different types of μ-CHP units: heat-led with electricity rejection and electricity-led with heat rejection are considered. It has to be remembered that a μ-CHP generated power is typically independent of its heat. However, the main aim of using a μ-CHP unit has to be primarily specified. The range of μ-CHP is considered between 1 and 3 kWe. As most μ-CHPs are heat led, the ratio between heat led μ-CHP and an electricity led μ-CHP assumed to be 2.5 in μ-CHP/PV study of this study. Moreover, present percentages of PV, electricity led μ-CHP and heat led μ-CHP along studied distribution feeder assumed to be equal to 30%, 20%, and 50% respectively.

Different penetration levels are considered to check the impact of μ-CHP/PV units on presented smart grid-based VVO: normal, high, and very high. At each penetration level, ECOE strives for optimizing the grid and finding the optimal response of each VVCC. The active power generation profile of microgeneration comprised of PVs, heat-led and electricity-led μ-CHP units in the case study for all quasi real-time intervals resulting from the AMI data.

5.3.1.2 Community Energy Storage Systems

Community energy storage (CES) is one of the recent advanced smart grid technologies that provide distribution grids with lots of benefits in terms of stability, reliability, quality, and control. As it benefits both customers and utilities, this technology has become a crucial element of recent microgrids. CES is typically located at the edge of the grid (rather than distribution substation), close to customers and DERs to smooth the impacts of intermittent DERs such as PVs or other smart grid components such as EVs and help integration of these sources to the smart grid. The main concept of CES initially used in HV/MV substations equipped with a CES battery can give advantages much more than conventional substation batteries. This type of CES could provide regional control benefits at transformer/feeder levels. New types of substation CES units could provide voltage regulation though their four quadrant inverters without need to have OLTC in substation anymore. In recent years, this concept became more applicable on the customer side and DERs. The main tasks of CES in distribution networks can be summarized as peak-shaving, smooth DER's intermittencies (output shifting/leveling), power quality improvements (voltage and reactive power supports), islanding during outages, frequency regulation etc. Typically, a CES consists of a battery (mostly lithium ion in a battery box), a four-quadrant inverter (in grid inverter panel), and a measurement/control system (in meter and breaker panel) that includes battery management system (BMS) and inverter monitoring/control. The main difference between CES and DGs lies in the fact that CES has a fully dispatchable four-quadrant inverter that enables it to bi-directionally exchange active and reactive powers. In other words, CES is able to regulate voltage, frequency, and inject/absorb reactive power to/from the grid. Figure presents the main structure of CES inspired from [24].

The operation of CES is very simple. CES systems store energy in their batteries and supply active or reactive power due to system active/reactive demand(s). CES can be considered as a backup power of a certain group of residential houses as it can supply power for a short period while discharging. Typically, the voltage rates of CES at LV side is 240/120V AC. Moreover, the rated power is typically 25–50 kW [25], and the rated energy for 25 kW is from 25 to 75 kWh and for 50 is from 50 to 150 kWh. In other words, the discharge time of most CES system is from 1 to 3 hours at each operating cycle. Typically, the discharge power is equal to the rated power but charging power has to be less than or equal to the rated power. By dividing rated energy and rated power, discharge time is obtained. Typically, the charge time has to be less than two times of the discharge time [24]. Fig. 5.5 shows the CES main structure.

Generally, CES such as any battery ESS has three modes of operation: discharge, standby, and charge. According to the four-quadrant inverter capability, CES discharge can be fully active power, active/reactive

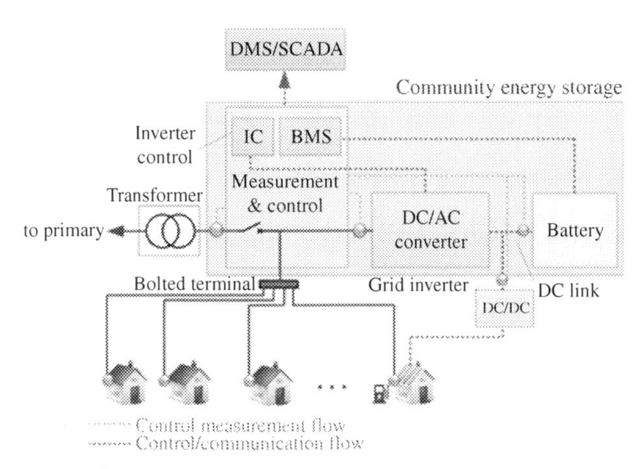

Figure 5.5 Community energy storage main structure.

(inductive), and active/reactive (capacitive). ECOE prefers CES systems more than switchable CBs as they can supply the active power. In addition, ECOE prefers CES systems more than DGs as well because, CES systems provide reactive power supports and other benefits to the grid. One of the most important tasks of CES can be peak shaving, energy conservation, and optimization engine, which can help CES to discharge during peak quasi real-time intervals. This could benefit the ECOE as well. The second mode of operation of CES systems is standby mode. In this operating mode, the CES systems are generally disconnected from the grid but they are in standby and ready to be integrated to the grid while system needs them. Due to the concept of four-quadrant inverters explained in Section 6.1, it is possible to say that these inverters may inject/absorb reactive power into the system while they are in standby mode ($P=0$, $Q<0$ or $Q>0$). This could create a new mode of operation for CES in future. For instance, VVO could send a control command to a standby CES to be connected to the grid and inject reactive power. This suggested mode of operation needs further technical/economic investigations.

The last CES mode of operation is charge mode. Typically, CES system charge in very light load conditions. ECOE can help CES to charge in light load quasi real-time intervals as well. As explained, the charge time has to be typically less than two times of discharge time. From power loss minimization point of view, as CES consumes active power in charge mode, active power loss increases slightly but as power loss level is very low in light load condition, the total impact of CES charge mode to the system is not significant. From CVR point of view, fewer active power consumption leads to more energy conservation. As such, one may argue that extending CES charge time very close to two times of discharge time and setting four-quadrant inverter to consume the fewest possible active power to charge batteries that comes with reactive power support can be more beneficial to smart grid-based energy conservation and optimization solution during light loads. Regarding this argument, it has to be reminded that the value of energy conservation and optimization objective function is at its lowest rate during very light load condition. Hence, this won't be much beneficial to the whole system if operating cost of CES is taken into account as well. Moreover, the impact of time-consuming charge on CESS battery efficiencies has to be investigated.

In brief, CES systems provide different intelligent layers in terms of energy management, feeder optimization, and local control to the grids. This could definitely help energy conservation and optimization solution objectives. Moreover, as most of new CES communication technologies are based on DNP3 and IEC 61850 protocols/standards, proposed engine in this research can be integrated with CES control system. For instance, ECOE would be able to find CES discharging control commands for peak shaving during peak time intervals.

5.3.2 Electric Vehicle Penetration

Penetration of dispatchable resources on the customer side or cogeneration loads such as EV has brought new opportunities as well as challenges to distribution networks. EV as one of microgrid sources has seen rapid growth in North America. Although distribution planners intend to provide command and controls strategies to adjust cogeneration impacts on loads but dispatchable energy sources can be a tool for system service optimization. As such, new studies have shown the fact that new inverter technologies could provide reactive power injection/ absorption opportunity for microgrids [26,27]. Using these types of technologies such as vehicle to grids (V2Gs) with reactive power support capabilities have considerable impacts on the optimization, reactive power need, the size and the number of switchable shunt CBs in distribution networks; as it will be possible to employ a percentage of EVs as reliable reactive power generating sources in V2G mode to reduce active and/or reactive power loss and conserve the energy consumption in the system.

In general, EVs could be classified based on three main charging levels [28] as well as on their technology shown in Table 5.1 Large penetration of EVs at different consumption levels and within various locations (house, workplace, shopping centers) could lead to salient changes in distribution network demand and daily load profile.

Many research studies have tried to evaluate the profile and the trajectory of these changes based on different EV penetration levels [29,30] which could be an expedient measuring factor. Generally, when an EV injects reactive power into the grid, it may raise distribution feeder capacity. It may improve voltage profile and power loss if there is no overcompensation of the reactive load. It may also lead to an increase in voltage, which results in an increase in loads. This increase in load may be either greater, or smaller than the reduction of losses. Hence, the overall input

Table 5.1 Different EV charging levels

AC Level 1	AC Level 2	AC Level 3 (TBD)	DC Level 1	DC Level 2	DC Level 3 (TBD)
120V, 1.4 Kw (12 A) 120V, 1.9 kW (16 A)	240V, up to 19.2 kW (80 A)	> 20 kW, single and 3-phase	200–450V DC, up to 36 kW (80 A)	200–450V DC, up to 90 kW (200 A)	200–600V DC, up to 240 kW (400 A)

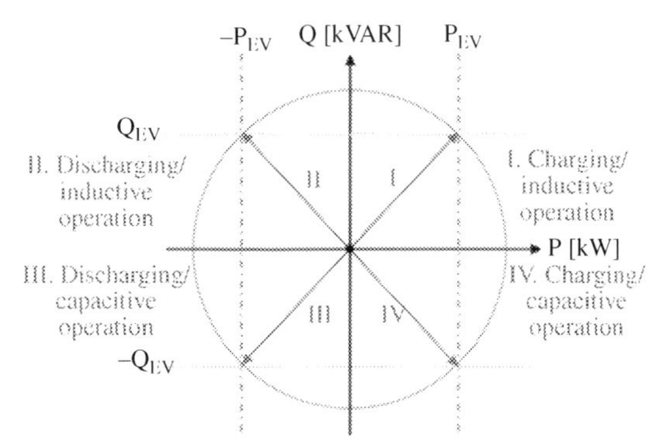

Figure 5.6 Active and reactive power generation of an EV with 4-quadrant inverter.

of kW in the system may either increase, or decrease. Power flow, i.e., run within energy conservation and optimization algorithm covers these following points such as system overcompensation, load rise and overall input of kW. EV charging stations may follow different operating modes [29–32]. Fig. 5.6 illustrates different operating points of AC/DC inverter of an EV. The most common EV operating scenario is $P > 0$, $Q = 0$ (on the border between Quadrant I and IV) which shows that EV consumes only the active power of the grid. Another possible operating mode of EV inverter enables EVs to inject reactive power into the grid (Quadrant IV) while still consuming active power of the grid for charging [26,27,33]. This leads to the rise in distribution feeder capacity, improvement in voltage profile, and power loss reduction.

The topology of the bidirectional charger needs minimal changes for it to be suitable for reactive support [34]. Single phase inverters have been tested for their ability to inject reactive power into the network in [35,36]. According to these research studies, if the voltage rate of the DC-link capacitor of the inverter increased at least 3% it would be possible to inject VAR. Moreover, the current ripple rate of the DC-link capacitor is strong enough for inverters to operate in capacitive mode. It should be mentioned that the total losses of the AC/DC converter could be slightly increased by reactive power support in normal operation of the charger [34] but the EV battery and the input inductor current are not affected by capacitive operation at all [27].

Moreover, studies have shown that the DC-link capacitor could be capable of supplying reactive power into the grid without engaging the EV battery [34]. Thus, VAR injection by EV inverters will not cause any degradation on EVs battery life [37]. This important study shows that EV charging stations could be one of the future candidates for reactive power injection into the grid even when the electric car is not connected to the charging station. As shown in Fig. 5.6, the amount of reactive power that an EV charging station can inject in the charging mode is confined by the charger power limit and the active power drawn from the grid. It is clear that if an EV is using the maximum active power from a charging station, this station cannot generate reactive power but as the charging stations are in idle mode most of the time, they could inject reactive power into the grid when required. Moreover, the charger can be rated at 10%–20% higher power than the maximum real power drawn during charging to make the charger capable of injecting VAR during all operating conditions [34]. Therefore, one might need to acknowledge this feature as one of EV's potential advantages.

5.4 IMPACT OF DER PENETRATION ON PROPOSED SMART GRID ADAPTIVE ENERGY CONSERVATION & OPTIMIZATION

Penetration of DERs throughout smart microgrids has raised grid complexities as well as uncertainties that affect conventional monitoring, optimization and control systems. Hence, electric power utilities are trying to verify the impact of DER penetration on their monitoring, control and optimization systems. As mentioned above, DERs such as μ-CHP units or PVs are used for different operational aims. PV modules supply residential buildings with clean active power while μ-CHP units can provide both heat and electricity. Prior to investigating the impact of different μ-CHP/PV penetration levels on different ECOE subparts, it is necessary to identify μ-CHP and PV types and penetration levels. This chapter takes two main types of μ-CHP units into account:heat-led with electricity rejection and electricity-led with heat rejection. As in real-field practices, μ-CHP units are considered 1, 3, and 5 kWe in this study. The average share of generated power of PV, electricity-led μ-CHP and heat-led μ-CHP along studied microgrid feeder nodes assumed to be equal to 30%, 20%, and 50% respectively. Moreover, as most μ-CHPs are heat-led, the ratio between heat-led μ-CHP and an electricity-led μ-CHP is assumed to be 2.5 in this study. Fig. 5.7 depicts the daily electricity generation of μ-CHP units and PVs in the last node of the system, i.e., node-33, produced by AMI data.

Three different penetration levels are taken into account to check the impact of μ-CHP/PV units on proposed smart microgrid-based ECOE:

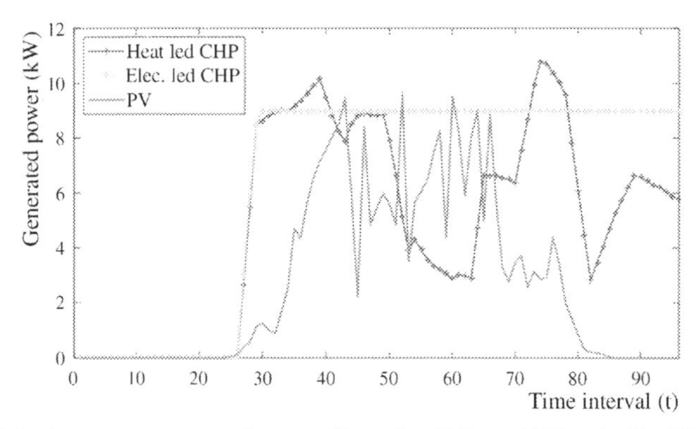

Figure 5.7 Active power generation profiles of μ-CHP and PV units for 96 quasi real-time operating intervals.

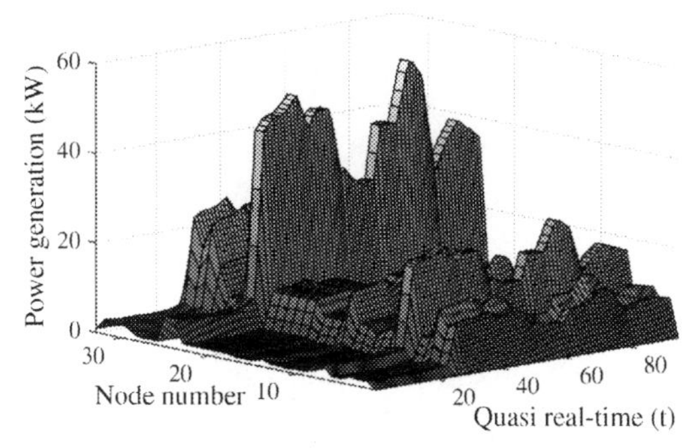

Figure 5.8 Active power generation profiles of cogeneration units collected by the AMI for 96 quasi real-time intervals.

normal (25%), high (50%), and very high (75%). At each penetration level, the proposed engine strives to optimize the grid and find the optimal response of each VVCC. The active power generation of microgeneration, comprising PVs, heat-led, and electricity-led μ-CHP units, for all quasi real-time intervals (i.e., every 15 minutes) resulting from the AMI data, is shown in Fig. 5.8.

5.4.1 Proposed Smart Grid-based Energy Conservation and Optimization Engine

This section of the chapter proposes a quasi-real-time ECOE, which uses real AMI data (e.g., active/reactive energy consumption records of the loads) as its inputs. Fig. 5.9 presents the main control-command topology of the presented solution in a typical North American distribution feeder

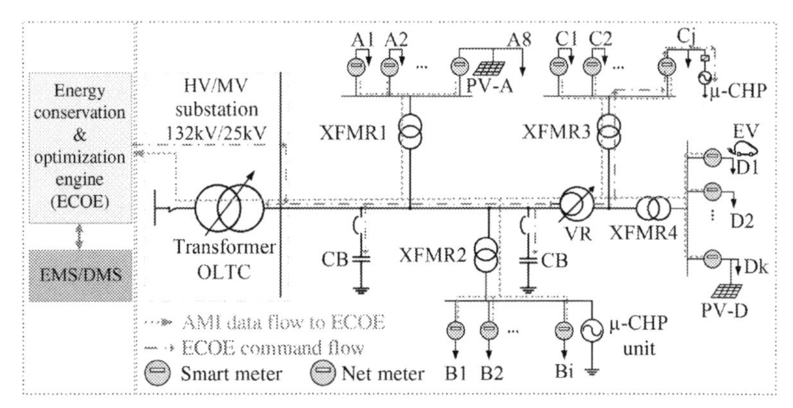

Figure 5.9 AMI-based ECOE main topology in a typical distribution feeder.

comprising microgeneration units such as μ-CHP and PV. The proposed system includes an optimization engine, which initially collects the load data of grid nodes using AMI. This engine optimizes the grid through its optimization algorithm to find the optimal settings of VVCCs. By finding the microgrid's optimal solution, the optimization engine sends control commands to VVCCs whether to keep their existing settings or to change them according to new optimal configurations (see command flow in Fig. 5.9).

The proposed smart microgrid adaptive ECOE is suitable for the distributed command and control topology of microgrids, as it could perform decentralized optimization and control over different microgrid assets from the MV substation downstream to its distribution feeder(s) using local AMI data. Further detail regarding the main architecture and tasks of the proposed solution can be found in [17,38,39].

5.4.2 Case Study: Data and Simulations

To test the accuracy and the applicability of presented smart microgrid-based ECOE in the presence of various DER penetrations, a 33-node microgrid [40] is used. Fig. 5.10 depicts a single line diagram and Table 5.2 presents general data of the case study. As depicted in Table 5.2, the studied microgrid comprises 33 nodes with 32 termination points, i.e., smart meters. The active/reactive power consumption of the nodes is collected by an ECOE designed in MATLAB. The energy storage model is based on Fig. 5.5. For the EV model, this chapter uses [39] model which employs ZIP model for EVs. According to [39], each EV type has

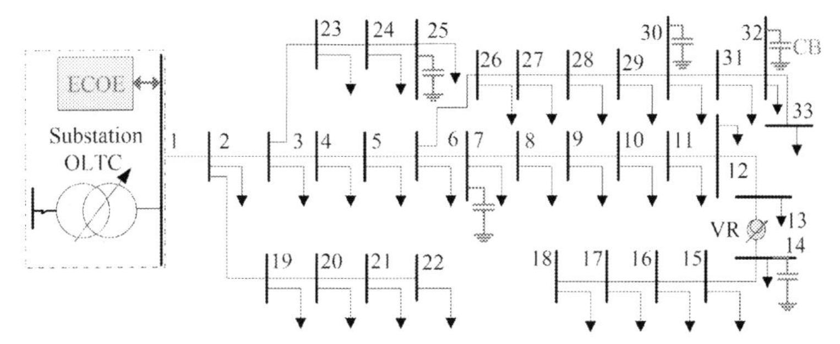

Figure 5.10 33-node microgrid case study comprising of ECOE control components such as OLTC, VR, and CBs.

Table 5.2 General data of 33-node microgrid

Node no.	Location(s)	Tap/reactive power (kVAr)	Tap/kVAr range	Tap/ switches
OLTC, VR	1, 14	− 5 to +5	16 + 1	LTC, CR
CBs	7, 14, 25, 30, 32	50 to 250	5 each	CBs
Optimization Technique:	Particle Swarm Optimization [41]			
Convergence Rate, Iteration	0.999, 50			

Table 5.3 Coefficient-factor setups for the case study

Load ZIP coefficients	Z	I	P
Active, reactive	0.418, 0.515	0.135, 0.023	0.447, 0.462
Costs	\$/kW, \$kW	\$/Tap	\$/kVAr-year
π_t, $C_{S.E,t}$	0.12	—	—
$C_{w,t}$, $C_{r,t}$, $C_{c,t}$	—	0.09	0.09
Fuzzification values	*dmax*		*dmin*
f_{Loss}, f_{CVR}	0.35 (P.U), 7 (%)		0.1 (P.U), 6 (%)
f_{OLTC}, f_{VR}, f_{CB}	0.1		0.09

its ZIP model that has to be integrated with other EV types in order to find the aggregated ZIP model for all EVs connected to the grid. Here, the ECOE is tested in quasi-real-time, i.e., every 15 minutes for a complete day (96 AMI time intervals). According to ECOE objectives Eqs. (5.1–5.15) and constraints Eqs. (5.16–5.27), the ECOE found optimal solutions for each quasi real-time interval and sent control commands

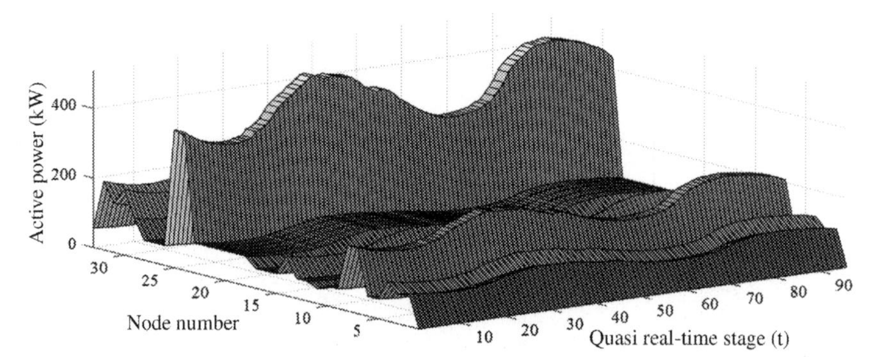

Figure 5.11 Active power consumptions of microgrid loads collected by the AMI for 96 quasi real-time intervals.

Table 5.4 Energy conservation and optimization engine results

Scenarios	Scenario-1	NP	HP	VHP
Obj. function ($)	17967.93	16525.74	15103.54	13654.36
Power loss (kW)	15679.72	13477.8	11558.48	9886.308
ΔE% by CVR	4.342111	4.773744	5.194733	5.581884
ΔV% by CVR	4.775398	5.228552	5.67508	6.087458
CVRf	0.907936	0.912401	0.914834	0.916583

to proper VVCCs. Table 5.3 gives ZIP coefficients and the different costs and fuzzification values used in the proposed engine during the study. Moreover, Fig. 5.11 shows the active power consumption of the load for all quasi real-time intervals collected from local AMI. To fully assess the impacts of different μ-CHP/PV penetration levels on the presented smart microgrid-based EOCE objective function subparts, four operating scenarios were studied: scenario-1: ECOE without μ-CHP/PV penetration, scenario-2: ECOE performs in normal penetration (NP) of μ-CHP/PVs, (i.e., 25% of nodes consist of μ-CHP/PV units), scenario-3: ECOE performs in high penetration (HP) of μ-CHP/PVs, (50% of nodes consist of μ-CHP/PV units), and scenario-4: ECOE performs in very high penetration (VHP) of μ-CHP/PVs (75%). Therefore, this chapter represents the percentages of penetration levels for combination of μ-CHP and PV units as key parameters of the optimization problem.

The main results of the study are shown in Table 5.4. Fig. 5.12 compares the objective function values for four different operating scenarios. Fig. 5.13 gives the power loss reduction achieved by the proposed ECOE.

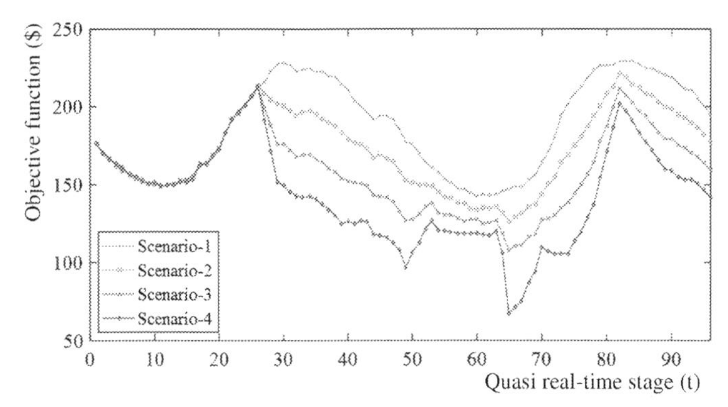

Figure 5.12 Objective function results by ECOE for different operating scenarios: (1) ECOE without penetration, (2) ECOE with normal penetration, (3) ECOE with high penetration, and (4) ECOE with very high DER penetration.

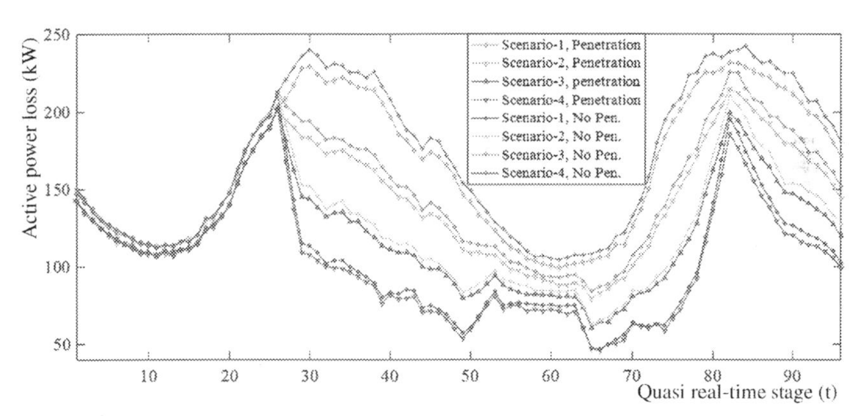

Figure 5.13 Total active power losses calculated by ECOE for different operating scenarios. First four scenarios: with μ-CHP/PV penetration; second four scenarios: without μ-CHP/PV penetration.

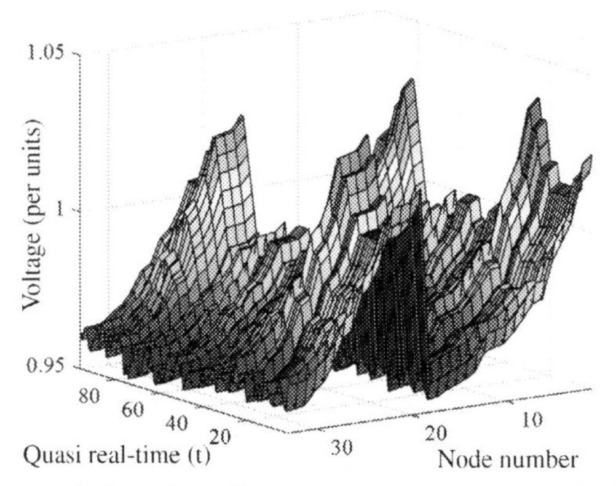

Figure 5.14 Voltage of all nodes of the system in scenario-4 (high penetration of μ-CHP/PV) found by ECOE showing all nodes are kept within the ANSI band.

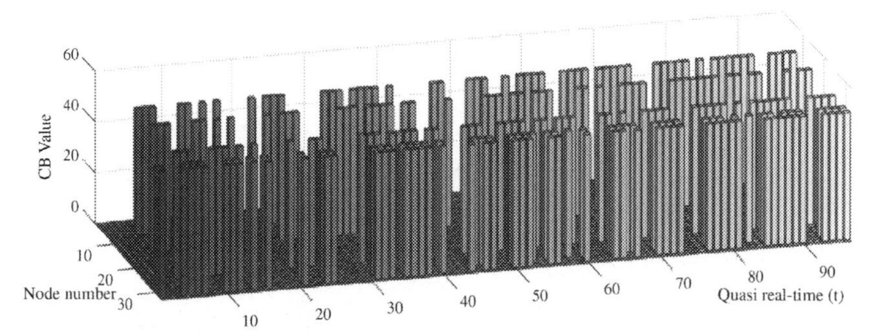

Figure 5.15 Optimal Values of shunt CB in scenario-4 (high penetration of μ-CHP/PV) achieved by ECOE.

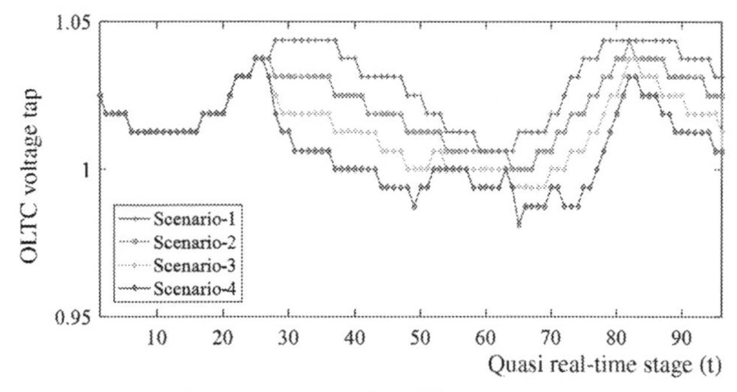

Figure 5.16 OLTC optimal tap positions for different operating scenarios: (1) ECOE without penetration, (2) ECOE with normal penetration, (3) ECOE with high penetration, and (4) ECOE with very high DER penetration.

Table 5.5 Number of bank and/or tap switching of VVCCs

Scenarios	Scenario-1	NP	HP	VHP
CB-7, 14	54, 56	50, 52	51, 51	37, 41
CB-25, 30, 32	44, 51, 52	53, 52, 54	39, 47, 48	39, 45, 41
Switching No.	261	257	236	203
Total kVAr	15500	15150	14200	13150
OLTC, VR	21, 6	22, 8	26, 2	33, 10

For each operating scenario, it is possible to check whether the voltage levels of all microgrid nodes are within ANSI range. Fig. 5.14 illustrates all microgrid node voltages in scenario-4. Fig. 5.15 gives CB's optimal kVAr and Fig. 5.16 depicts OLTC operations for all operating scenarios every 15 minutes as ECOE control command examples.

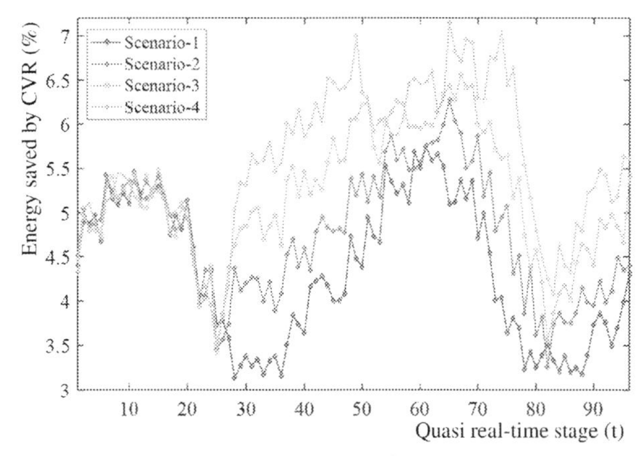

Figure 5.17 Saved energy resulted by CVR in different operating scenarios: (1) ECOE without penetration, (2) ECOE with normal penetration, (3) ECOE with high penetration, and (4) ECOE with very high DER penetration.

Table 5.5 summarizes the number of capacitor bank and/or tap switching of OLTC/VR for the studied day. Finally, Fig. 5.17 represents the percentage of energy saved from consumption by performing CVR using ECOE in the four penetration scenarios for each quasi real-time interval.

5.4.3 Result Analysis and Further Discussions

The correctness and the applicability of a smart microgrid-based ECOE were tested in four different operating scenarios in the case-study. From Table 5.4 and Fig. 5.12, it can be seen that the optimization engine accurately minimized the objective function at each quasi real-time interval. A comparison of scenario-1 with scenario-4 shows that the AMI-based ECOE could minimize the objective function as well as power loss better (by about 24%) in the last scenario. The reason for this lies in the fact that, by increasing penetration levels of μ-CHP/PV, DER generation will also increase. This can decrease the power losses of the system locally. Thus, Fig. 5.13 shows that the ECOE slightly decreased power loss in each scenario compared with a case in which the system has the same generation but does not use ECOE. Monitoring the voltage levels of all microgrid nodes in the operating scenario considered in Fig. 5.14 showed that the engine kept all voltages within ANSI range and performed CVR to conserve energy.

Fig. 5.15 illustrated that the operating cost minimization subpart of ECOE led the system to adopt less KVAR in most quasi real-time intervals (e.g., 50KVAR). Moreover, as seen in Table 5.5, by increasing the penetration level of μ-CHP/PVs, ECOE can minimize microgrid loss with less total KVAR injection and a lower number of CB switching. Fig. 5.16 showed that the OLTC located in node-1 of the case study had more impact on ECOE, as it covers more microgrid nodes. It has to be mentioned that this study uses BFS power flow technique which supports OLTC tap changes. As such, any time an OLTC tap changes, the power flow solution will run again to obtain the optimal solution precisely. It is clear that changing OLTC taps can affect voltage levels of downstream nodes. However, the ECOE is capable of keeping all node voltages within the ANSI standard and avoid unacceptable voltage solutions. By increasing the penetration of μ-CHP/PV units, OLTC reduced the voltage of the system better with a larger number of tap operations.

Regarding the impact of μ-CHP/PV on the CVR subpart of ECOE, it is possible to conclude from Table 5.4 and Fig. 5.17 that increasing the level of μ-CHP/PV penetration a case which could lead to more effective energy conservation (CVR). There are two main reasons for this. First, by increasing the level of μ-CHP/PV penetration, DER generation will also increase. Thus, it would be easier for the ECOE to lower the voltage levels of nodes to perform energy conservation. In other words, by increasing local generation due to μ-CHP/PV penetration, voltage reduction of CVR will be increased. Second, CVR led to a greater reduction in energy consumption by increasing μ-CHP/PV penetration, as the characteristics of the load remained the same in all scenarios. Table 5.3 showed the ZIP coefficients of system loads. Some smart grid components, such as EVs could change the characteristics of loads, which can cause changes in ZIP coefficient values and mixes. The results showed that if the load characteristics are not fully constant power, it is possible to expect that increasing μ-CHP/PV penetration could save more energy through CVR. As both energy consumption and voltage reduction were improved by increasing μ-CHP/PV penetration, the CVR factor of feeder shown in Table 5.4 slightly improved as well.

It can be realized from Table 5.3 that fuzzification and cost values defined in this study led ECOE to focus more on the CVR subpart rather than on the loss reduction subpart. That further loss reduction could be achieved by giving more weight to the ECOE loss minimization subpart rather than to the CVR subpart. This could lead ECOE to focus more on loss minimization by injecting more KVAr into the system while performing less CVR. Thus, using fuzzification gives grid operators a significant flexibility to set their optimization priorities based on microgrid operational requirements.

As a result, the study of the impact of μ-CHP/PV penetration led the smart microgrid-based ECOE to optimize microgrids with more accuracy. Fuzzification technique could assist grid operators to correctly give weights to ECOE objective function subparts at each quasi real-time interval according to microgrid technical/economic needs.

5.5 CONCLUSION

This chapter presented a smart microgrid adaptive ECOE (i.e., ECOE) capable of minimizing grid loss, microgrid voltage, and reactive power control assets and energy conservation (CVR) costs using local AMI data. Moreover, to assess the performance of ECOE in the presence of different μ-CHP/PV penetration levels, (i.e., normal, high, and very high), a 33-node microgrid is used. The test results showed considerable changes in the operating results of the proposed smart microgrid ECOE as a result of changes in μ-CHP/PV penetration levels. The results of this study showed that advanced smart microgrid energy optimization and conservation solutions can be used to optimize microgrids in the presence of different DER penetration levels. The proposed ECOE minimized the costs of active power loss effectively. It injected less reactive power into the system with a lower number of CB switching as the local grid generation level was elevated by increasing the μ-CHP/PV penetration level. Moreover, CVR effectiveness can be improved by increasing the μ-CHP/PV penetration level due to the enhancement of microgrid local active power generation using fuzzification. Hence, OLTC could reduce system voltage levels more by increasing the μ-CHP/PV penetration level.

In summary, an effective weighting strategy for smart microgrid-based ECOE subsystems, based on microgeneration penetration levels, would be critically important for the voltage and reactive power optimization of future smart microgrid. Fuzzification technique, used in this study, led to the better performance of the smart microgrid-based ECOE in the presence of different μ-CHP/PV penetration levels. The results showed that fuzzification values of the ECOE that are set by the grid operator can define ECOE key strategy and flow of each grid operating time interval. Considering DER penetration on smart microgrid-based energy conservation and optimization solutions could help smart microgrid planners and operators to adopt effective technologies according to their impacts on microgrid feeders. Moreover, studying the impact of DER penetration could result in more accurate energy conservation and optimization solutions as, typically, conventional approaches do not take DER impact on energy conservation into account. Applying predictive algorithms to find weighting factors for each quasi real-time interval for each ECOE subpart could be a novel future area of study that might help ECOE to perform more effectively in the presence of different microgeneration penetrations and load conditions.

In conclusion, the results of this chapter can help new smart microgrid-based energy optimization solutions, including energy conservation solutions, to use AMI data to increase efficiency and accuracy in the presence of different microgeneration penetration levels.

REFERENCES

[1] H. Farhangi, A road map to integration, IEEE Power Energy Magazine 12 (3) (May/Jun. 2014) 52–66.
[2] Electrical Power Systems and Equipment-Voltage Ratings, ANSI Standard C 84.1, 1995.
[3] R.F. Preiss, V.J. Warnock, Impact of voltage reduction on energy and demand, IEEE Trans. Power Apparatus Systems 97 (5) (Sep. 1978) 1665–1671.
[4] R.F. Preiss, V.J. Warnock, Impact of voltage reduction on energy and demand: phase II, IEEE Trans. Power Systems (2) (May 1986) 92–95.
[5] D.M. Lauria, Conservation voltage reduction (CVR) at northeast utilities, IEEE Trans. Power Delivery 2 (4) (Oct. 1987) 1186–1191.
[6] B.W. Kennedy, R.H. Fletcher, Conservation voltage reduction (CVR) at SNOHOMISH county PUD, IEEE Trans. Power Systems 6 (3) (Aug. 1991) 986–998.
[7] D. Kirshner, Implementation of conservation voltage reduction at commonwealth Edison, IEEE Trans. Power Systems 5 (4) (Nov. 1990) 1178–1182.
[8] S. Auchariyamet, S. Sirisumrannukul, Volt/VAr Control in Distribution Systems by Fuzzy Multiobjective and Particle Swarm, Proceedings of the 6th International conference on Electrical Engineering/ Electronics, Computer, Telecom. and Info. Tech., ECTI-CON, Pattaya, Thailand, May 2009.
[9] A. Rahideh, M. Gitizadeh, A. Rahideh, Fuzzy Logic in real time voltage/reactive power control in FARS regional electric network, Electr. Power System Res. 76 (Feb. 2006) 996–1002.
[10] N.S. Markushevich, I.C. Herejk, R.E. Nielsen, Functional requirements and cost-benefit study for distribution automation at B.C. Hydro, IEEE Trans. Power Systems 9 (2) (May 1994) 772–781.
[11] T.L. Wilson, Conservation energy with voltage reduction-fact or fantasy, in: Proceedings of the IEEE Rural Electric Power Conference, Colorado Springs, Co, May 2002.
[12] Global Energy Partners LLC, Utility Distribution System Efficiency (DEI), Northwest Energy Efficiency Alliance Report #08-192, Portland, Oregon, USA, Jun. 27, 2008.

[13] L. Schwartz, Is it smart if it is not clean? Strategies for Utility Distribution Systems, Part One, RAP, Montpelier, Vermont, USA, May 2010.
[14] K.P. Schneider, F.K. Tuffner, J.C. Fuller, R. Singh, Evaluation of Conservation Voltage Reduction (CVR) on a National Level, Pacific Northwest National Laboratory, United States Department of Energy, DE-AC05-76RL01830, Jul. 2010.
[15] M. Wakefield, EPRI Smart Grid Demonstration Initiative & Early Results, EPRI Clean Energy-Session 7, GreenGov Symposium, Washington DC, USA, Nov. 2011.
[16] Electric Authority, Te Mana Hiko, Available for: <https://www.ea.govt.nz/>
[17] M. Manbachi, M. Nasri, B. Shahabi, H. Farhangi, A. Palizban, S. Arzanpour, et al., Real-time adaptive VVO/CVR topology using multi-agent system and IEC 61850-based communication protocol, IEEE Trans. Sustain. Energy 5 (2) (April 2014) 587−597. Apr. 2014.
[18] M. Manbachi, H. Farhangi, A. Palizban, S. Arzanpour, Smartgrid adaptive volt-var optimization: challenges for sustainable future grids, Sustain. Cit. Soc. J. 28 (Jan. 2017) 242−255.
[19] Communication networks and systems for power utility automation-Part 7-420: basic communication structure-distributed energy resources logical nodes, IEC 61850-7-420-2009 standard, Oct. 2009.
[20] IEC 61850, Power Utility Automation, Relevant Application: EMS, DMS, DA, SA, DER, AMI, Storage, EV, Available for: <http://www.iec.ch/smartgrid/standards/>.
[21] M. Manbachi, H. Farhangi, A. Palizban, S. Arzanpour, Smart grid adaptive energy conservation and optimization engine utilizing Particle Swarm Optimization and Fuzzification, Appl. Energy J. 174 (Jul. 2016) 69−79.
[22] IEEE 802-2014, Revision to IEEE std. 802-2001 IEEE Standard for Local and Metropolitan Area Networks: Overview and Architecture Sponsored by the LAN/MAN Standards Committee, 12 June 2014 IEEE-SA Standards Board. Available for: <https://standards.ieee.org/getieee802/download/802-2014.pdf>.
[23] IEEE Std. 2030, IEEE Guide for Smart Grid Interoperability of Energy Technology and Information Technology Operation with the Electric Power System (EPS), End-Use Applications, and Loads, Sep. 2011.
[24] Functional Specifications for Community Energy Storage (CES) Unit, Revision 2.2, American Electric Power (AEP), 12 Sep. 2009.
[25] H. Asgeirsson, DTE energy: energy storage demonstration projects, in: IEEE PES General Meeting - Energy Storage Super Session, 2011.
[26] M.C. Kisacikoglu, B. Ozpineci, L.M. Tolbert, Effects of V2G Reactive Power Compensation on the Component Selection in an EV or PHEV Bidirectional Charger, Proceedings of the IEEE Energy Conversion Congress and Exposition (ECCE), Atlanta, Georgia, USA, Sep. 2010, pp. 870−876.
[27] M.C. Kisacikoglu, B. Ozpineci, L.M. Tolbert, Reactive Power Operation Analysis of a Single-Phase EV/PHEV Bidirectional Battery Charger, Proceedings of the 2011 8th International Conference on Power Electronics (ECCE), Jeju, Korea, May/Jun. 2011, pp. 585−592.
[28] R.C. Green, L. Wang, M. Alam, The Impact of Plug-in Hybrid Electric Vehicles on Distribution Networks: a Review and Outlook, Proceedings of the IEEE Power and Energy Society General Meeting, Minneapolis, USA, Jul. 2010.
[29] S. Shao, T. Zhang, M. Ipattanasomporn, S. Rahman, Impact of TOU Rates on Distribution Load Shapes in a Smart Grid with PHEV Penetration, Proceedings of the IEEE PES Transmission and Distribution Conference and Exposition, New Orleans, LA, USA, Apr. 2010.
[30] K. Clement, H. Haesen, J. Driesen, Coordinated Charging of Multiple Plug-In Hybrid Electric Vehicles in Residential Distribution Grids, Proceedings of the IEEE PES Power Systems Conference and Exposition, Seattle, USA, Mar. 2009.

[31] S.S. Raghavan, A. Khaligh, Impact of Plug-in Hybrid Electric Vehicle Charging on a Distribution Network in a Smart Grid Environment, Proceedings of the IEEE PES on Innovative Smart Grid Technology (ISGT), Washington DC, Jan. 2012.
[32] C. Weiller, Plug-in hybrid electric vehicle impacts on hourly electricity demand in the United States, Energy Policy Journal 39 (2011) 3766–3778.
[33] M.C. Kisacikoglu, B. Ozpineci, L.M. Tolbert, EV/PHEV bidirectional charger assessment for V2G reactive power operation, IEEE Trans. Power Electron. 28 (12) (Dec. 2013) 5717–5727.
[34] M. Ehsani, M. Falahi, S. Lotfifard, Vehicle to grid services: potential and applications, Energies J. 5 (2012) 4076–4090.
[35] M.C. Kisacikoglu, B. Ozpineci, L.M. Tolbert, F. Wang, Single-phase inverter design for V2G reactive power compensation, in: Proceedings of the 26th Annual IEEE Applied Power Electronics Conference and Exposition (APEC), Fort Worth, TX, USA, Mar. 2011.
[36] L.J. Borle, Zero Average Current Error Control Methods for Bidirectional AC-DC Converters, PhD thesis, Curtin University of Technology, Perth, Australia, 1999.
[37] M.C. Kisacikoglu, B. Ozpineci, L.M. Tolbert, Examination of a PHEV Bidirectional Charger System for V2G Reactive Power Compensation, in: Proceedings of the 26th Annual IEEE Applied Power Electronics Conf. and Exposition (APEC), Fort Worth, USA, Mar. 2011.
[38] M. Manbachi, A. Sadu, H. Farhangi, A. Monti, A. Palizban, F. Ponci, et al., Real-time co-simulation platform for smart grid volt-var optimization using IEC 61850, IEEE Trans. Ind. Informat. 12 (4) (Aug. 2016) 1392–1402.
[39] M. Manbachi, A. Sadu, H. Farhangi, A. Monti, A. Palizban, F. Ponci, et al., Impact of EV penetration on volt-var optimization of distribution networks using real-time co-simulation monitoring platform, Appl. Energy J. 169 (4) (May. 2016) 28–39.
[40] M.E. Baran, F.F. Wu, Network reconfiguration in distribution systems for loss reduction and load balancing, IEEE Trans. Power Delivery 4 (3) (Apr. 1989) 1401–1407.
[41] M. Manbachi, M. Ordonez, AMI-based energy management for islanded AC/DC microgrids utilizing energy conservation and optimization, IEEE Trans. Smart Grid (Aug. 2017). Available for: <https://doi.org/10.1109/TSG.2017.2737946>.

FURTHER READING

SAE J1772, SAE Charging Configurations and Ratings Terminology, SAE International, Hybrid Committee, Version. 031611, 2011.

CHAPTER 6

Short-term Scheduling of Future Distribution Network in High Penetration of Electric Vehicles in Deregulated Energy Market

Mehrdad Ghahramani, Sayyad Nojavan, Kazem Zare and Behnam Mohammadi-ivatloo
University of Tabriz, Tabriz, Iran

6.1 INTRODUCTION

Over the recent decade, renewable energy units, mainly wind turbines (WTs), due to their environmental advantages (namely no carbon, no SOx and NOx emissions) became increasingly attractive [1]. In addition, advancement of technology and moving toward smart grids especially at distribution level facilitated participation of customers in providing required energy and reserve of distribution systems through demand response (DR) programs [2]. Moreover, increasing penetration of plug-in electric vehicles (PEVs) as one of the main loads of future distribution networks (FDN) is inevitable [3]. It is believed that optimal scheduling of renewable energy resources, PEVs and DR programs will decrease the total costs of FDNs [4].

6.1.1 Problem Definition

The presence of one or more of aforementioned elements in day-ahead scheduling of FDNs because of their planning complexity will lead DNO toward more complex and challenging tasks (i.e., jeopardizing power balance, reducing reliability, and increasing operation costs). Therefore, in order to appropriate and optimal economic decisions in the presence of various energy units and their complexities a proper operation scheduling is needed [5].

Operation of Distributed Energy Resources in Smart Distribution Networks
DOI: https://doi.org/10.1016/B978-0-12-814891-4.00006-0

6.1.2 Literature Review

Determining the optimal amount of energy and reserve through optimal day-ahead scheduling (ODAS) of renewable resources as the key feature of FDNs has attracted high attention and various studies have been published in this context. In [6], an operational scheduling of distribution networks based on stochastic method is proposed to determine the optimal amount of energy and reserve provided by different energy resources while considering the errors of wind and load forecast. In [7], through the optimal operation of distribution networks the idea of integrating energy storage systems to the distribution networks is investigated and its effectiveness on efficiency of these networks is studied. In [8], in order to deal with the problems of PEVs uncertainty, the storage and distributed generation scheduling is utilized. In addition, Tabu Search and Simulated Annealing algorithms are utilized for minimizing the total operation cost of network. The optimization method which is proposed in [9] uses a multiobjective cost-emissions stochastic method in order to schedule distributed energy units while considering the alternative nature of load, wind, and solar generation. In [10], probability distribution functions are used for modeling the uncertainties of wind and solar generation while a two-stage stochastic objective function is produced in order to minimize the total operational cost. A robust stochastic optimization method is employed in [11] to build a stochastic model for wind power generation while price-based and incentive-based DR programs are utilized for improving network operation. The authors of [12] propose a robust optimization scheduling based on concepts of optimal unit commitment which wind power, DR programs, and bulk energy storages is used in cooptimization of energy and reserve scheduling. In [13] a multiobjective optimal operation problem is solved using fuzzy satisfying method in the presence of fuel cell power. An optimal planning of storage systems based on the optimal operational power flow of distribution networks is presented in [14], which the uncertainties of wind power are considered. In [15], a fuzzy method is employed in order to analyze the effects of DG units operation on network losses, and the ability of distribution system in demand supply while the uncertainties of demand and installed capacity of DG units are considered. In [16], an optimal short-term scheduling is used for a smart active distribution system which utilizes each hour network reconfiguration with the aim of minimizing operational costs. In [17], an optimal power flow method is utilized for the 1-hour-ahead operation of distribution networks in which wind power uncertainties are addressed through a two-stage stochastic method. In [18], in order to determine optimal amounts of energy and reserve scheduling of smart distribution system in the presence of high penetration of wind power, a stochastic multiobjective short-term scheduling based on augmented e-constraint method is presented. A new multiobjective model is proposed in [19] to determine an optimal active and reactive power scheduling for minimizing both the cost and voltage magnitude difference in the presence of capacitor banks, distributed generation, and EVs. In [20], an interval optimization is utilized for modeling the uncertainties of load and renewable energy generation in (ODAS) distribution systems. In [21], a multiobjective optimization method is utilized for optimal scheduling of the single-phase PEVs charging/discharging through unbalanced three-phase distribution networks. A bi-level information-gap decision theory is presented in [22] to hedge DNO against uncertainties of wholesale market prices.

6.1.3 Procedure and Contributions

In this chapter an ODAS model for FDNs is proposed in which PEVs and responsive loads can take part in optimal scheduling of both energy and reserve while the goal is minimizing total costs of distribution network operator (DNO). The proposed model determines optimal generation of DG units, PEVs and WTs in FDNs. Also, the proposed model ascertains transactions with the upstream market, load reduction of responsive loads, and PEVs charge/discharge, while the active and reactive power balance of the system is considered. The initiative contributions of this study are highlighted as follows:

- proposing a novel model for ODAS of FDNs while considering technical constraints of distribution networks, diesel generators (DGs), PEVs and DR programs;
- integrating PEVs, intermittent renewable generation, and DR programs in operational scheduling of FDNs while EV battery's stored energy and load flexibility takes part in both energy and reserve scheduling.

6.1.4 Chapter Organization

The next sections of this study are organized as follows. Problem formulation and modeling for deterministic ODAS of the FDNs are given in

Section 6.2. Numerical results are detailed in Section 6.3. In Section 6.4, the chapter closes with some important findings through the conclusion.

6.2 PROBLEM FORMULATION

The deterministic formulation of the objective function and constraints for ODAS of FDNs is completely detailed in this section. Furthermore, the mathematical models of distribution network, WTs and DR programs are described in this section. The following assumptions are considered in proposed operational scheduling.

- Decision making about behavior of market participants are handled by DNO.
- All of the needed information including the bids of responsive loads and forecasting data of wind speed, solar irradiance and loads demand are received by DNO.
- The required data of DG units, distribution system model and upstream grid prices are available for DNO.

The mathematical model of the proposed ODAS of distribution system is provided in the following.

6.2.1 Objective Function

The objective function of this chapter is minimizing total cost of DNO with determining optimal amounts of energy and reserve required for FDNs. The total operation cost of distribution network is the cost of providing required energy and reserve through different resources.

$$Min \sum_{t=1}^{N_T} \left[\begin{array}{l} \varsigma_{ug}(t) \times P_{ug}^{DAS}(t) \\ + \sum_{j=1}^{N_{DG}} \left\{ cost_{DG}^{DAS,E}(j,t) + cost_{DG}^{DAS,SU}(j,t) + cost_{DG}^{DAS,R}(j,t) \right\} \\ + \sum_{i=1}^{N_{LL}} \left\{ cost_{LL}^{DAS,E}(i,t) + cost_{LL}^{DAS,R}(i,t) \right\} \\ + \sum_{d=1}^{N_{DRA}} \left\{ cost_{DRA}^{DAS,E}(d,t) + cost_{DRA}^{DAS,R}(d,t) \right\} \end{array} \right] \quad (6.1)$$

The proposed objective function has four terms. The costs of purchasing power from upstream grid are proposed in first term. DG operation costs including the energy cost, startup cost, and reserve cost are represented in second term. The costs of providing energy and reserve by demand response aggregators (DRAs) and large loads (LLs), are represented respectively.

6.2.2 Constraints and Mathematical Modeling

A series of equality and inequality constraints which should be taken into account in ODAS of FDNs are proposed in the following:

6.2.2.1 Power Flow Constraints

Power flow equations show that in all of the buses, the supplied active and reactive powers should be equal to consumed active and reactive powers. The balance of supplied and consumed power for nth bus at time t can be formulated as follows [23]:

$$\begin{aligned} &P_{ug}^{DAS}(t) + \sum_{j\in n} P_{DG}^{DAS}(j,t) + \sum_{w\in n} P_{Wind}^{DAS}(w,t) \\ &+ \sum_{v} P_{ch}(t,v) - \sum_{v} P_{dis}(t,v) \\ &+ \sum_{i\in n} P_{LL}^{DAS}(i,t) + \sum_{d\in n} P_{DRA}^{DAS}(d,t) - P_{load}^{DAS}(n,t) \\ &= V_{i,h}\sum_{j} V_{j,h}(G_{ij}\cos\delta_{i,h} + B_{ij}\sin\delta_{j,h}) \end{aligned} \quad \forall m,n,t \tag{6.2}$$

$$\begin{aligned} &Q_{ug}^{DAS}(t) + \sum_{j\in n} Q_{DG}^{DAS}(j,t) + \sum_{w\in n} Q_{Wind}^{DAS}(w,t) \\ &+ \sum_{d\in n} Q_{DRA}^{DAS}(d,t) + \sum_{i\in n} Q_{LL}^{DAS}(i,t) - Q_{load}^{DAS}(n,t) \\ &= V_{i,h}\sum_{j} V_{j,h}(G_{ij}\cos\delta_{i,h} - B_{ij}\sin\delta_{j,h}) \end{aligned} \quad \forall m,n,t \tag{6.3}$$

6.2.2.2 Distribution Network Constraints

In order to have technical safe operation, the level of voltage of all buses should be in a specific interval based on Eq. (6.4).

$$V_{\min} \leq V(n,t) \leq V_{\max} \tag{6.4}$$

Due to the limitation of substation in power interaction with upstream grid, the active and reactive power of upstream-grid should be controlled [24]. The active and reactive powers of the substation are restricted as Eqs. (6.5) and (6.6).

$$P_{ug}^{\min} \leq P_{ug}^{DAS}(t) \leq P_{ug}^{\max} \tag{6.5}$$

$$Q_{ug}^{\min} \leq Q_{ug}^{DAS}(t) \leq Q_{ug}^{\max} \tag{6.6}$$

6.2.2.3 DG Unit Constraints

The performance cost of DG units can be calculated as a quadratic function as shown in Eq. (6.7) [17]:

$$\text{cost}_{DG}^{DAS,E}(j,t) = \alpha_{3,j} \times (P_{DG}^{DAS}(j,t))^2 + \alpha_{2,j} \times P_{DG}^{DAS}(j,t) + \alpha_{1,j} \quad \forall j,t \tag{6.7}$$

The startup cost of DG units obtained from Eqs. (6.8) and (6.9) specifies that it should be positive.

$$cost_{DG}^{DAS,su}(j,t) = \varsigma_{su}(j) \times (a(j,t) - a(j,t-1)) \tag{6.8}$$

$$cost_{DG}^{DAS,su}(j,t) \geq 0 \tag{6.9}$$

The cost of DGs participation in providing required reserve of distribution networks can be considered as 20% of the marginal price of their participation in power production. Thus, the reserve cost could be calculated through Eq. (6.10).

$$cost_{DG}^{DAS,R}(j,t) = 0.2 \times (\alpha_{2,j} + 2 \times \alpha_{3,j} \times P_{DG}^{\max}(j,t)) \times R_{DG}^{DAS}(j,t) \quad \forall j,t \tag{6.10}$$

In addition, DGs should operate with considering the upper and lower limits of maximum installed capacity as Eqs. (6.11) and (6.12).

$$P_{DG}^{\min}(j) \times a(j,t) \leq P_{DG}^{DAS}(j,t) \leq P_{DG}^{\max}(j) \times a(j,t) \quad \forall j,t \tag{6.11}$$

$$P_{DG}^{DAS}(j,t) + R_{DG}^{DAS}(j,t) \leq P_{DG}^{\max}(j) \times a(j,t) \quad \forall j,t \tag{6.12}$$

The technical constraints do not allow DG units to ramp up and ramp down their production more than a determined ratio [25], which can be stated as Eqs. (6.13) and (6.14).

$$P_{DG}^{DAS}(j,t) - P_{DG}^{DAS}(j,t-1) \leq P_{RU}(j) \times (1 - b(j,t)) + P_{DG}^{DAS,\min}(j) \times b(j,t) \quad \forall j,t \tag{6.13}$$

$$P_{DG}^{DAS}(j,t-1) - P_{DG}^{DAS}(j,t) \leq P_{RD}(j) \times (1 - c(j,t)) + P_{DG}^{DAS,\min}(j) \times c(j,t) \quad \forall j,t \tag{6.14}$$

In addition, as Eqs. (6.15) and (6.16), when the conventional DG units is turned on, it should maintain current mode for specified hours before shutting-down and also after shutting-down it should be in this condition for determined hours before restarting again [26].

$$\sum_{h=t}^{t+T_{ut}(j)-1} a(j,h) \geq T_{ut}(j) \times b(j,t) \quad \forall j,t \tag{6.15}$$

$$\sum_{h=t}^{t+T_{DT}(j)-1} (1 - a(j,h)) \geq T_{DT}(j) \times c(j,t) \quad \forall j,t \tag{6.16}$$

The following constraints defining the hourly mode of DGs do not let a DG unit be in both the operating and turning off conditions at the same time.

$$b(j,t) - c(j,t) = a(j,t) - a(j,t-1) \quad \forall j,t \tag{6.17}$$

$$b(j,t) + c(j,t) \leq 1 \quad \forall j,t \tag{6.18}$$

6.2.2.4 Wind Turbine Model

The power produced by WTs can be formulated as Eq. (6.19) which has been obtained from power curve of WTs [27].

$$P_{wind}(v) = \begin{cases} P_{rated} \times \dfrac{(v - v_{cut-in})}{(v_{rated} - v_{cut-in})} & v_{cut-in} \leq v \leq v_{rated} \\ P_{rated} & v_{rated} \leq v \leq v_{cut-out} \\ 0 & otherwise \end{cases} \tag{6.19}$$

In every time horizon, the accessible wind power should be bigger than scheduled power of WTs as (6.20).

$$P_{wind}^{DAS}(w,t) \leqslant P_{wind}(v(t)); \quad \forall w,t \tag{6.20}$$

6.2.2.5 Demand Response Model

With progression of smart grid facilities, the ability of taking part in DR programs is made more than before for various kinds of customers [28]. In this study, it is considered that LLs and DRAs have the possibility of taking part in providing both energy and reserve of FDN. The DRA, has the role of a middleman between the DNO and small scale consumers including commercial and residential consumers and enables the ability of taking part for enormous volumes of small scale customers in DR programs [29]. DRAs offer their load reduction as a step-by-step biding model to the DNO as presented in [30] and if the submitted offers are accepted, the DRAs are called to act their proposal. The DR programs which are studied in this chapter are modeled as the following [10]:

$$\hbar_{\min}^{d} \leq H_1^d \leq \hbar_1^d \tag{6.21}$$

$$0 \leq H_k^d \leq (\hbar_{k+1}^d - \hbar_k^d) \quad \forall\, k = 2, 3, \ldots, K \tag{6.22}$$

$$P_{DRA}^{DAS}(d,t) = \sum_k H_k^d \tag{6.23}$$

$$cost_{DRA}^{DAS,E}(d,t) = \sum_k \omega_{DRA}^{k,d} \times H_k^d \tag{6.24}$$

Eq. (6.21) restricts the accepted offer of consumer d at hour t between minimum amount of the possible interruption and submitted the amount of interruption in step one. According Eq. (6.22), in the rest steps, the accepted offers of consumer d at hour t could be among zero and their submitted amount of interruption in the related steps. As mentioned in Eq. (6.23), the total amount of curtailable demand for consumer d at hour t is equal to summation of accepted offers over all steps. Furthermore, the cost of curtailed energy can be obtained from Eq. (6.24).

In addition, the amount of decreased energy of every DRA which is not accepted in the energy scheduling can be utilized in the reserve scheduling. However, as displayed in Eq. (6.25), the whole amount of energy and reserve decrease of every DRA should be lower than its maximum decrease offer for load consumption. The cost of providing reserve with dth DRA in tth period and at a cost equal to $\Omega_{DRA}(d,t)$ is calculated in (6.26).

$$P_{DRA}^{DAS}(d,t) + R_{DRA}^{DAS}(d,t) \leq P_{DRA}^{\max}(d,t) \tag{6.25}$$

$$cost_{DRA}^{DAS,R}(d,t) = R_{DRA}^{DAS}(d,t) \times \Omega_{DRA}(d,t) \tag{6.26}$$

LLs bid their maximum amount of energy which they can decrease at a desirable cost. Eqs. (6.27)−(6.29) can model the taking part of the ith LL in both scheduling energy decrease and providing required reserve [31]:

$$P_{LL}^{DAS}(i,t) + R_{LL}^{DAS}(i,t) \leq P_{LL}^{\max}(i,t) \tag{6.27}$$

$$cost_{LL}^{DAS,E}(i,t) = P_{LL}^{DAS}(i,t) \times \omega_{LL}(i,t) \tag{6.28}$$

$$cost_{LL}^{DAS,R}(d,t) = R_{LL}^{DAS}(i,t) \times \Omega_{LL}(i,t) \tag{6.29}$$

6.2.2.6 Plug-In Electrical Vehicle model

The short-term modeling for PEV is described through (6.30)−(6.36) as follows [32]. The Eqs. (6.30) and (6.31) express that state of charge of vth vehicle at time t depends on its state of charge at time t-1 and also depends on charging/discharging and the state of traveling. The Eq. (6.32) guarantees that state of charge will be in operating limits. The required energy for traveling is the function of traveling distance and the efficiency according to (6.33). Based Eqs. (6.34) and (6.35) for every PEV the charging and discharging rate are restricted to technical characteristics of operating state. Every PEV have three states as charging mode ($Uch(i,t,v) = 1$) discharging mode $Udis(i,t,v) = 1$ and traveling mode $Uch(i,t,v) + Udis(i,t,v) = 0$ which is limited in Eq. (6.36).

$$SOC(t,v) = SOC(t-1,v) + \eta_v^{ch} P_{ch}(t,v) - \frac{P_{dis}(t,v)}{\eta_v^{dis}} - P_{tra}(t,v) \tag{6.30}$$

$$SOC(t,v) = E_v^0 + \eta_v^{ch} P_{ch}(t,v) - \frac{P_{dis}(t,v)}{\eta_v^{dis}} - P_{tra}(t,v) \tag{6.31}$$

$$SOC_{\min}^{v} \leq SOC(t,v) \leq SOC_{\max}^{v} \tag{6.32}$$

$$P_{tra}(t,v) = \Delta D(t,v) \times \Omega_v \tag{6.33}$$

$$\underline{P_{ch}} \times Uch(t,v) \leq P_{ch}(t,v) \leq \overline{P_{ch}} \times Uch(t,v) \tag{6.34}$$

$$\underline{P_{dis}} \times Udis(t,v) \leq P_{dis}(t,v) \leq \overline{P_{dis}} \times Udis(t,v) \tag{6.35}$$

$$Uch(t,v) + Udis(t,v) \leq 1 \tag{6.36}$$

6.3 CASE STUDIES AND NUMERICAL RESULTS

6.3.1 System Data

In order to study the proposed method, an IEEE 33-bus test system is utilized in which DG placement of it is obtained from [33] expresses DG units placed in proper buses. The utilized test case is shown in Fig. 6.1.

In utilized case study three WTs are located in buses 14, 16, and 31 which their technical data obtained from [5]. The rated power, rated speed, cut-in speed, and cut-out speed are equal to 3 MW, 13 m/s, 3 m/s, and 25 m/s respectively. In addition four DGs are connected to buses 8, 13, 16, and 25. Table 6.1 tabulates the cost function coefficients of DGs which are obtained from [34].

The technical data of DGs including start-up cost, minimum up time, minimum down time, ramp up and ramp down ratio, upper and lower bounds of power production are provided in Table 6.2 [34].

In addition, the cost of provided reserve capacity by DG units considered 25% of marginal price of their energy production. The hourly step-by-step offers of LLs and DRAs for energy decline are presented through Tables 6.3 and 6.4, respectively. In this chapter it is assumed that the reserve capacity provided by all responsive loads has a cost equal to 20% of maximum decline offers.

NYISOs PJM on Wednesday, July 17, 2013 [35] is utilized in this study to use the appropriate numerical data for forecasted energy prices of load and upstream grid.

Fig. 6.2 provides the hourly forecast energy consumption of the FDN. The power factor of all loads is supposed in a constant value equal to 0.95 lagging.

The forecast of hourly prices of energy in upstream grid market are given in Fig. 6.3. Furthermore, Fig. 6.4 presents hourly forecasted amounts for wind speed [36].

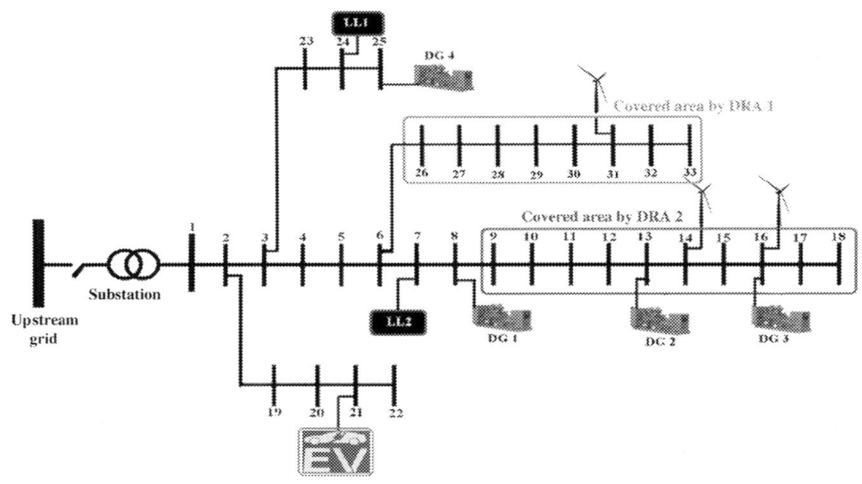

Figure 6.1 IEEE 33-bus test system.

Table 6.1 Cost coefficient of DG units

Cost coefficients

Unit	$\alpha_{1,j}$ ($)	$\alpha_{2,j}$ ($/MW)	$\alpha_{3,j}$ ($/MWh^2)
DG1	27	87	0.0025
DG2	25	87	0.0035
DG3	28	92	0.0035
DG4	26	81	0.184

Table 6.2 Technical data of DG units

Technical data

Unit	SUT ($)	MUT/MDT (h)	RU/ RD (MW/h)	Pmax (MW)	Pmin (MW)
DG1	15	2	1.8	3.5	1
DG2	25	1	1.5	3	0.75
DG3	28	1	1.5	3	0.75
DG4	26	2	1.8	4.1	1

Table 6.3 Bid-quantity energy decline offers of LLs

Hour	LL1		LL 2	
	Maximum decline (MW)	Price ($/MWh)	Maximum decline (MW)	Price ($/MWh)
10	0.85	43	0.40	18
11	0.90	77	0.40	30
12	0.90	122	0.45	53
13	0.95	108	0.45	43
14	1	273	0.45	79
15	1	122	0.45	43
16	1	404	0.50	79
17	1	304	0.50	73
18	1	126	0.50	67
19	1	118	0.45	47
20	0.95	84	0.45	40
21	0.90	104	0.45	33
22	0.90	318	0.40	32
23	0.85	72	0.40	16

Table 6.4 Bid-quantity energy decline offers of DRAs

DRA	DRA1				DRA2			
Covered Buses	26, 27, 28, 29, 30, 31, 32,33				9, 10, 11, 12, 13, 14, 15, 16, 17,18			
Quantity (MW) Price ($/MWh)	0–0.1 12	0.1–0.7 90	0.7–1.2 154	1.2–1.5 192	0–0.3 38	0.3–0.8 102	0.8–1.3 167	1.3–1.8 231

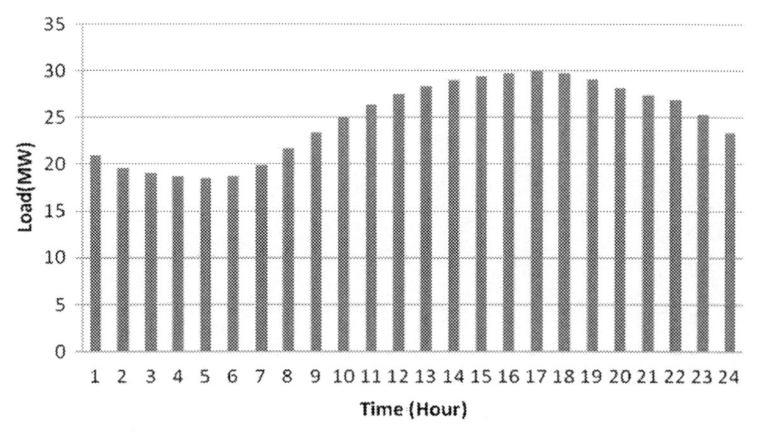

Figure 6.2 Hourly consumption forecast of FDN.

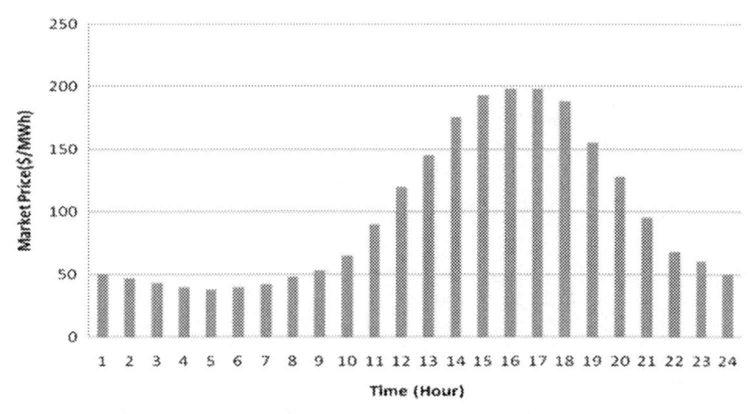

Figure 6.3 Hourly energy prices forecast of upstream grid.

In this chapter it is assumed that a little urban parking lot with the capacity of 10 EVs is located in bus 21. The travel patterns of PEVs are tabulated in Table 6.5 as following [37]:

The optimization of MILP problem which is presented in this study is carried out in GAMS software by CPLEX solver while termination threshold is adjusted in 0.01% [38].

6.3.2 Effectiveness of Proposed Method

In order to show the efficiency of the proposed framework, two case studies are considered separately as follows:

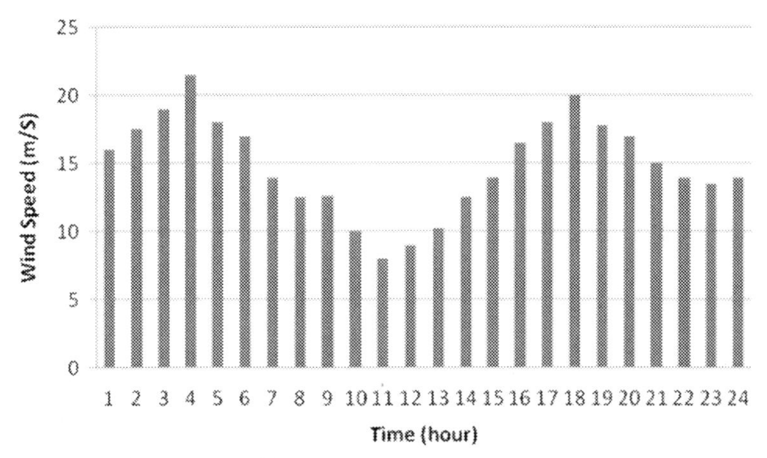

Figure 6.4 Hourly wind speed forecast.

Table 6.5 The travel patterns of PEVs (km)

Time (h)	V1	V2	V3	V4	V5	V6	V7	V8	V9	V10
T1	0	0	0	4.6	0	0	0	4	0	0
T2	0	3.6	0	1.8	0	2	0	0	0	0
T3	0	5	0	0	0	2.2	0	0.6	0	2
T4	0	0	0	0	3.6	0	0	1.2	0	0
T5	2.4	0	0	0	1.8	4.2	2.8	3.6	0	4.4
T6	0	4.8	0	0	1.4	1.8	0	0	0	0
T7	0	0	0	0	1.6	2.6	0	0	0	0
T8	4.8	0	1	2	0	3.8	0	0	0	2
T9	0	0	0	0	1.2	3	1.2	0.8	0	1.2
T10	0	2.4	0	4	0	0	3.4	0	0	1
T11	0	0	0	4.6	2.4	0	0	4.4	0	0.4
T12	4	0	0	0	4.2	3	0	1.2	0	0
T13	0	0	0	0	2	0	3.4	0	4.2	0
T14	0	0	0	0	3	0	0	4	0	0
T15	0	0	0	0	0	1.4	0	0	3.8	0
T16	3.6	0	0	4.6	0	0	3.8	0	0	4
T17	0	0	3.6	0	1.6	0	0	3	0	4
T18	0	0	0	4	0	0	0	1.8	0	0
T19	0	0	4	0	2.2	2.6	0	0	2	4
T20	0	0	0	0	3	0	4.2	3.2	2.2	0
T21	0	0	4.8	3.8	0	0	0	2.6	1	1
T22	0	0	0	0	3.8	0	0	0.4	0	0
T23	0	4.8	0	0	0	0	0	0	0	2.2
T24	0	0	0	0	2.2	0	3.2	0	0.4	0

- Case 1: In the first one, regarding to the MILP model, PEVs are not considered in short-term operation of FDN.
- Case 2: In the second case, in order to show the effectiveness of utilizing PEVs, regarding to the MILP model, EVs are considered in the short-term scheduling of FDN.

6.3.2.1 Case 1: Without the Presence of PEVs

Regarding to the obtained results, in hours which the price of upstream grid market is more than the cost of production of DG units, the DG units start to produce power and therefore the purchased power from upstream grid decreases. In addition according to this point that short-term operation of WTs has no cost therefore this units according to the wind speed work in their maximum power. Furthermore, in high price hours in order to decrease the purchasing power from upstream grid and reducing costs DR programs start to work. In this case, DG units and DR programs supply the required reserve and therefore according to limitations of DG units in producing power, in some hours the DG units lose the opportunity of power production and reducing the costs of FDN.

6.3.2.2 Case 2: In the Presence of PEVs

In this case the set of EVs which are parked in parking lot according their travel pattern, work as an energy storage. In hours which the price of upstream grid is low, the set of parking lot starts to charge the EVs and in hours which the price is high according to travel pattern the stored energy is returned to the network and therefore the purchased energy decreases. In addition the presence of parking lot and its participation in providing required reserve releases the capacity of DG units. Therefore, according to Fig. 6.5 the contribution of DG units in producing power through high price hours increases and the cost of FDN decreases.

In addition according to Figs. 6.6 and 6.7 in hours 3, 4, 5, 11, 12, and 13 which the consumption and the price are low, the purchasing power from upstream gird increases for charging EVs; and through hours 7, 8, 9, 15, 16, and 17 which the both of consumption and the price are high with discharging EVs the purchasing power from upstream grid decreases.

According to Fig. 6.7 in hours 3, 4, 5, 10, 11, and 12 urban parking lot treat as load and absorbs energy and through hours 7, 8, 9, 15, 16, and 17 urban parking lot treat as generation unit and release the absorbed energy.

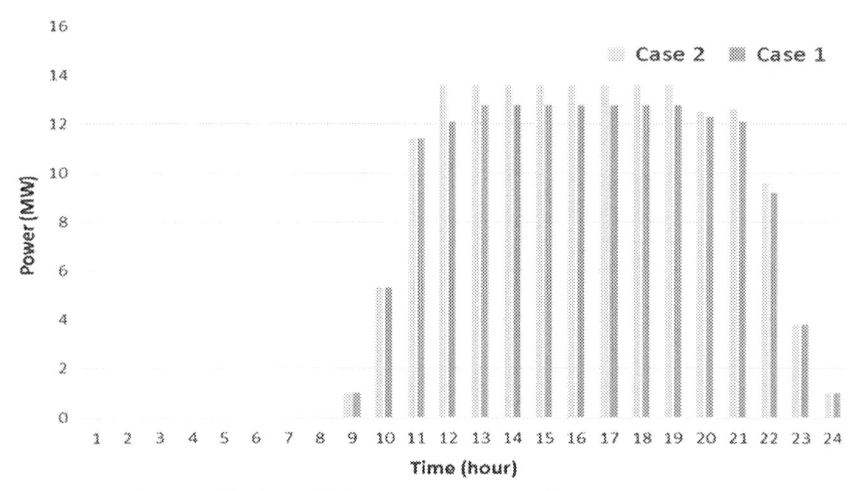

Figure 6.5 The contribution of DG units in power production.

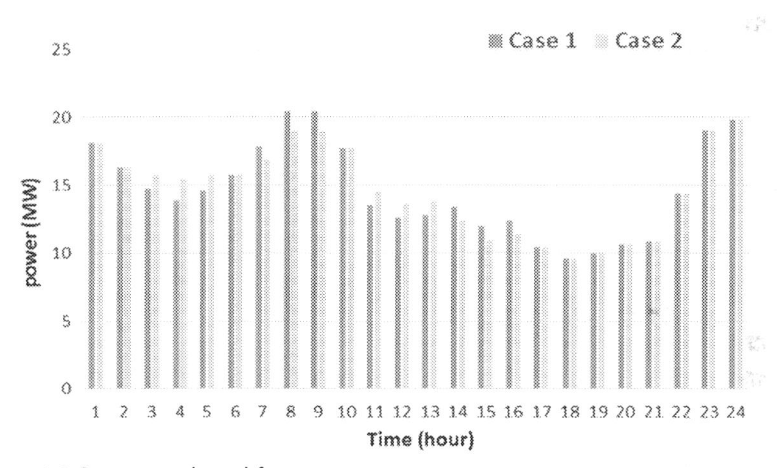

Figure 6.6 Power purchased from upstream.

Fig. 6.8 shows the hourly energy of different energy sources in case 2 separately. In addition as shown in Fig. 6.9 according to this point that WTs day-ahead operation has no cost therefore in both case 1 and case 2 according to wind speed WTs work in their maximum power production.

In order to compare case 1 and case 2 and show the effect of utilizing urban parking lot, the cost of network operation through the aforementioned cases is presented in Table 6.6.

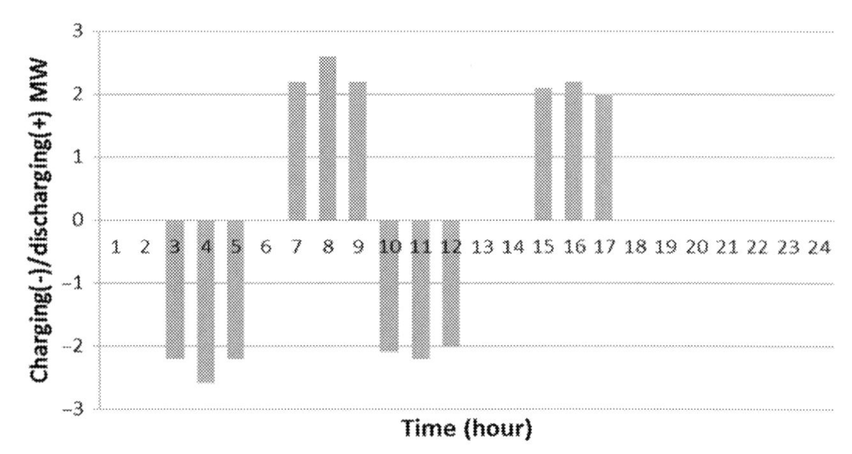

Figure 6.7 Electrical vehicles charge/discharge scheduling in case 2.

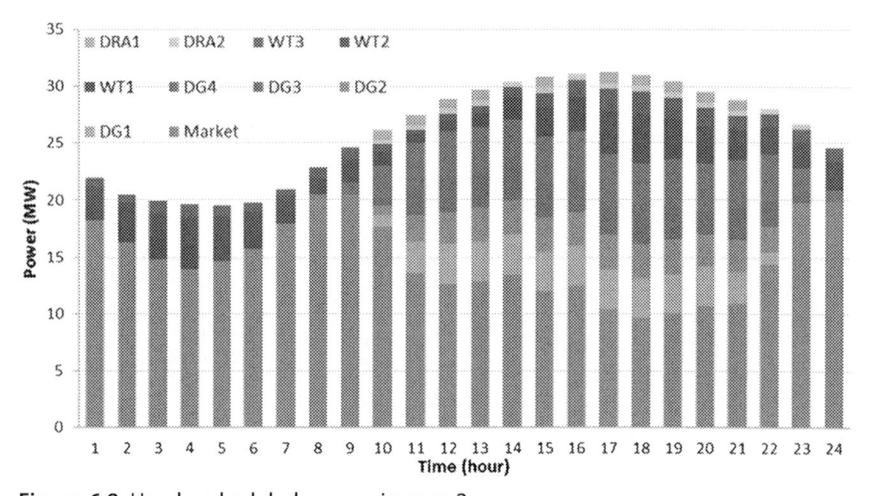

Figure 6.8 Hourly scheduled energy in case 2.

As shown in this table there is a huge difference between case 1 and case 2. Therefore a proper operational scheduling for incorporation of EVs in FDNs can reduce the costs of day-ahead scheduling of distribution network.

6.4 CONCLUSION

Optimal day-ahead scheduling of distribution networks in the presence of renewable sources and DR programs attracted more attention through

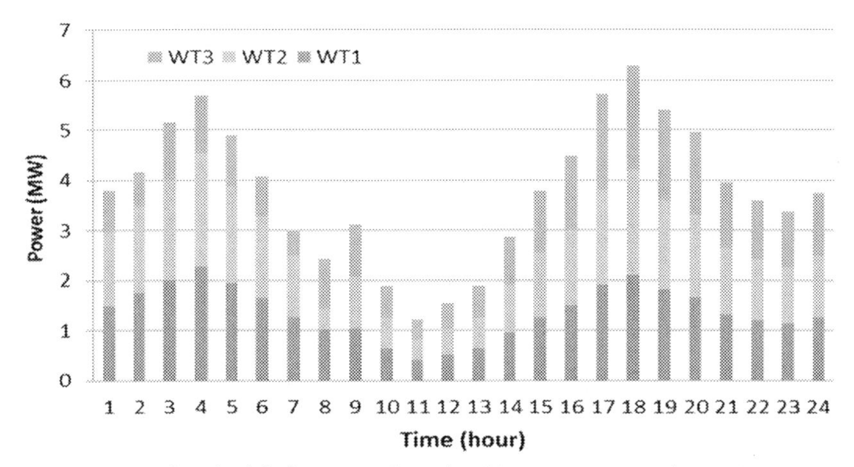

Figure 6.9 Hourly scheduled energy of wind turbines in case 1 and case 2.

Table 6.6 The operation cost of case 1 and case 2

Case	Network operation cost
Case 1	63572 $
Case 2	57394 $

recent years. In this chapter the effect of incorporating EVs in ODAS of distribution networks has been studied. In addition an IEEE 33 bus test system with an urban parking lot with 10 EVs is used to show the effectiveness of the proposed method. The obtained results show that in hours in which the energy price of upstream grid is high the purchase from this market decreases and the required energy supplies more contribution of urban parking lot, distributed energy resources, and DR programs. It can be seen that the proposed model has the ability of scheduling the FDNs. In addition this chapter shows utilizing EVs in ODAS of FDNs reduces the operation costs.

REFERENCES

[1] F. Jabari, et al., Optimal short-term scheduling of a novel tri-generation system in the presence of demand response programs and battery storage system, Energy Convers. Manage. 122 (2016) 95–108.

[2] M. Majidi, et al., A multi-objective model for optimal operation of a battery/PV/fuel cell/grid hybrid energy system using weighted sum technique and fuzzy satisfying approach considering responsible load management, Sol. Energy 144 (2017) 79–89.

[3] H. Fathabadi, Novel solar powered electric vehicle charging station with the capability of vehicle-to-grid, Sol. Energy 142 (2017) 136–143.
[4] G. Mokryani, Active distribution networks planning with integration of demand response, Sol. Energy 122 (2015) 1362–1370.
[5] M. Mazidi, et al., Robust day-ahead scheduling of smart distribution networks considering demand response programs, Appl. Energy 178 (2016) 929–942.
[6] A. Zakariazadeh, et al., Stochastic operational scheduling of smart distribution system considering wind generation and demand response programs, Int. J. Electr. Power Energy Systems 63 (2014) 218–225.
[7] M. Farrokhifar, Optimal operation of energy storage devices with RESs to improve efficiency of distribution grids; technical and economical assessment, Int. J. Electr. Power Energy Systems 74 (2016) 153–161.
[8] A. Ahmadian, et al., Fuzzy load modeling of plug-in electric vehicles for optimal storage and DG planning in active distribution network, IEEE Trans. Vehicular Technol. (2016).
[9] A. Zakariazadeh, et al., Stochastic multi-objective operational planning of smart distribution systems considering demand response programs, Electr. Power Systems Res. 111 (2014) 156–168.
[10] M. Mazidi, et al., Integrated scheduling of renewable generation and demand response programs in a microgrid, Energy Convers. Manage. 86 (2014) 1118–1127.
[11] Z. Tan, et al., The optimization model for multi-type customers assisting wind power consumptive considering uncertainty and demand response based on robust stochastic theory, Energy Convers. Manage. 105 (2015) 1070–1081.
[12] E. Heydarian-Forushani, et al., Robust scheduling of variable wind generation by coordination of bulk energy storages and demand response, Energy Convers. Manage. 106 (2015) 941–950.
[13] T. Niknam, et al., An efficient algorithm for multi-objective optimal operation management of distribution network considering fuel cell power plants, Energy 36 (1) (2011) 119–132.
[14] M. Sedghi, et al., Optimal storage planning in active distribution network considering uncertainty of wind power distributed generation, IEEE Trans. Power Systems 31 (1) (2016) 304–316.
[15] A. Soroudi, et al., Possibilistic evaluation of distributed generations impacts on distribution networks, IEEE Trans. Power Systems 26 (4) (2011) 2293–2301.
[16] S. Golshannavaz, et al., Smart distribution grid: optimal day-ahead scheduling with reconfigurable topology, IEEE Trans. Smart Grid 5 (5) (2014) 2402–2411.
[17] Y. Tan, et al., A two-stage stochastic programming approach considering risk level for distribution networks operation with wind power, IEEE Systems J. 10 (1) (2016) 117–126.
[18] A. Zakariazadeh, et al., Economic-environmental energy and reserve scheduling of smart distribution systems: A multiobjective mathematical programming approach, Energy Convers. Manage. 78 (2014) 151–164.
[19] T. Sousa, et al., A multi-objective optimization of the active and reactive resource scheduling at a distribution level in a smart grid context, Energy 85 (2015) 236–250.
[20] C. Chen, et al., An interval optimization based day-ahead scheduling scheme for renewable energy management in smart distribution systems, Energy Convers. Manage. 106 (2015) 584–596.
[21] M. Esmaili, et al., Multi-objective optimal charging of plug-in electric vehicles in unbalanced distribution networks, Int. J. Electr. Power Energy Systems 73 (2015) 644–652.

[22] M. Mazidi, et al., Incorporating price-responsive customers in day-ahead scheduling of smart distribution networks, Energy Convers. Manage. 115 (2016) 103–116.
[23] X. Dong, et al., Heuristic planning method of EV fast charging station on a freeway considering the power flow constraints of the distribution network, Energy Proc. 105 (2017) 2422–2428.
[24] Z. Yang, et al., A novel network model for optimal power flow with reactive power and network losses, Electr. Power Systems Res. 144 (2017) 63–71.
[25] S. Nojavan, et al., Optimal stochastic energy management of retailer based on selling price determination under smart grid environment in the presence of demand response program, Appl. Energy 187 (2017) 449–464.
[26] E. Mahboubi-Moghaddam, et al., Reliability constrained decision model for energy service provider incorporating demand response programs, Appl. Energy 183 (2016) 552–565.
[27] Y. Jiang, et al., Day-ahead stochastic economic dispatch of wind integrated power system considering demand response of residential hybrid energy system, Appl. Energy 190 (2017) 1126–1137.
[28] A. Keshtkar, et al., An adaptive fuzzy logic system for residential energy management in smart grid environments, Appl. Energy 186 (2017) 68–81.
[29] M. Asensio, et al., Risk-constrained optimal bidding strategy for pairing of wind and demand response resources, IEEE Trans. Smart Grid 8 (1) (2017) 200–208.
[30] J. Aghaei, et al., Contribution of emergency demand response programs in power system reliability, Energy 103 (2016) 688–696.
[31] M.H. Amini, et al., Simultaneous allocation of electric vehicles' parking lots and distributed renewable resources in smart power distribution networks, Sustain. Cities Soc. 28 (2017) 332–342.
[32] M.E. Baran, et al., Network reconfiguration in distribution systems for loss reduction and load balancing, IEEE Trans. Power Deliv. 4 (2) (1989) 1401–1407.
[33] S. Wong, et al., Electric power distribution system design and planning in a deregulated environment, IET Gen. Transm. Distribut. 3 (12) (2009) 1061–1078.
[34] Diesel generators specification sheets, Kohler Power Systems Company, online available at: <http://www.yestramski.com/industrial/generatorsdiesel/industrial-diesel-generators-all.htm>.
[35] (2013, July. 17). New York Independent System Operator, online available at: http://www.nyiso.com/public/markets_operations/index.jsp.
[36] Willy Online Pty Ltd, Online available at: http://wind.willyweather.com.au/.
[37] A. Soroudi, et al., Risk averse energy hub management considering plug-in electric vehicles using information gap decision theory, Plug In Electric Vehicles in Smart Grids, Springer, Singapore, 2015, pp. 107–127.
[38] The GAMS/CPLEX manual, online available at: http://www.gams.com/dd/docs/solvers/cplex/index.html.

CHAPTER 7

Application of Load Shifting Programs in Next Day Operation of Distribution Networks

Mehrdad Ghahramani, Sayyad Nojavan, Kazem Zare and Behnam Mohammadi-ivatloo
University of Tabriz, Tabriz, Iran

7.1 INTRODUCTION

As mentioned in Chapter 6, in the last few decades, air pollution and critical environmental issues have been some of the drivers to move toward the use of renewable energy sources (RES). In addition, development of low carbon technologies, containing wind turbine (WT) and battery energy storage systems (BESS) facilitated the goal of reaching to the clean energy generation [1]. Also, with developing toward smart grids, the requisite infrastructures for communication to the customers which is necessary for implementation of demand response (DR) programs are obtained [2]. In Chapter 6 the effect of electrical vehicles in distribution power networks is studied. Through this chapter DR programs according to their importance in determining the operational costs of distribution networks are studied separately. In addition the focus of this chapter is on the effect of DR programs on the interactions of network with upstream grid and other sources of network. DR is defined as a tariff or programs with the aim of decreasing demand when market prices are high or when the balance between product and consumption is threatened during unpredictable fluctuations in customers load or power generation [3]. It is believed that optimal scheduling of renewable generators and DR programs will improve the reliability and reduce the total operation cost of smart grids [4,5]. As a result of [6,7], by employing DR programs, because of shifting demands from high price hours (peak) to low price hours, decrease in procurement costs of electricity and having a flatten demand profile is expected. Therefore, renewable power producers and

Operation of Distributed Energy Resources in Smart Distribution Networks
DOI: https://doi.org/10.1016/B978-0-12-814891-4.00007-2

DR programs are asked to contribute in power generation and their competition with deterministic energy production facilities.

7.1.1 Problem Definition

The concurrent connection of renewable generation, DR programs, diesel generators, and energy storage systems to electrical networks may cause technical challenges for independent system operator (ISO). In other words, the incorrect contribution of energy supply sources may impair power balance and leads to higher operation cost and reliability endangerment of distribution networks. Accordingly, an appropriate operation scheduling is essential for optimal and proper financial decisions in the presence of various alternative energy utilizations spatially DR programs [8].

7.1.2 Literature Review

Determining the amount of energy and reserve by optimal day-ahead scheduling (ODAS) with considering contribution of various sources of energy as the key feature of smart grids specifically at the distribution level have attracted high regard, and various literatures have been devoted in this context. The survey in [9] is determining a stochastic model for ODAS of smart distribution network (SDN). The study presented in [10] proposes an ODAS for optimizing both emission and operation costs of microgrid using a multiobjective generation management model. In [11], authors institute a microgrid system with various types of distributed generators considering cost of both operation and the pollutant treatment. Also, the effect of uncertainty agents such as load and renewable resources variation are determined and the objective function solved with the combination of particle swarm optimization (PSO) with Monte Carlo simulation. In [12] an optimization model for ODAS of distribution network using vehicle to grid characteristics of electric vehicles (EVs) is presented for reducing the operational costs of the SDN. An ODAS for a storage integrated in an active distribution network is presented in [13] in which the goal is minimizing the cost of network operation. Also according to the results of this study, using batteries in distribution network has some other benefits such as voltage adjustment, peak shaving, and reliability improvement. The research proposed in [14] offers a fuzzy model for considering uncertainties of operation scheduling of distribution network associated with diesel generator (DG) units. In [15] a model was presented for ODAS of active distribution networks while minimizing the operational cost considering distributed generation and network topology. A two stage optimization model for one hour scheduling of distribution networks has been conducted in [16]. The first stage of the proposed model considers the operational risk of wind power while the second stage addresses the problems of over load and power violation. A hierarchical method for operation scheduling of SDNs integrated with EVs are investigated in [17]. The results show that the demand curve can be flattened with optimal scheduling of EVs. A novel multiobjective method is presented in [18,19] for an optimal scheduling of SDN with the goal of minimizing operational costs considering EVs. In [20] a multiobjective model for optimal energy and reserve scheduling of SDNs is presented in order to minimize both operation costs and emissions considering stochastic characteristic of wind power. A bi-level information-gap decision theory (IGDT) for operational scheduling of SDNs under the presence of wholesale price uncertainty and DG units has been provided in [21] which the purpose is profit maximizing of distribution system. The studies of [22–24] proposed a multiobjective fuzzy PSO for ODAS of SDNs considering fuel cell power plants with the aims of minimizing cost, emission, losses, and voltage deviation. The authors of [25,26] have investigated a stochastic multiobjective model for scheduling of distributed energy resources with the purpose of minimizing operation costs and pollutant emissions considering wind and solar power fluctuations.

7.1.3 Procedure

This chapter suggests an ODAS model for SDN in which responsive loads can take part in both energy and reserve scheduling. The goal of the suggested model is minimizing the total operation cost of SDNs with considering the contribution of various energy resources. This model specifies generation of DG units and WTs in SDN schedule. Also, the proposed model determines transactions with the wholesale market, accepted load reduction, and BESS charge/discharge, while the load balance of the system is satisfied.

7.1.4 Contributions

This chapter proposes a complete model for day-ahead scheduling of SDNs considering contribution of renewable and nonrenewable distributed energy sources. Furthermore, in this study a complete model of DR programs is proposed in which load flexibility takes part in both energy and reserve scheduling.

7.1.5 Chapter Organization

The reminder of this chapter is organized as follows. In Section 7.2, mathematical modeling and formulation for deterministic day-ahead scheduling of the SDN are described. Numerical results are given in Section 7.3. The chapter is concluded with some important findings in Section 7.4.

7.2 PROBLEM FORMULATION

In this section a deterministic formulation for ODAS of SDNs, containing the objective function and constraints, is completely described. Moreover, the models of distribution network, WTs and DR programs are presented in this section. The proposed operation scheduling is based on the following assumptions.

- The optimization process and decisions on the behavior of participants of the market are handled by ISO.
- To accomplish day-ahead scheduling of the network, the bids of responsive loads and forecasting data which contain wind speed and forecasted loads are received by ISO.
- The required data of DG units, distribution network model and wholesale market prices are available for ISO.

The mathematical formulation of the studied ODAS of distribution network is provided in the following.

7.2.1 Objective Function

The energy and reserve requirements of the SDN are scheduled by the ISO with the objective to minimize the total operation cost during the scheduling horizon. The total operation cost is the cost of providing energy and reserve from different resources.

$$Min\sum_{t=1}^{N_T}\left[\begin{array}{l}\varsigma_{ug}(t)\times P_{ug}^{DAS}(t)\\ +\sum_{j=1}^{N_{DG}}\left\{cost_{DG}^{DAS,E}(j,t)+cost_{DG}^{DAS,SU}(j,t)+cost_{DG}^{DAS,R}(j,t)\right\}\\ +\sum_{i=1}^{N_{LL}}\left\{cost_{LL}^{DAS,E}(i,t)+cost_{LL}^{DAS,R}(i,t)\right\}\\ +\sum_{d=1}^{N_{DRA}}\left\{cost_{DRA}^{DAS,E}(d,t)+cost_{DRA}^{DAS,R}(d,t)\right\}\end{array}\right] \quad (7.1)$$

The first term of (7.1) is the cost of purchasing energy from upstream grid, i.e., purchasing from wholesale market. The second term respectively represents the energy cost, startup cost, and reserve cost of DG units. The third and fourth terms denote the costs due to load reduction and reserve provided by large loads (LLs) and demand response aggregators (DRAs), respectively.

7.2.2 Constraints and Mathematical Formulation

ODAS of SDNs should be solved taking into account a series of equality and inequality constraints, which are prepared in the following.

7.2.2.1 Power Flow Constraints

Power flow equations show that the active and reactive powers are satisfied in all of the buses of the network. The balance of active and reactive power should be taken into consideration for nth bus at time t, which can be formulated as Eqs. (7.2)–(7.5) [27].

$$\begin{aligned}&\sum_{j\in n}P_{DG}^{DAS}(j,t)+\sum_{w\in n}P_{Wind}^{DAS}(w,t)+\sum_{d\in n}P_{DRA}^{DAS}(d,t)\\&+\sum_{b\in n}(\eta_{ch}\times P_{BESS}^{DAS,ch}(b,t)+\eta_{dis}\times P_{BESS}^{DAS,dis}(b,t))\\&+\sum_{i\in n}P_{LL}^{DAS}(i,t)+P_{flow}(m,n,t)\\&=\sum_{k\in(n,k)}P_{flow}(n,k,t)+Re(Z)(m,n)\times\ell(m,n,t)+P_{load}^{DAS}(n,t)\end{aligned}\quad \forall m,n,t \quad (7.2)$$

$$\begin{aligned}&\sum_{j\in n} Q_{DG}^{DAS}(j,t)+\sum_{w\in n} Q_{Wind}^{DAS}(w,t)+\sum_{d\in n} Q_{DRA}^{DAS}(d,t)\\&+\sum_{b\in n}(\eta_{ch}\times Q_{BESS}^{DAS,ch}(b,t)+\eta_{dis}\times Q_{BESS}^{DAS,dis}(b,t))\\&+\sum_{i\in n} Q_{LL}^{DAS}(i,t)+Q_{flow}(m,n,t)\\&=\sum_{k\in(n,k)} Q_{flow}(n,k,t)+Im(Z)(m,n)\times\ell(m,n,t)+Q_{load}^{DAS}(n,t)\end{aligned}\quad\forall m,n,t \tag{7.3}$$

$$\begin{aligned}&v(n,t)-v(m,t)=(Re(Z)^2(n,m)+Im(Z)^2(n,m))\times\ell(n,m,t)\\&-2\times(Re(Z)(n,m)\times P_{flow}(n,m,t)+Im(Z)(n,m)\times Q_{flow}(n,m,t))\end{aligned}\quad\forall m,n,t \tag{7.4}$$

$$\ell(n,m)=\frac{P_{flow}^2(n,m)+Q_{flow}^2(n,m)}{v(m)}\quad\forall m,n,t \tag{7.5}$$

where, $v(n)=\left|V(n)\right|^2$ *and* $\ell(n,m)=\left|I(n,m)\right|^2$

For applying a linear programming method, Eq. (7.5) can be improved to linear format taking into account v $(m,\ t)$ equal to one. In addition, v and ℓ are two auxiliary variables introduced for linearization of the AC power flow equations. Accordingly, the formulations of quadratic active and reactive power transfer between buses can be revised to linear format by utilizing piecewise linear approximation concept [28].

7.2.2.2 Distribution Network Constraints

Technical constraints of distribution network should be considered for warranting safe operation of the network. Voltage level for all buses of the network and feeder current limits should be guaranteed in (7.6) and (7.7).

$$V(n)_{\min}^2\leqslant v(n,t)\leqslant V(n)_{\max}^2\quad\forall n,t \tag{7.6}$$

$$\ell(n,m,t)\leq I_{substation}^{\max}(n,m)^2\quad\forall n,m,t \tag{7.7}$$

Because of the limitation of substation transformers' capacity in importing and exporting power from upstream grid, the voltage and current of substation should be controlled [29]. Therefore, when a substation is connected to bus 1 of the SDN Eqs. (7.8) and (7.9) ensure the safe act of the network.

$$v(n, m, t) = constant, \quad n = substaion\ bus \quad \forall t \tag{7.8}$$

$$\ell(n, m, t) \leq (I_{substation}^{\max})^2, \quad m = 1, \quad n = substaion\ bus \quad \forall t \tag{7.9}$$

7.2.2.3 Battery Energy Storage System

There are three various performance states for BESS including charging, discharging, and idle. In charging state, BESS operates as a consumer and attracts energy from network. In discharging state, BESS operates as a DG unit and delivers the attracted energy to the network. In idle state, BESS neither attracts energy nor delivers the attracted energy to the network [30]. The following equation is utilized to model the performance states of the BESS:

$$BESS_{ch}(\beta, t) + BESS_{dis}(\beta, t) \leq 1; \quad BESS_{ch}, BESS_{dis} \in \{0, 1\}, \forall t \tag{7.10}$$

According to Eq. (7.10), when $BESS_{ch}(\beta, t)$ is equal to one BESS is in charging state and BESS is in discharging state when $BESS_{dis}(\beta, t)$ is equal to one. When both of the $BESS_{dis}(\beta, t)$ and $BESS_{ch}(\beta, t)$ are zero, BESS is in idle state. For the proper functioning of the BESS the following constraint should be considered [31]:

$$SOC(\beta, t) = SOC(\beta, t-1) + \eta^{ch} \times P_{BESS,ch}^{DAS}(\beta, t) - \eta^{dis} \times P_{BESS,dis}^{DAS}(\beta, t) \tag{7.11}$$

$$SOC^{\min}(\beta) \leq SOC(\beta, t) \leq SOC^{\max}(\beta) \tag{7.12}$$

$$0 \leq P_{BESS}^{DAS}(\beta, t) \leq P_{BESS,ch}^{\max} \times BESS_{ch}(\beta, t) \tag{7.13}$$

$$0 \leq P_{BESS,dis}^{DAS}(\beta, t) \leq P_{BESS,dis}^{\max} \times BESS_{dis}(\beta, t) \tag{7.14}$$

The amount of stored energy in the BESS called the state of charge is calculated in Eq. (7.11) and limited in Eq. (7.12). Eqs. (7.13) and (7.14) represent the ramp rate limits of the battery charge and discharge, respectively.

7.2.2.4 DG Unit Constraints

The operational cost of DG units can be considered as a quadratic function of generated power. Similar to Chapter 6, it can be linearized using the piecewise linear approximation method [32,33]. The startup (SU) cost of DG units can be stated similar to Chapter 6. The reserve cost of the DG units can be assumed as a 20% of the highest marginal price of the power production of the unit according to Chapter 6.

The lower and upper bounds of power generation of the DG units with their respective reserve should be considered in solving the problem according to Chapter 6. The ramp up (RU) and ramp down (RD) ratio of the DG power units should be taken into account [4] similar to Chapter 6.

The conventional DG unit should produce power for determined hours before shutting-down based on Chapter 6. Moreover, the unit should be in shut-down condition for specified hours before starting up according to Chapter 6 [34]. For the startup and shut-down condition of the units are defined according to Chapter 6.

7.2.2.5 Wind Turbine Model

The available power from WTs is formulated similar to Chapter 6 which has been obtained from power curve of WTs [35]. In every time horizon, the accessible wind power should be bigger than scheduled power of WTs.

7.2.2.6 Demand Response Model

With progression in smart grid infrastructures the involvement of different kinds of customers in DR programs is made possible [36]. In this study, LLs and DRAs can take part in both energy and reserve scheduling. The DRA, as a medium between the ISO and small customers, such as commercial and residential customers, enables the contribution of enormous volumes of customers in DR programs [37]. Because the DRAs represent the aggregated DR of end-users, they should offer their load reduction as a step-by-step biding model as illustrated and presented in [38]. The DRAs submit the offers of load reduction to the ISO and if the offers are accepted, the DRAs are called to decrease the proposed quantity. The studied DR programs are modeled according to Chapter 6 [9].

The energy decline of every DRA that is not utilized in the energy scheduling can be committed in the reserve scheduling. However, as displayed in Chapter 6, the summation of the energy decline and reserve capacity of every DRA should be limited to its maximum load decline offer. The reserve cost of *d*th DRA in *t*th period and at a cost equal to $\Omega_{DRA}(d, t)$ is calculated in Chapter 6. LLs bid their maximum amount of energy decline at a desirable cost. The presence of the *i*th LL in both scheduling of energy decline and reserve amount is modeled in Chapter 6 [39].

7.3 CASE STUDIES AND NUMERICAL RESULTS

7.3.1 System Data

As discussed in Chapter 6, in order to evaluate the proposed model, an IEEE 33-bus test system is studied in this chapter [40]. Considering the results which are obtained from [41], DG units are placed in proper buses. The test case which is studied in this chapter is shown in Fig. 7.1.

Three WTs are used in the case study, which are located in buses 13, 15, and 30. The WT data are adopted from [42]. The rated power of the WTs is equal to 3 MW. The respective cut-in and cut-out speeds of the WTs are selected as 3 m/s and 25 m/s. Rated speed of the WTs are equal to 13 m/s. As discussed in Chapter 6 four DGs are used in the case study, which are connected to buses 7, 12, 15, and 24. The cost function coefficients of DGs are tabulated in Chapter 6 which are obtained from [43]. The technical data of DGs including startup (SU) cost, minimum up (MU) time, minimum down (MD) time, ramp up (RU) and ramp down (RD) ratio, upper and lower bounds of power production are provided in

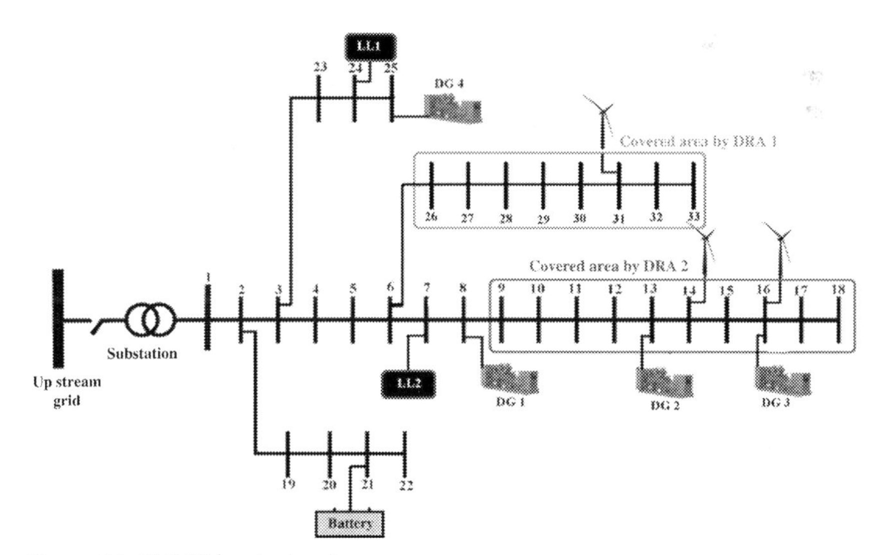

Figure 7.1 IEEE 33-bus test system.

Chapter 6 [43]. In addition, the cost of reserve capacity which is provided by DG units is 25% of their marginal price of the energy production.

The capacity of BESS is 0.5. It is assumed that the lowest and highest storable energy of the BESS respectively is 20% and 80% of its rated capacity; also the ramp rates of BESS charging and discharging are restricted to 0.1 MW for each hour. The hourly offers of step-by-step bid-quantity energy decline of LLs and DRAs are presented in Chapter 6. It is assumed that the cost of reserve capacity which is provided by all responsive loads is 20% of their maximum energy decline offers.

In order to use the appropriate numerical data for forecasted load and upstream grid energy prices, NYISOs PJM on Wednesday, July 17, 2013 [44] have been utilized.

The hourly consumption forecast of the SDN is provided in Chapter 6. The power factor of all loads is supposed in a constant value equal to 0.95 lagging.

The hourly energy prices of upstream grid are given in Chapter 6. Furthermore, the hourly forecasted amounts for wind speed are provided in Chapter 6.

The optimization of MILP problem which is proposed in this chapter is carried out in GAMS software by utilizing CPLEX solver while termination threshold is adjusted in 0.01% [45].

7.3.2 Effectiveness of Proposed Method

For the purpose of investigating the efficiency of the proposed framework, two various case studies are considered as follows:

- Case I: In the first case, regarding to the MILP model, DR programs are not considered in ODAS of the SDN.
- Case II: In the second case, in order to understand the effectiveness of DR programs, regarding to the MILP model, DR programs are considered in the ODAS of SDN.

The results of purchased power from upstream grid which are obtained from ODAS are shown in Fig. 7.2.

Case I: As illustrated in Fig. 7.2, within the hours in which the energy prices of upstream grid are low, particularly in 24:00 and among the period of 1:00–9:00, the required energy of the distribution network is purchased from the market of upstream grid. Therefore, when the energy prices of the upstream grid is higher than the costs of energy production

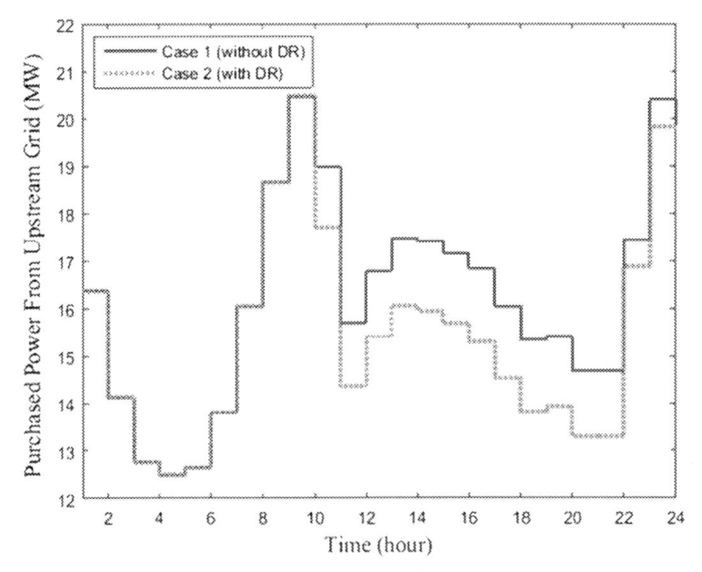

Figure 7.2 Power interactions with upstream grid.

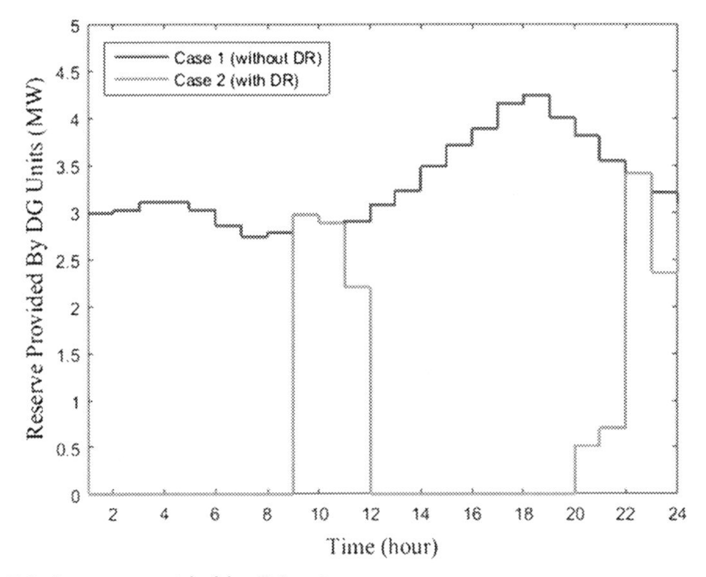

Figure 7.3 Reserve provided by DG units.

in DG units, specifically between the period of 10:00–23:00, the purchased energy from the energy market of upstream grid decreases. Accordingly, using DG units the purpose of addressing the SDN consumption will be less costly than purchasing energy from the energy

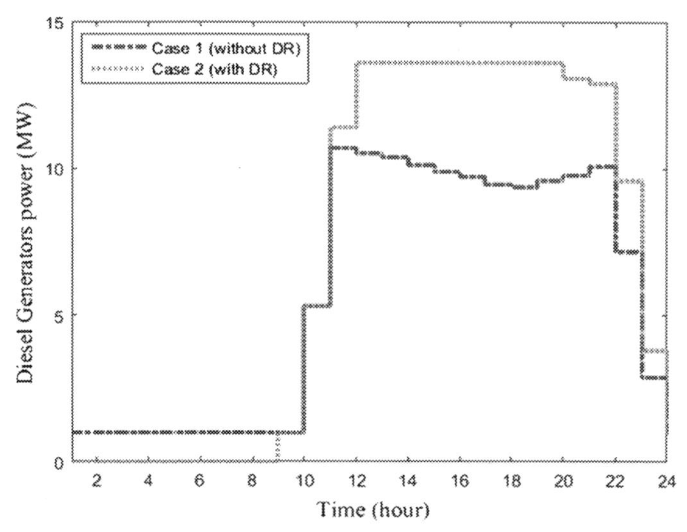

Figure 7.4 Power produced by DG units.

market of upstream grid. This proposed MILP scheduling model leads to a less operating cost and it shows the proposed model is more economical.

But as provided in Figs. 7.3 and 7.4, in case I, all of the needed reserve of SDN is covered by DG units. In addition, it is observable that in peak period, hours 14:00–21:00 in order to attain the goal of providing reserve capacity, one or more of the DG units should be in the mode of operation. Although, the period of 10:00–23:00, in which the energy prices of upstream grid are high, it is the best time of energy selling for DG units, but in case I providing reserve forcing ISO to buy energy from the upstream grid with higher prices. In addition, between the period of 1:00–9:00 and 24:00 in which the prices of upstream grid are low, there is a need to providing reserve forcing DG units to be in operation mode with a higher cost. As can be seen, procurement of reserve capacity is costly and accordingly the operation cost of SDN increases. This increased cost is the cost of addressing the uncertainties of load consumption and wind generation in ODAS of SDN.

Case II: In order to understand the effectiveness of DR programs, in the second case the ODAS of SDN has been simulated in the presence of DR programs and the results of scheduled reserve which are provided by DRAs and LLs have been shown in Fig. 7.5.

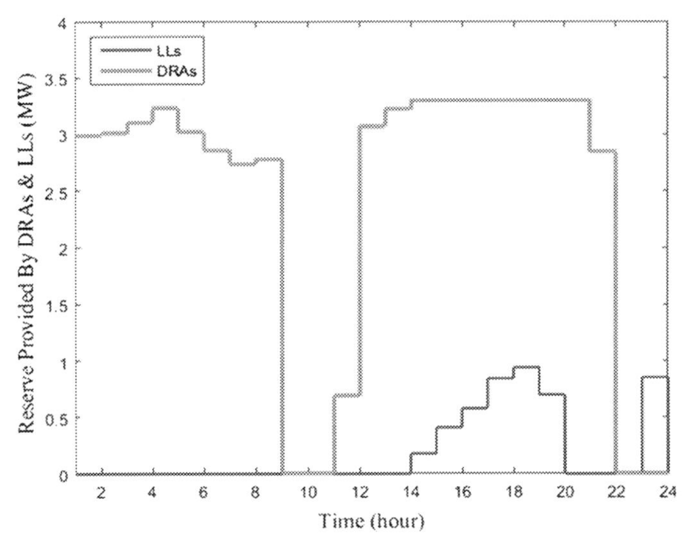

Figure 7.5 Reserve provided by DRAs and LLs.

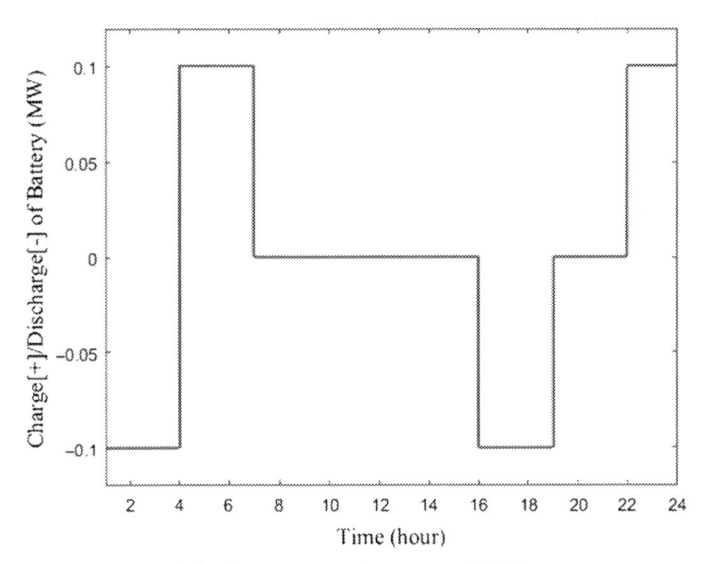

Figure 7.6 Charging and discharging time horizons of BESS.

As provided in Fig. 7.5, within the hours 10:00–23:00 when upstream grid prices are relatively high, the load decline of responsive loads has been scheduled by the ISO. Simultaneously, as illustrated in this figure, DRAs and LLs are scheduled to prepare the amount of required

Table 7.1 Comparison between the cost of case I and case II

	Energy providing cost ($)	Reserve providing cost ($)	Total operation cost ($)
Case I	62794.470	2571.768	65366.238
Case II	57048.852	1695.028	58743.880

reserve capacity and therefore as shown in Fig. 7.4, the DG units can be fully used to provide the network load. Accordingly, as illustrated in Fig. 7.3, providing reserve is not employing the capacity of DG units and they can be fully used for accommodating network consumption with lower operation cost in case II.

The BESS charging and discharging time horizons are illustrated in Fig. 7.6. It can be seen that in peak hours when the prices of upstream grid are high, the BESS is in discharge state and in off-peak hours when the prices of upstream grid are low the BESS is in charged state. Accordingly, BESS leads the distribution network to a lower operating cost by limiting the purchased energy from the upstream grid market.

Table 7.1 provides a comparison between case I and case II that shows the cost of ODAS with and without considering the role of DR programs. As shown in Table 7.1, with considering responsive loads, the operating cost is decreased of about 10.1%. This difference in values of operating cost is because in case I, all the needed amount of energy and reserve capacity should be prepared by upstream grid or by DG units. Therefore, at some of peak time horizons, providing reserve employs the capacity of DG units and as a result of this their presence in energy producing is reduced. But in case II, the load decline of responsive loads acts as an economical option within higher prices of upstream grid.

7.4 CONCLUSION

In this chapter a novel formulation and solution method is proposed to solve ODAS of the SDN considering DR programs. The proposed method incorporates DR programs in ODAS of SDN in order to minimize total cost of energy and reserve requirement. The proposed model is formulated as trustable mixed integer linear program that can be efficiently solved via commercial software. The results demonstrate that the proposed method is able to attain economic and reliable energy and reserve scheduling of distribution network with reasonable computational effort. Compared with alternative scheduling approaches, the proposed method is more applicable in that, it entirely takes into account the contribution of various kinds of energy sources in day-ahead scheduling of distribution networks. Moreover, it was shown that considering DR programs can decrease total operation costs and provide a more efficient use of DG units. As expected, utilizing DR programs decreases the values of operation cost of SDN by freeing up the capacity of DG units. In addition, it was shown that the proposed method is useful for addressing the contribution of DR programs in ODAS of distribution network.

REFERENCES

[1] V. Krakowski, et al., Feasible path toward 40–100% renewable energy shares for power supply in France by 2050: a prospective analysis, Appl. Energy 171 (2016) 501–522.
[2] P. Siano, et al., Assessing the benefits of residential demand response in a real time distribution energy market, Appl. Energy 161 (2016) 533–551.
[3] M.A.F. Ghazvini, et al., A multi-objective model for scheduling of short-term incentive-based demand response programs offered by electricity retailers, Appl. Energy 151 (2015) 102–118.
[4] E. Mahboubi-Moghaddam, et al., Reliability constrained decision model for energy service provider incorporating demand response programs, Appl. Energy 183 (2016) 552–565.
[5] S. Behboodi, et al., Renewable resources portfolio optimization in the presence of demand response, Appl. Energy 162 (2016) 139–148.
[6] S. Nojavan, et al., Optimal bidding strategy of electricity retailers using robust optimisation approach considering time-of-use rate demand response programs under market price uncertainties, IET Gen. Trans. Distribut. 9 (4) (2015) 328–338.
[7] G. Gutiérrez-Alcaraz, et al., Effects of demand response programs on distribution system operation, Int. J. Electr. Power Energy Systems 74 (2016) 230–237.
[8] C. Chen, et al., An interval optimization based day-ahead scheduling scheme for renewable energy management in smart distribution systems, Energy Convers. Manage. 106 (2015) 584–596.
[9] A. Zakariazadeh, et al., Stochastic operational scheduling of smart distribution system considering wind generation and demand response programs, Int. J. Electr. Power Energy Systems 63 (2014) 218–225.
[10] V. Hosseinnezhad, et al., Optimal day-ahead operational planning of microgrids, Energy Convers. Manage. 126 (2016) 142–157.
[11] H. Wu, et al., Dynamic economic dispatch of a microgrid: mathematical models and solution algorithm, Int. J. Electr. Power Energy Systems 63 (2014) 336–346.
[12] X. Lin, et al., Distribution network planning integrating charging stations of electric vehicle with V2G, Int. J. Electr. Power Energy Systems 63 (2014) 507–512.
[13] M. Sedghi, A. Ahmadian, M. Aliakbar-Golkar, Optimal storage planning in active distribution network considering uncertainty of wind power distributed generation, IEEE Trans. Power Systems 31 (1) (2016) 304–316.
[14] A. Soroudi, et al., Possibilistic evaluation of distributed generations impacts on distribution networks, IEEE Trans. Power Systems 26 (4) (2011) 2293–2301.

[15] S. Golshannavaz, et al., Smart distribution grid: optimal day-ahead scheduling with reconfigurable topology, IEEE Trans. Smart Grid 5 (5) (2014) 2402–2411.
[16] Y. Tan, et al., A two-stage stochastic programming approach considering risk level for distribution networks operation with wind power, IEEE Systems J. 10 (1) (2016) 117–126.
[17] A. Zakariazadeh, et al., Integrated operation of electric vehicles and renewable generation in a smart distribution system, Energy Convers. Manage. 89 (2015) 99–110.
[18] T. Sousa, et al., A multi-objective optimization of the active and reactive resource scheduling at a distribution level in a smart grid context, Energy 85 (2015) 236–250.
[19] M. Esmaili, et al., Multi-objective optimal charging of plug-in electric vehicles in unbalanced distribution networks, Int. J. Electr. Power Energy Systems 73 (2015) 644–652.
[20] A. Zakariazadeh, et al., Economic-environmental energy and reserve scheduling of smart distribution systems: a multiobjective mathematical programming approach, Energy Convers. Manage. 78 (2014) 151–164.
[21] M. Mazidi, et al., Incorporating price-responsive customers in day-ahead scheduling of SDNs, Energy Convers. Manage. 115 (2016) 103–116.
[22] T. Niknam, et al., Multi-objective daily operation management of distribution network considering fuel cell power plants, IET Renew. Power Gen. 5 (5) (2011) 356–367.
[23] T. Niknam, et al., A practical multi-objective PSO algorithm for optimal operation management of distribution network with regard to fuel cell power plants, Renew. Energy 36 (5) (2011) 1529–1544.
[24] T. Niknam, et al., An efficient algorithm for multi-objective optimal operation management of distribution network considering fuel cell power plants, Energy 36 (1) (2011) 119–132.
[25] A. Zakariazadeh, et al., Stochastic multi-objective operational planning of smart distribution systems considering demand response programs, Electr. Power Systems Res. 111 (2014) 156–168.
[26] S.M. Mohseni-Bonab, et al., A two-point estimate method for uncertainty modeling in multi-objective optimal reactive power dispatch problem, Int. J. Electr. Power Energy Systems 75 (2016) 194–204.
[27] A.R. Baran, et al., A three-phase optimal power flow applied to the planning of unbalanced distribution networks, Int. J. Electr. Power Energy Systems 74 (2016) 301–309.
[28] M.R. Sarker, et al., Optimal coordination and scheduling of demand response via monetary incentives, IEEE Trans. Smart Grid 6 (3) (2015) 1341–1352.
[29] Y.-Y. Hsu, et al., A combined artificial neural network-fuzzy dynamic programming approach to reactive power/voltage control in a distribution substation, IEEE Trans. Power Systems 13 (4) (1998) 1265–1271.
[30] F. Jabari, et al., Optimal short-term scheduling of a novel tri-generation system in the presence of demand response programs and battery storage system, Energy Convers. Manage. 122 (2016) 95–108.
[31] M.E. Baran, et al., Network reconfiguration in distribution systems for loss reduction and load balancing, IEEE Trans. Power Delivery 4 (2) (1989) 1401–1407.
[32] A.A. Cárdenas, et al., A polyhedral-based approach applied to quadratic cost curves in the unit commitment problem, IEEE Trans. Power Systems 31 (5) (2016) 3674–3683.
[33] S. Nojavan, et al., Optimal stochastic energy management of retailer based on selling price determination under smart grid environment in the presence of demand response program, Appl. Energy 187 (2017) 449–464.

[34] Y. Jiang, et al., Day-ahead stochastic economic dispatch of wind integrated power system considering demand response of residential hybrid energy system, Appl. Energy 190 (2017) 1126–1137.
[35] A. Keshtkar, et al., An adaptive fuzzy logic system for residential energy management in smart grid environments, Appl. Energy 186 (2017) 68–81.
[36] M. Asensio, et al., Risk-constrained optimal bidding strategy for pairing of wind and demand response resources, IEEE Trans. Smart Grid 8 (1) (2017) 200–208.
[37] J. Aghaei, et al., Contribution of emergency demand response programs in power system reliability, Energy 103 (2016) 688–696.
[38] M. Mazidi, et al., Integrated scheduling of renewable generation and demand response programs in a microgrid, Energy Convers. Manage. 86 (2014) 1118–1127.
[39] M. Majidi, et al., A multi-objective model for optimal operation of a battery/PV/ fuel cell/grid hybrid energy system using weighted sum technique and fuzzy satisfying approach considering responsible load management, Sol. Energy 144 (2017) 79–89.
[40] S. Wong, et al., Electric power distribution system design and planning in a deregulated environment, IET Gen. Trans. Distribut. 3 (12) (2009) 1061–1078.
[41] M. Mazidi, et al., Robust day-ahead scheduling of smart distribution networks considering demand response programs, Appl. Energy 178 (2016) 929–942.
[42] Diesel generators specification sheets, Kohler Power Systems Company, online available at: <http://www.yestramski.com/industrial/generatorsdiesel/industrial-diesel-generators-all.htm>.
[43] (2013, July. 17). New York Independent System Operator, online available at: http://www.nyiso.com/public/markets_operations/index.jsp.
[44] Willy Online Pty Ltd, Online available at: http://wind.willyweather.com.au/.
[45] The GAMS/CPLEX manual, online available at: http://www.gams.com/dd/docs/solvers/cplex/index.html.

CHAPTER 8

Impacts of Solar Parks and Wind Farms on Controlled Islanding of Radial Distribution Networks

Farkhondeh Jabari and Behnam Mohammadi-ivatloo
University of Tabriz, Tabriz, Iran

8.1 LITERATURE REVIEW

In the literature, several remarkable efforts have been carried out on optimal islanding of electricity grids. In this context, Jabari et al., developed a novel backward elimination method for defensive splitting of interconnected and distribution systems [1–5]. A two-stage linear stochastic programming problem is solved in Refs. [6,7] to model the intermittent nature of wind generations in splitting of radial distribution systems with the aim of minimizing total load generation mismatch, transmission congestion, and improving the voltage profile considering energy storage devices.

Ref. [8] developed a comprehensive learning particle swarm optimization method for minimization of total real and reactive power procurement cost taking into account optimal power flow constraints within islands and controlled load shedding in each candidate solution. Caldon et al. [9] proposed an automatic adaptive identification islanding configuration approach to determine the best splitting. In this study, a centralized controller predicts the active and reactive values of all loads and investigates the probability of successful islanding operation of created sections. In Ref. [10], the genetic algorithm is used to improve the reliability of legacy radial distribution systems by adding some interconnections between different feeders in islanding operating mode. Cooperative control strategy of small-scale power generation and storage technologies and its impacts on stabilization of islanded distribution systems is experimentally studied in Ref. [11]. Hemmatpour et al. [12], applied an adaptive multiobjective harmony search algorithm to split increase the loadability index in islanding operating mode. In Ref. [13], a serial of tree knapsack

Operation of Distributed Energy Resources in Smart Distribution Networks
DOI: https://doi.org/10.1016/B978-0-12-814891-4.00008-4

based islanding scenario generation algorithm is addressed taking into account some technical criteria such as load priority and controllability, power balance, voltage security margin, and feeder capacity constraints.

Ref. [14] introduced a control architecture algorithm based on islanding security region (ISR) in a way that ISR makes it possible to detect the uncontrolled splitting phenomenon before occurring the upstream grid loss. Control measures are also implemented to pull system back into ISR and make an intentional defensive partitioning decision in time. A mixed-integer second-order conic programming problem is solved by López et al. [15] to design islanding process of current distribution grids, minimize transmission active power losses, maximize average interruption frequency index, average interruption duration index, and energy not supplied. A bioinspired metaheuristic artificial immune systems based optimal islanding framework is provided by Oliveira et al. [16] to minimize total transmission lines losses with respect to load and wind production uncertainties. Other islanding search algorithms such as three-layer artificial neural network [17], hybrid big bang-big crunch algorithm [18], and decimal coded quantum particle swarm optimization [19] have widely been used by scholars.

In the last decade, planning and operation of renewable energy resources based energy systems are attracting world attention due to major concerns about energy crisis and global warming [20–25]. As mentioned, some scholars have focused on defensive splitting of meshed and radial distribution systems. However, impacts of renewable based power generation technologies such as solar parks and wind farms have not been discussed, yet. Therefore, this chapter provides a stochastic splitting algorithm for partitioning the radial distribution systems and evaluating the impact of solar and wind generations' uncertainties.

The remainder of this chapter can be organized as follows: Section 8.2 presents the problem formulation consisting of islanding search algorithm, objective function and constraints, and the proposed stochastic programming approach. Simulation result and discussions are drawn in Section 8.3. Finally, concluding remarks appear in Section 8.4.

8.2 PROBLEM FORMULATION

8.2.1 Islanding Search Algorithm

As mentioned before, if interconnections between different feeders and upstream network are disrupted, the controlled islanding strategy should be implemented in order to minimize total load generation mismatch and improve system reliability and efficiency. For this purpose, an interface probabilistic area consisting of boundary branches and nodes should be determined. As it is obvious from Fig. 8.1, coherent groups of generating units are specified based on electrical distance between them. For example, if we consider that three independent islands are formed with following coherent generating units:

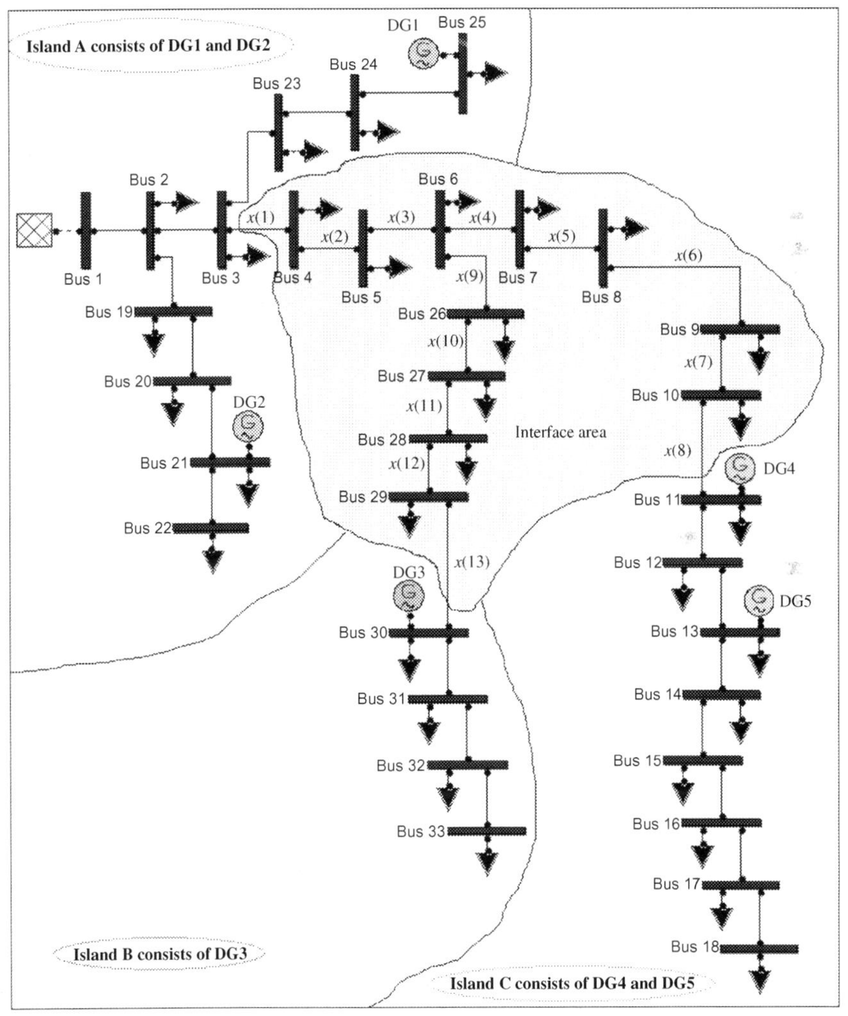

Figure 8.1 Coherent generating units and boundary region in a test distribution grid.

- DG1 and DG2,
- DG3,
- DG4 and DG5.

Obviously, first and second generating units can be considered as two roots of island A. Similarly, the third distributed generator (DG) is a root of island B. Others are as two source nodes for expanding area C. All lines that connect two coherent DGs are not belonging to boundary region. These lines are connected together. Other branches belong to interface network as boundary lines. All nodes belong to interface region can be called as probabilistic buses. Primary branch and node information matrices of three islands are formed according to interface network. As illustrated in Fig. 8.2, islands A, B, and C have a maximum of 20, 15, and 19 buses, respectively.

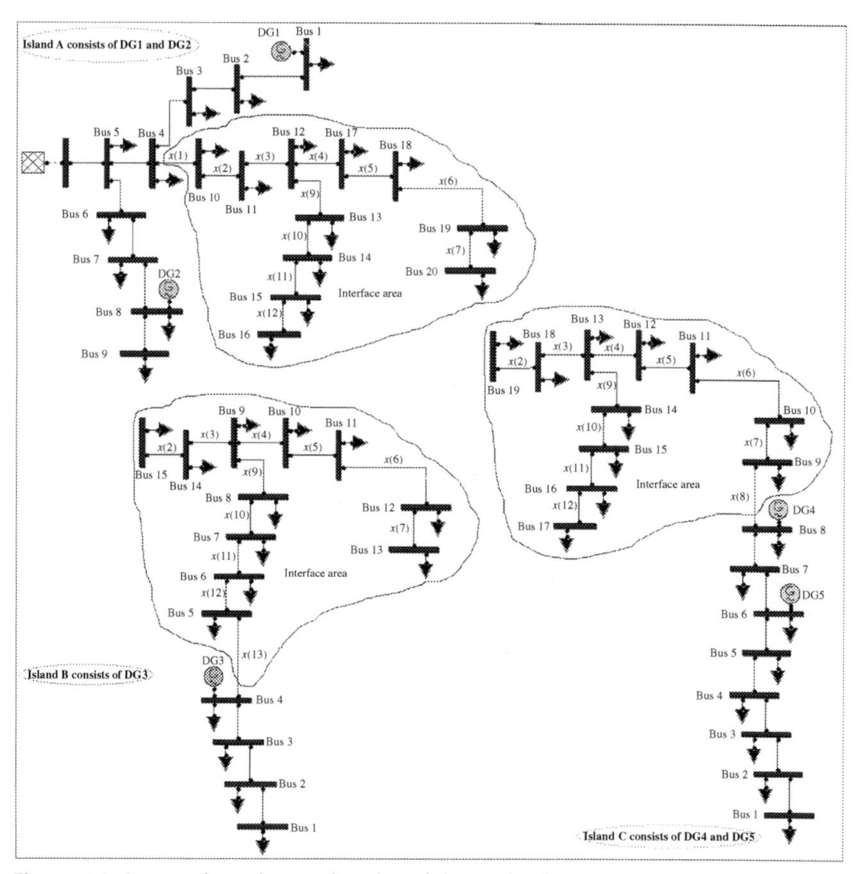

Figure 8.2 Primary branches and nodes of three islands.

The impedance of kth probabilistic branch can be formulated as following relations:

$$Z(i,j) = M \times (1 - x(k)) + Z_0(i,j) \quad (8.1)$$

$$x(k) = \begin{cases} 0 & \text{if } k^{th} \text{ probabilistic line is disconnected} \\ 1 & \text{if } k^{th} \text{ probabilistic line is connected} \end{cases} \quad (8.2)$$

$$Z_0(i,j) = R(i,j) + jX(i,j) \quad (8.3)$$

where,

$Z(i,j)$: impedance of line i to j

M: large constant factor

$Z_0(i,j)$, $R(i,j)$and $X(i,j)$: Impedance, resistance and reactance of line i to j in terms of per unit

$x(k)$: Binary state variable of probabilistic line k

If $x(k) = 0$, k^{th}probabilistic line is disconnected and $Z(i,j) \cong M$. If $x(k) = 1$, k^{th}probabilistic line is opened and $Z(i,j) = Z_0(i,j)$.

The primary node data matrices are then updated as follows: If i^{th}bus of area A is transmitted from island A to area B, i^{th}node of island A will be removed from its primary node data matrix as follows:

- $NEM_A(i) = 1$: If i^{th} bus of area A is not belong to interface network, i^{th} element of NEM_A is considered to be equal to 1.
- Otherwise, i^{th} element of $NEM_A$is equal to multiplication of binary state variables of all probabilistic lines that connect it to roots. As shown in Fig. 8.2, 5th and 16th buses of area A are deterministic and probabilistic, respectively. Hence, 5th and 16th elements of $NEM_A$are equal to 1 and $x(1) \times x(2) \times x(3) \times x(9) \times x(10) \times x(11) \times x(12)$, respectively.
- When,$NEM_A(i) = 0$, i^{th} node of area A is transmitted to other adjacent islands. Otherwise, $NEM_A(i) \neq 0$.

8.2.2 Objective Function and Constraints

In this chapter, total real power losses and voltage variations are minimized as Relation (8.4):

$$\text{Objective} = \frac{\sum_{k=1}^{N_{island}} \sum_{i=1}^{N_B^k - 1} R_{i_k} I_{i_k}^2}{\sum_{j=1}^{N_B} R_j I_j^2} + \sum_{k=1}^{N_{island}} \sum_{n=1}^{N_B^k} \left| V_{n_k} - 1 \right|^2 \tag{8.4}$$

where,

R_{i_k}: resistance of feeder j in island k

I_{i_k}: current of feeder j in island k

R_j: resistance of feeder j in preislanding distribution system

I_j: current of feeder j in preislanding distribution system

N_B^k: number of nodes in island k

N_{Island}: number of islands

N_B: number of nodes in preislanding distribution system

V_{n_k}: voltage magnitude of bus n in island k in per unit

Subject to:

- voltage limits

$$V_i^{min} \leq V_i \leq V_i^{max} \tag{8.5}$$

where,

V_i^{min}: minimum voltage magnitude of bus I

V_i^{max}: maximum voltage magnitude of bus i

- Feeder capacity

$$\sqrt{P_k^2 + Q_k^2} \leq S_k^{max} \tag{8.6}$$

where,

P_k: active power flow in feeder k

Q_k: reactive power flow in feeder k

S_k^{max}: power flow capacity of feeder k

8.2.3 Wind Production Uncertainty

The wind speed constantly changes. Its variation is generally modeled with Weibull probability distribution function as Eq. (8.7) [22]:

$$f(u_w) = \frac{k_w}{A_w} \left(\frac{u_w}{A_w} \right)^{k_w - 1} \exp\left(-\left(\frac{u_w}{A_w} \right)^{k_w} \right) \tag{8.7}$$

where,

u_w: wind speed

$A_w$and k_w: scale and shape factors, respectively

The output mechanical power of a wind turbine (P_m) is approximated by Eq. (8.8) [23]:

$$P_m = \begin{cases} 0 & u_w < u_{in};\, u_{out} \le u_w \\ P_r \dfrac{u_w - u_{in}}{u_r - u_{in}} & u_{in} \le u_w < u_r \\ P_r & u_r \le u_w < u_{out} \end{cases} \tag{8.8}$$

where,

u_{in}, $u_{out}$and u_r: wind turbine generator cut-in, cut-out and rated speeds in m/sec

P_r: wind turbine generator rated power in MW

8.2.4 Solar Photovoltaic Cells

In the last decade, use of solar photovoltaic cells to generate electricity is rapidly gaining popularity due to zero carbon footprints [22,26]. This chapter aims to present optimal short-term dispatching of solar photovoltaic based multichiller plants in the presence of time-of-use cooling-demand response programs. Use of solar irradiance as primary energy source during extremelyhot summer days not only mitigates total carbon footprints, but also reduces total energy consumptions of electrical chillers from fossil fuels based nonrenewable energy sources, especially by applying peak clipping and valley filling demand response strategies on cooling demand. The power output of a photovoltaic module can be calculated from Eq. (8.9) [27].

$$P_t^{pv} = \eta\, S\, \Phi_t \left[1 - 0.005 \times (T_t^a - 25)\right] \tag{8.9}$$

where,

P_t^{pv}: lower output of a photovoltaic panel

η: conversion coefficient of a photovoltaic panel

S: array area of a photovoltaic module

Φ_t: solar irradiance

T_t^a: ambient temperature at time t

The flowchart of the proposed probabilistic splitting approach is illustrated in Fig. 8.3. As obvious from this figure, the uncertainties of solar parks and wind farms are considered in each generated scenario. This improves the reliability of the obtained optimum separating points.

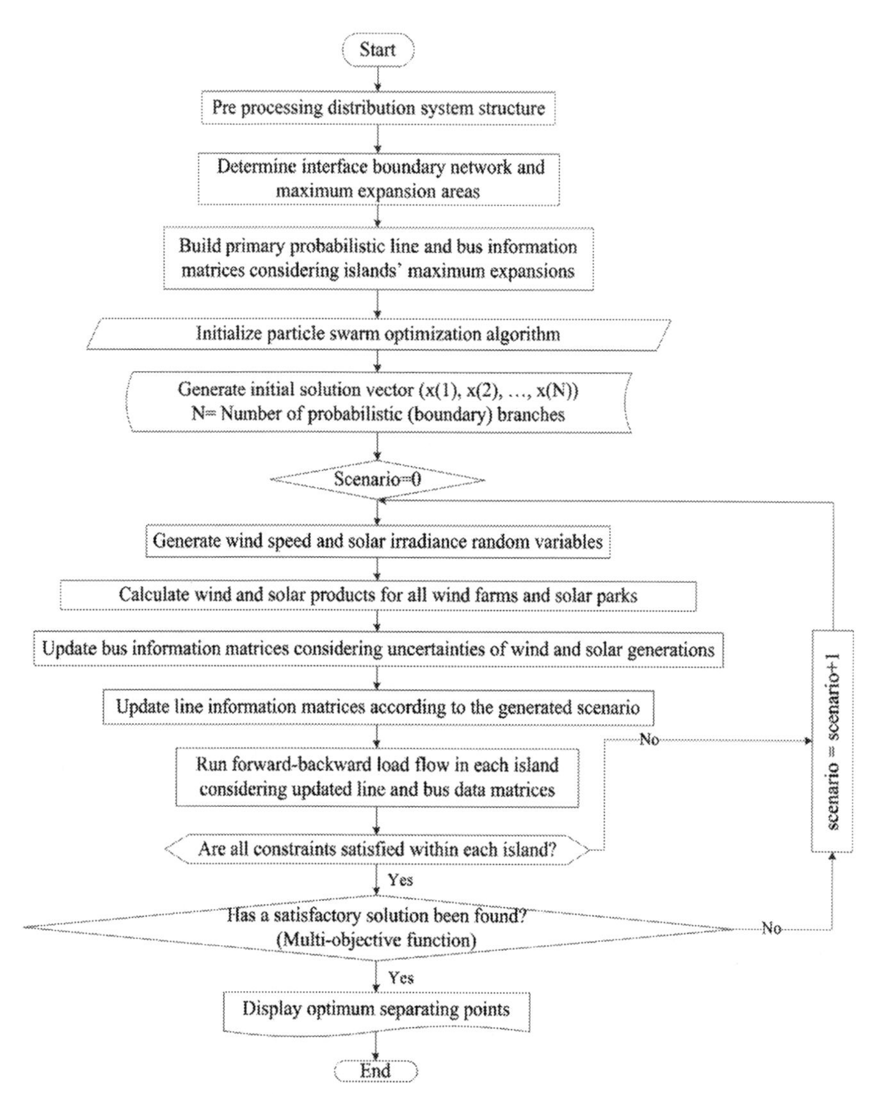

Figure 8.3 Proposed probabilistic islanding algorithm considering wind and solar generation uncertainties.

8.3 SIMULATION RESULT AND DISCUSSIONS

To prove the feasibility and the applicability of the proposed algorithm in finding the best solution vector, a 12.66 kV and 33-bus radial distribution grid [28] is selected as the test system. Fig. 8.4 depicts the structure of this benchmark network with 32 branches and 33 nodes.

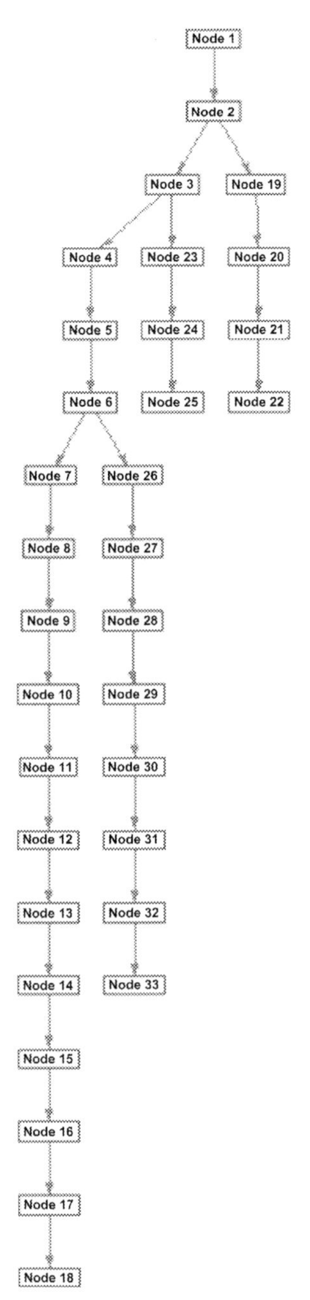

Figure 8.4 Schematic presentation of 12.66 kV, 33-bus, 32-line radial distribution system.

Tables 8.1 and 8.2 summarize two lines and bus data of 33-bus radial distribution system, respectively. In this research, minimum and maximum values of bus voltage magnitude are considered to be equal to 0.9 and 1.05, respectively.

This test network is connected to upstream system via node 1. Its initial operating point obtained from backward-forward sweep power flow without considering solar photovoltaic cells and wind turbines is reported in Table 8.3 and Fig. 8.5.

Table 8.1 Line data for 33-bus radial test grid

From bus	To bus	$R(\Omega)$	$X(\Omega)$
1	2	0.0922	0.0470
2	3	0.4930	0.2511
3	4	0.3660	0.1864
4	5	0.3811	0.1941
5	6	0.8191	0.7070
6	7	0.1872	0.6188
7	8	0.7114	0.2351
8	9	1.0300	0.7400
9	10	1.0440	0.7400
10	11	0.1966	0.0650
11	12	0.3744	0.1238
12	13	1.4680	1.1550
13	14	0.5416	0.7129
14	15	0.5910	0.5260
15	16	0.7463	0.5450
16	17	1.2890	1.7210
17	18	0.7320	0.5740
2	19	0.1640	0.1565
19	20	1.5042	1.3554
20	21	0.4095	0.4784
21	22	0.7089	0.9373
3	23	0.4512	0.3083
23	24	0.8980	0.7091
24	25	0.8960	0.7011
6	26	0.2030	0.1034
26	27	0.2842	0.1447
27	28	1.0590	0.9377
28	29	0.8042	0.7006
29	30	0.5075	0.2585
30	31	0.9744	0.9630
31	32	0.3105	0.3619
32	33	0.3410	0.5302

Table 8.2 Bus data for 33-bus radial test grid

Bus number	*P*(kW)	*Q*(kW)
1	0	0
2	120	72
3	108	48
4	144	96
5	72	36
6	72	24
7	240	120
8	240	120
9	72	24
10	72	24
11	54	36
12	72	42
13	72	42
14	144	96
15	72	12
16	72	24
17	72	24
18	108	48
19	108	48
20	108	48
21	108	48
22	108	48
23	108	60
24	504	240
25	504	240
26	72	30
27	72	30
28	72	24
29	144	84
30	240	720
31	180	84
32	252	120

It is assumed that three solar parks and three wind farms have been located at nodes 5, 27, and 9 with following characteristics:

Solar irradiance: 1860 W/m^2

Wind speed: 1 m/s

Number of PV modules in 1st solar park: 1000

Number of PV modules in 2nd solar park: 1200

Number of PV modules in 3rd solar park: 1500

Cut-in speed in 1st wind farm: 3 m/s

Table 8.3 Optimum operating point of test system without considering solar and wind generations

Optimum operating point of test system	Without DGs
Active losses (kW)	253.9667
Reactive losses (kVAR)	169.0311
System active load (kW)	3715
System reactive load (kVAR)	2300

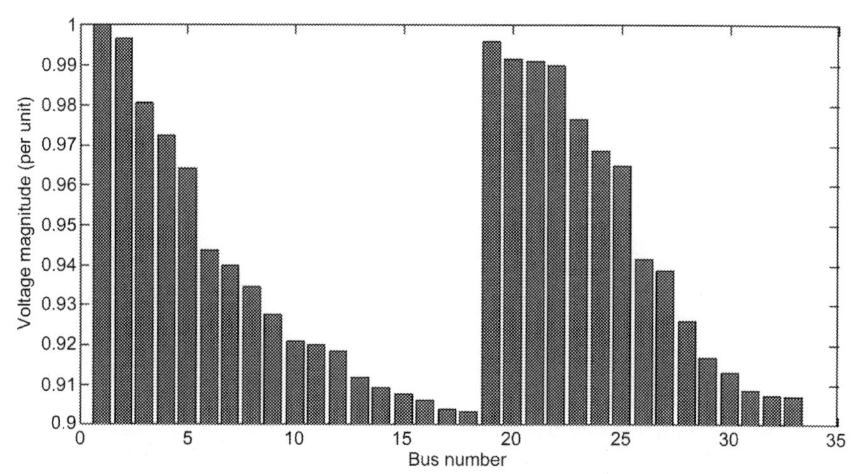

Figure 8.5 Voltage profile of 33-bus radial distribution system without considering solar and wind generations.

Cut-in speed in 2nd wind farm: 4 m/s
Cut-in speed in 3rd wind farm: 5 m/s
Cut-out speed in 1st wind farm: 55 m/s
Cut-out speed in 2nd wind farm: 65 m/s
Cut-out speed in 3rd wind farm: 80 m/s
Rated speed in 1st wind farm: 35 m/s
Rated speed in 2nd wind farm: 37 m/s
Rated speed in 3rd wind farm: 40 m/s
Rated power in 1st wind farm: 120 kW
Rated power in 2nd wind farm: 280 kW
Rated power in 3rd wind farm: 360 kW

It is supposed that the electrical connection between main power system and downstream network is lost. Hence, the proposed splitting method is implemented in order to prevent from partial blackout of radial distribution grid while minimizing total active power losses and voltage drops and considering all mentioned operational constraints. Simulations are performed on a Lenovo with 2.10 GHz CPU, 4 GB RAM in Matlab software package. Firstly, it is assumed that all coherent generators are clustered as follows:

- Coherent cluster A: G1 and G2
- Coherent cluster B: G3
- Coherent cluster C: G4, G5, and G6

The interface network between three mentioned primary islands is shown in Fig. 8.6.

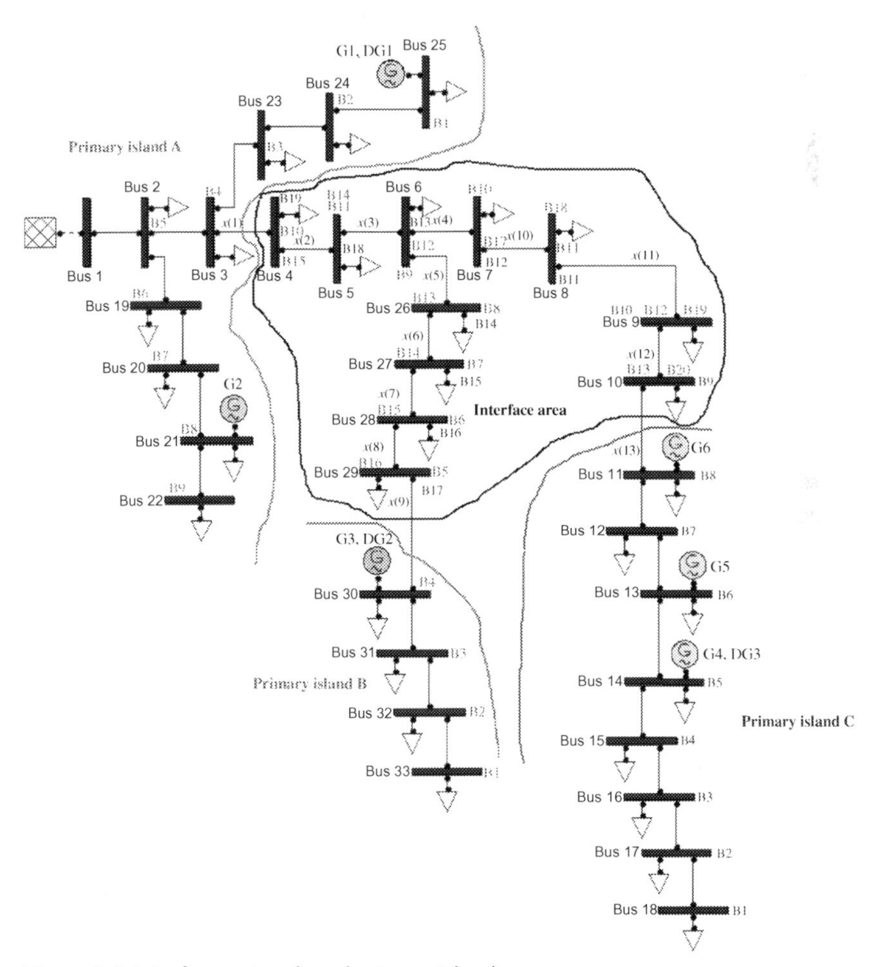

Figure 8.6 Interface network and primary islands.

The primary line and bus data matrices of three mentioned islands with and without DGs can be considered as follows.

- Bus data of island A without/with solar parks and wind farms

Bus number	Active power (kW)	Reactive power (kVAR)
1	504	240
2	504	240
3	108	60
4	108	48
5	120	72
6	108	48
7	108	48
8	108	48
9	108	48
10	144	96
11	72-output power of 1st solar park and wind farm	36
12	72	24
13	72	30
14	72-output power of 2nd solar park and wind farm	30
15	72	24
16	144	84
17	240	120
18	240	120
19	72-output power of 3rd solar park and wind farm	24
20	72	24

- Line data of island A

In bus	Out bus	Resistance (Ω)	Reactance (Ω)
1	2	0.8960	0.7011
2	3	0.8980	0.7091
3	4	0.4512	0.3083
4	5	0.4930	0.2511
4	10	$0.3660 + M \times (1\text{-}x(1))$	$0.1864 + M \times (1\text{-}x(1))$
5	6	0.1640	0.1565
6	7	1.5042	1.3554
7	8	0.4095	0.4784
8	9	0.7089	0.9373
10	11	$0.3811 + M \times (1\text{-}x(2))$	$0.1941 + M \times (1\text{-}x(2))$
11	12	$0.8191 + M \times (1\text{-}x(3))$	$0.7070 + M \times (1\text{-}x(3))$

(Continued)

In bus	Out bus	Resistance (Ω)	Reactance (Ω)
12	13	0.2030 + $M \times$ (1-x(5))	0.1034 + $M \times$ (1-x(5))
12	17	0.1872 + $M \times$ (1-x(4))	0.6188 + $M \times$ (1-x(4))
13	14	0.2842 + $M \times$ (1-x(6))	0.1447 + $M \times$ (1-x(6))
14	15	1.0590 + $M \times$ (1-x(7))	0.9377 + $M \times$ (1-x(7))
15	16	0.8042 + $M \times$ (1-x(8))	0.7006 + $M \times$ (1-x(8))
17	18	0.7114 + $M \times$ (1-x(10))	0.2351 + $M \times$ (1-x(10))
18	19	1.0300 + $M \times$ (1-x(11))	0.7400 + $M \times$ (1-x(11))
19	20	1.0440 + $M \times$ (1-x(12))	0.7400 + $M \times$ (1-x(12))

- Bus data of island B without/with solar parks and wind farms

Bus number	Active power (kW)	Reactive power (kVAR)
1	72	48
2	252	120
3	180	84
4	240	720
5	144	84
6	72	24
7	72-output power of 2nd solar park and wind farm	30
8	72	24
9	72	120
10	240	120
11	240	24
12	72-output power of 3rd solar park and wind farm	24
13	72	24
14	72-output power of 1st solar park and wind farm	36
15	144	96

- Line data of island B

In bus	Out bus	Resistance (Ω)	Reactance (Ω)
1	2	0.3410	0.5302
2	3	0.3105	0.3619
3	4	0.9744	0.9630
4	5	0.5075 + $M \times$ (1-x(9))	0.2585 + $M \times$ (1-x(9))
5	6	0.8042 + $M \times$ (1-x(8))	0.7006 + $M \times$ (1-x(8))
6	7	1.0590 + $M \times$ (1-x(7))	0.9377 + $M \times$ (1-x(7))
7	8	0.2842 + $M \times$ (1-x(6))	0.1447 + $M \times$ (1-x(6))

(Continued)

In bus	Out bus	Resistance (Ω)	Reactance (Ω)
8	9	$0.2030 + M \times (1\text{-}x(5))$	$0.1034 + M \times (1\text{-}x(5))$
9	10	$0.1872 + M \times (1\text{-}x(4))$	$0.6188 + M \times (1\text{-}x(4))$
9	14	$0.8191 + M \times (1\text{-}x(3))$	$0.7070 + M \times (1\text{-}x(3))$
10	11	$0.7114 + M \times (1\text{-}x(10))$	$0.2351 + M \times (1\text{-}x(10))$
11	12	$1.0300 + M \times (1\text{-}x(11))$	$0.7400 + M \times (1\text{-}x(11))$
12	13	$1.0440 + M \times (1\text{-}x(12))$	$0.7400 + M \times (1\text{-}x(12))$
14	15	$0.3811 + M \times (1\text{-}x(2))$	$0.1941 + M \times (1\text{-}x(2))$

- Bus data of island C without/with solar parks and wind farms

Bus number	Active power (kW)	Reactive power (kVAR)
1	108	48
2	72	24
3	72	24
4	72	12
5	144	96
6	72	42
7	72	42
8	54	36
9	72	24
10	72-output power of 3rd solar park and wind farm	24
11	240	120
12	240	120
13	72	24
14	72	30
15	72-output power of 2nd solar park and wind farm	30
16	72	24
17	144	84
18	72-output power of 1st solar park and wind farm	36
19	144	96

- Line data of island C

In bus	Out bus	Resistance (Ω)	Reactance (Ω)
1	2	0.7320	0.5740
2	3	1.2890	1.7210
3	4	0.7463	0.5450
4	5	0.5910	0.5260

(*Continued*)

In bus	Out bus	Resistance (Ω)	Reactance (Ω)
5	6	0.5416	0.7129
6	7	1.4680	1.1550
7	8	0.3744	0.1238
8	9	$0.1966 + M \times (1\text{-}x(13))$	$0.0650 + M \times (1\text{-}x(13))$
9	10	$1.0440 + M \times (1\text{-}x(12))$	$0.7400 + M \times (1\text{-}x(12))$
10	11	$1.0300 + M \times (1\text{-}x(11))$	$0.7400 + M \times (1\text{-}x(11))$
11	12	$0.7114 + M \times (1\text{-}x(10))$	$0.2351 + M \times (1\text{-}x(10))$
12	13	$0.1872 + M \times (1\text{-}x(4))$	$0.6188 + M \times (1\text{-}x(4))$
13	14	$0.2030 + M \times (1\text{-}x(5))$	$0.1034 + M \times (1\text{-}x(5))$
13	18	$0.8191 + M \times (1\text{-}x(3))$	$0.7070 + M \times (1\text{-}x(3))$
14	15	$0.2842 + M \times (1\text{-}x(6))$	$0.1447 + M \times (1\text{-}x(6))$
15	16	$1.0590 + M \times (1\text{-}x(7))$	$0.9377 + M \times (1\text{-}x(7))$
16	17	$0.8042 + M \times (1\text{-}x(8))$	$0.7006 + M \times (1\text{-}x(8))$
18	19	$0.3811 + M \times (1\text{-}x(2))$	$0.1941 + M \times (1\text{-}x(2))]$

Secondly, binary particle swarm optimization algorithm (BPSO) is used to obtain the best islanding solution among $2^{13} = 8192$ scenarios. Moreover, bus data matrices can be updated using three following matrices:

$$
\begin{aligned}
\mathrm{NEM\_}A = [&1\\
&1\\
&1\\
&1\\
&1\\
&1\\
&1\\
&1\\
&1\\
&x(1)\\
&x(1) \times x(2)\\
&x(1) \times x(2) \times x(3)\\
&x(1) \times x(2) \times x(3) \times x(5)\\
&x(1) \times x(2) \times x(3) \times x(5) \times x(6)\\
&x(1) \times x(2) \times x(3) \times x(5) \times x(6) \times x(7)\\
&x(1) \times x(2) \times x(3) \times x(5) \times x(6) \times x(7) \times x(8)\\
&x(1) \times x(2) \times x(3) \times x(4)\\
&x(1) \times x(2) \times x(3) \times x(4) \times x(10)\\
&x(1) \times x(2) \times x(3) \times x(4) \times x(10) \times x(11)\\
&x(1) \times x(2) \times x(3) \times x(4) \times x(10) \times x(11) \times x(12)]
\end{aligned}
$$

$$
\begin{aligned}
\mathrm{NEM\_}B = [&1\\
&1\\
&1\\
&1\\
&x(9)\\
&x(8)\times x(9)\\
&x(7)\times x(8)\times x(9)\\
&x(6)\times x(7)\times x(8)\times x(9)\\
&x(5)\times x(6)\times x(7)\times x(8)\times x(9)\\
&x(4)\times x(5)\times x(6)\times x(7)\times x(8)\times x(9)\\
&x(4)\times x(5)\times x(6)\times x(7)\times x(8)\times x(9)\times x(10)\\
&x(4)\times x(5)\times x(6)\times x(7)\times x(8)\times x(9)\times x(10)\times x(11)\\
&x(4)\times x(5)\times x(6)\times x(7)\times x(8)\times x(9)\times x(10)\times x(11)\times x(12)\\
&x(3)\times x(5)\times x(6)\times x(7)\times x(8)\times x(9)\\
&x(2)\times x(3)\times x(5)\times x(6)\times x(7)\times x(8)\times x(9)]
\end{aligned}
$$

$$
\begin{aligned}
\mathrm{NEM\_}C = [&1\\
&1\\
&1\\
&1\\
&1\\
&1\\
&1\\
&1\\
&x(13)\\
&x(12)\times x(13)\\
&x(11)\times x(12)\times x(13)\\
&x(10)\times x(11)\times x(12)\times x(13)\\
&x(4)\times x(10)\times x(11)\times x(12)\times x(13)\\
&x(5)\times x(4)\times x(10)\times x(11)\times x(12)\times x(13)\\
&x(6)\times x(5)\times x(4)\times x(10)\times x(11)\times x(12)\times x(13)\\
&x(7)\times x(6)\times x(5)\times x(4)\times x(10)\times x(11)\times x(12)\times x(13)\\
&x(8)\times x(7)\times x(6)\times x(5)\times x(4)\times x(10)\times x(11)\times x(12)\times x(13)\\
&x(3)\times x(4)\times x(10)\times x(11)\times x(12)\times x(13)\\
&x(2)\times x(3)\times x(4)\times x(10)\times x(11)\times x(12)\times x(13)]
\end{aligned}
$$

According to Fig. 8.7, this test system is optimally separated into three stable islands by tripping 3rd and 12th boundary branches. Fig. 8.8 and

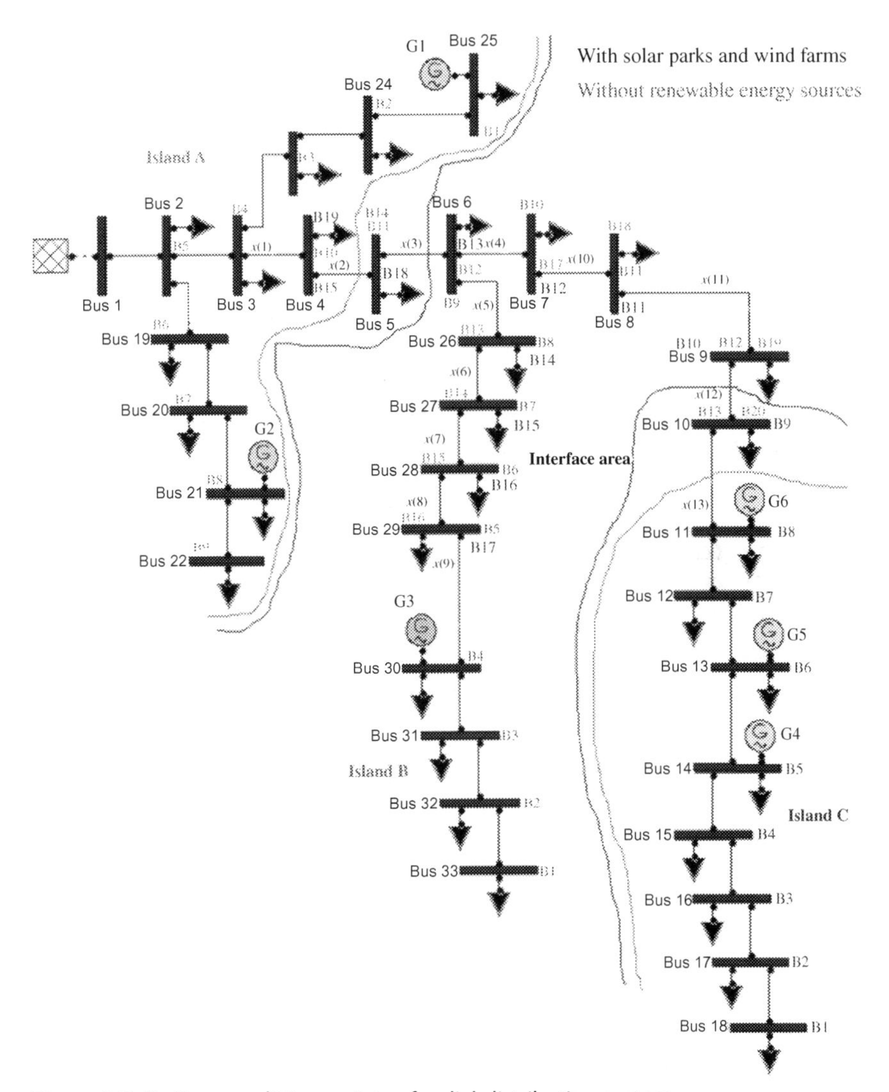

Figure 8.7 Optimum splitting points of radial distribution system.

Table 8.4 show the voltage profile and the total active power losses in two cases without and with renewable energy resources, respectively. It is found that the optimum separating points of an isolated radial distribution system changes in the presence of solar parks and wind farms. Moreover, the presented algorithm is able to find a good solution vector which reduces the total real power losses and improves the voltage profiles within all islands, significantly.

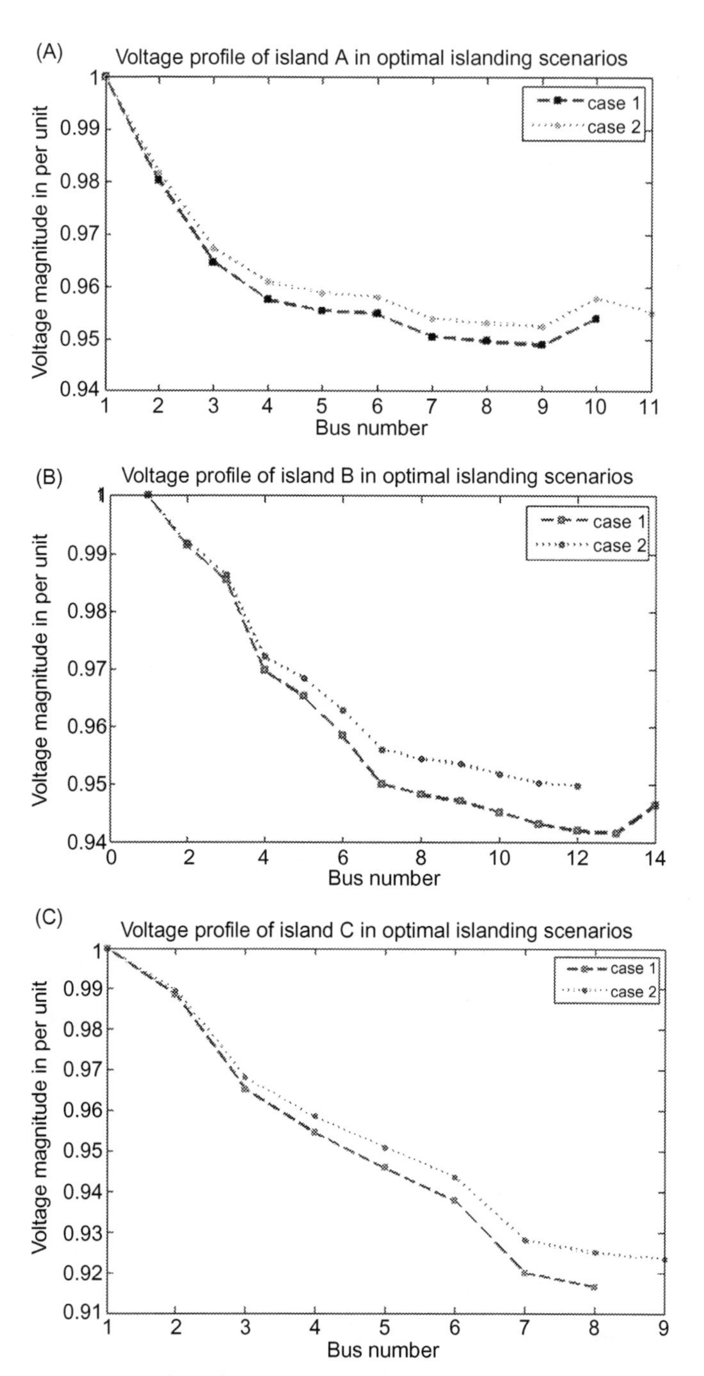

Figure 8.8 Voltage profile of islanded distribution system without (case 1: $x(2) = x(13) = 0$) and with (case 2: $x(3) = x(12) = 0$) renewable energies. (a). Voltage profile in island A. (b). Voltage profile in island B. (c). Voltage profile in island C.

Table 8.4 Total real power losses in two cases: Case 1: $x(2) = x(13) = 0$, Case 2: $x(3) = x(12) = 0$

Islands	Real power losses (kW)			
	Case 1: $x(2) = x(13) = 0$		Case 2: $x(3) = x(12) = 0$	Loss reduction
	Without solar and wind products	With solar and wind products		
A	102.8228	84.5159	82.0328	84.5159−82.0328 = 2.4831
B	61.2054	46.9267	46.8791	46.9267−46.8791 = 0.0476
C	142.9473	110.3567	103.7450	110.3567−103.7450 = 6.6117
Total	306.9755	241.7993	232.6569	241.7993−232.6569 = 9.1424

8.4 CONCLUSION

This chapter presented a backward-forward sweep based strategy for islanding of isolated radial distribution systems. The proposed approach is able to find a good splitting scenario within a huge initial search space under two cases with and without considering the output power of renewable energy resources. The numerical result demonstrated that the total active power losses and the voltage drops considerably reduces in the presence of solar parks and wind farms, as expected. By making an optimal islanding decision in the presence of renewable power generation technologies (case b), the amount of total active losses will be decreased. In the future, load-splitting based defensive islanding strategy can be applied on faulted distribution systems with the aim of minimizing islands' power mismatches.

REFERENCES

[1] F. Jabari, H. Seyedi, S.N. Ravadanegh, Large-scale power system controlled islanding based on backward elimination method and primary maximum expansion areas considering static voltage stability, Int. J. Electr. Power Energy Systems 67 (2015) 368–380.

[2] F. Jabari, H. Seyedia, S.N. Ravadanegh, B.M. Ivatloo, Stochastic contingency analysis based on voltage stability assessment in islanded power system considering load uncertainty using MCS and k-PEM, Handbook of Research on Emerging Technologies for Electrical Power Planning, Analysis, and Optimization, IGI Global, 2016, pp. 12–36.

[3] F. Jabari, B. Mohammadi-Ivatloo, Backward-forward sweep-based islanding scenario generation algorithm for defensive splitting of radial distribution systems, Power Quality in Future Electrical Power Systems, Institution of Engineering and Technology, 2017, pp. 283–304.

[4] F. Jabari, H. Seyedi, S.N. Ravadanegh, Online aggregation of coherent generators based on electrical parameters of synchronous generators, in: Power System Conference (PSC), 2015 30th International, 2015, pp. 8–13.
[5] F. Jabari, B. Mohammadi-Ivatloo, M. Rasouli, Optimal planning of a micro-combined cooling, heating and power system using air-source heat pumps for residential buildings, Energy Harvesting and Energy Efficiency, Springer, New York, 2017, pp. 423–455.
[6] P.M. de Quevedo, J. Allahdadian, J. Contreras, G. Chicco, Islanding in distribution systems considering wind power and storage, Sustain. Energy Grids Networks 5 (2016) 156–166. 2016/03/01/.
[7] J.P. Lopes, C. Moreira, A. Madureira, Defining control strategies for microgrids islanded operation, IEEE Trans. Power Systems 21 (2006) 916–924.
[8] A. El-Zonkoly, M. Saad, R. Khalil, New algorithm based on CLPSO for controlled islanding of distribution systems, Int. J. Electr. Power Energy Systems 45 (2013) 391–403.
[9] R. Caldon, A. Stocco, R. Turri, Feasibility of adaptive intentional islanding operation of electric utility systems with distributed generation, Electric Power Systems Res. 78 (2008) 2017–2023.
[10] H.E. Brown, S. Suryanarayanan, S.A. Natarajan, S. Rajopadhye, Improving reliability of islanded distribution systems with distributed renewable energy resources, IEEE Trans. Smart Grid 3 (2012) 2028–2038.
[11] J.Y. Kim, J.H. Jeon, S.K. Kim, C. Cho, J.H. Park, H.M. Kim, et al., Cooperative control strategy of energy storage system and microsources for stabilizing the microgrid during islanded operation, IEEE Trans. Power Electr. 25 (2010) 3037–3048.
[12] M.H. Hemmatpour, M. Mohammadian, A.A. Gharaveisi, Optimum islanded microgrid reconfiguration based on maximization of system loadability and minimization of power losses, Int. J. Electr. Power Energy Systems 78 (2016) 343–355.
[13] L. Jikeng, W. Xudong, W. Peng, L. Shengwen, S. Guang-hui, M. Xin, et al., Two-stage method for optimal island partition of distribution system with distributed generations, IET Gener. Transm. Distrib. 6 (2012) 218–225.
[14] Y. Chen, Z. Xu, J. Østergaard, Islanding Control Architecture in future smart grid with both demand and wind turbine control, Electric Power Systems Research 95 (2013) 214–224. 2//.
[15] J.C. López, M. Lavorato, M.J. Rider, Optimal reconfiguration of electrical distribution systems considering reliability indices improvement, Int. J. Electr. Power Energy Systems 78 (2016) 837–845. 6//.
[16] L.W. de Oliveira, F. d S. Seta, E.J. de Oliveira, Optimal reconfiguration of distribution systems with representation of uncertainties through interval analysis, Int. J. Electr. Power Energy Systems 83 (2016) 382–391. 12//.
[17] H. Fathabadi, Power distribution network reconfiguration for power loss minimization using novel dynamic fuzzy c-means (dFCM) clustering based ANN approach, Int. J. Electr. Power Energy Systems 78 (2016) 96–107. 6//.
[18] M. Esmaeili, M. Sedighizadeh, M. Esmaili, Multi-objective optimal reconfiguration and DG (Distributed Generation) power allocation in distribution networks using Big Bang-Big Crunch algorithm considering load uncertainty, Energy 103 (2016) 86–99. 5/15/.
[19] W. Guan, Y. Tan, H. Zhang, J. Song, Distribution system feeder reconfiguration considering different model of DG sources, Int. J. Electr. Power Energy Systems 68 (2015) 210–221. 6//.
[20] F. Jabari, S. Nojavan, B.M. Ivatloo, M.B. Sharifian, Optimal short-term scheduling of a novel tri-generation system in the presence of demand response programs and battery storage system, Energy Conv. Manage. 122 (2016) 95–108.

[21] F. Jabari, S. Nojavan, B.M. Ivatloo, Designing and optimizing a novel advanced adiabatic compressed air energy storage and air source heat pump based μ-Combined Cooling, heating and power system, Energy 116 (2016) 64–77.
[22] F. Jabari, S. Nojavan, B. Mohammadi-ivatloo, H. Ghaebi, H. Mehrjerdi, Risk-constrained scheduling of solar Stirling engine based industrial continuous heat treatment furnace, Appl. Thermal Eng. 128 (2018) 940–955. 2018/01/05/.
[23] F. Jabari, B. Mohammadi-ivatloo, G. Li, H. Mehrjerdi, Design and performance investigation of a novel absorption ice-making system using waste heat recovery from flue gases of air to air heat pump, Appl. Thermal Eng. 119 (2017) 108–118.
[24] F. Jabari, B. Mohammadi-ivatloo, M.B.B. Sharifian, S. Nojavan, Design and robust optimization of a novel industrial continuous heat treatment furnace, Energy (2017).
[25] F. Jabari, S. Nojavan, B. Mohammadi-ivatloo, Energy-exergy based analysis of a novel μ-cogeneration system in the presence of demand response programs, in: 3rd International Conference of IEA Technology and Energy Management, pp. 28–29 Feb 2017.
[26] A.M.F. Jabari, B. Mohammadi-ivatloo, Long-term solar irradiance forecasting using feed-forward back-propagation neural network, in: 3rd International Conference of IEA, Tehran, Iran, 28th February-1st March 2017.
[27] D.T. Nguyen, L.B. Le, Optimal bidding strategy for microgrids considering renewable energy and building thermal dynamics, IEEE Trans. Smart Grid 5 (2014) 1608–1620.
[28] Z. Boor, S.M. Hosseini, Optimal placement of DG to improve the reliability of distribution systems considering time varying loads using genetic algorithm, Majlesi J. Electr. Eng. 7 (2012).

CHAPTER 9

Reliability-Based Scheduling of Active Distribution System With the Integration of Wind Power Generation

Saeed Abapour and Kazem Zare
University of Tabriz, Tabriz, Iran

9.1 INTRODUCTION

It is important for distribution network operator (DNO) to supply reliable electricity to customers and improve the efficiency of electric system and optimal scheduling of distributed generation (DG) units. Accordingly, design and model distribution systems properly by DNO has been defined as an important issue in such systems. The reliability study can be used to design and assess the performance of the distribution network based on the availability of suitable input component data and the configuration of the system. The reliability assessment can be also used to recommend the numbers of new components that should be incorporated to improve the system's reliability [1].

Distribution networks are deployed as a radially form. Therefore, the integration of renewable energy resources (RERs) into distribution network will change the power flow path in feeders [2]. To achieve an appropriate status, an appropriate management system is required, as well as a suitable scheduling model to integrate renewable energy sources such as wind power generation (WPG). This paper proposes active management (AM) of distribution network. AM is an operative method to reinforce distribution networks for connection and operation of renewable energies [3]. Optimal operation of distribution networks with implementation of AM programs reduces the operation cost of DNO dramatically. Therefore, applying the AM can help to reduce the losses, control the voltage profile, modify the peak load, decrease the clean energy curtailment, and defer the need to reinforce the distribution network [4].

Operation of Distributed Energy Resources in Smart Distribution Networks
DOI: https://doi.org/10.1016/B978-0-12-814891-4.00009-6

A number of studies have addressed various aspects of AM. In Ref. [5], the concept and principle of AM are introduced, and several indices are proposed to assess both economic and technical impacts of AM on the distribution network. In Ref. [6], energy storage system (ESS) is presented as one of the AM options. The networks with ESS can be used to actively control the network in order to increase the generation capacity value. In Ref. [7], the network with AM is studied, which monitors and controls the generation value, and reduces wind curtailment and minimizes system losses by flexible demand scheduling. The presented case study shows that an AM scheme can increase the connected wind capacity to network by flexible demand management. In Ref. [8], the authors explore the options to reduce the curtailment of wind generation employing AM. A curtailment instruction will be sent to different wind generation units when network constraints occur. In Ref. [9], authors have presented the voltage constraint management in distribution system that increases the amount of DG units. In this paper, a centralized optimization approach is proposed that minimizes the curtailment value of renewable energy. In Ref. [10], the authors present an algorithm to manage an active distribution network. The main objective of the proposed method is minimizing the operation cost of system, which is expressed in terms of cost of energy losses, cost of energy curtailment, and cost of load shedding, and cost of reactive power. In Ref. [11], a method is proposed to assess maximum wind energy output in active distribution networks, which is based on a multiperiod optimal power flow analysis. AM schemes such as coordinated voltage and power factor control and the energy curtailment of renewable resources are integrated in the AM method.

In this work, the authors also focus on developing a reliability evaluation framework for a conventional and active distribution network in the presence of WTG and ESS to reduce the power outage. One of the worst disadvantages of radial networks is their low reliability issues. Therefore distribution network will have a high rate of interruption [12]. Regarding the importance of such an issue, several schemes have been proposed to improve the reliability of electric networks. A number of works have addressed the reliability evaluation of power system. In Ref. [13], optimal planning of distribution network is considered as a proper scheme to the network reliability improvement. In Ref. [14], the authors proposed the network expansion planning to improve the distribution network reliability. In the proposed approach, an expansion planning is presented considering heat and electrical energy distribution network. In Ref. [15], the authors have presented a dynamic probabilistic expansion planning of distribution network components and DG unit considering reliability. Several options such as the installation of feeders, transformers, and DG unit are considered. In Ref. [16], a new model for reliability assessment of active distribution networks is presented that considers the impact of islanding dynamics. In order to accomplish the reliability assessment, the effects of the impact of component failures in the islanding process as well as voltage and frequency variations are considered. In Ref. [17], the impacts of demand response on the service reliability in a residential distribution network are measured. In Ref. [18], the authors have proposed an analytical approach to assess the reliability of distribution networks. The studies are implemented within energy hubs with integration of renewable energies. The reliability of costumers is modeled as a function of reliability index of input resources in energy hub. In Ref. [19], the author models the system reliability based on different feeder's failure rate models which are a function of the values of reactive component of currents flowing through branches.

To achieve an accurate solution and realistic reliability evaluation of the distribution systems, a multistate model that describes the stochastic characteristics of different components of the WTG and ESS units in their respective states has been introduced [20,21]. This chapter addresses the reliability improvement of a passive and active distribution system with the integration of WTG and ESS units. The reliability of the WTG and ESS is estimated by applying a Markov model based on the respective states of the major components of the system. The proposed objective function maximizes the benefits of DNO in the short-term scheduling model. In this model, DNO has been considered the battery charging/discharging decisions, on load tap changer (OLTC) of substation transformer and control of reactive power compensators (RPCs) as three options of AM.

The rest of this chapter is organized as follows:

In Section 9.2, different distribution network configuration and concept of active network management are proposed. In Section 9.3, reliability models for the wind system and ESS are defined. In Section 9.4, mathematical formulation is presented. Test system data and problem assumptions are in Section 9.5. In Section 9.6, the proposed model is applied on the 9-bus radial distribution system and the simulation results are given and discussed. Finally, Section 9.7 summarizes the findings of this work.

9.2 DISTRIBUTION NETWORK CONFIGURATION

The reliability of the distribution network depends on the energy resource and configuration of the system. In distribution network, the reliability of components must be frequently assessed to reflect the impact of the frequency and duration of power interruption on the reliability of the system [22].

9.2.1 Active Distribution Network

Nowadays, distribution networks are affected by high penetration of the renewable DG units, and it will change planning and operation of the network. This problem will create a mismatch between the required network load value and the installed capacity of WPGs [23,24]. Therefore reinforcement of network is needed. To avoid additional costs, DNO should use a novel method for optimal scheduling of the network. In active distribution network, AM will create conditions for the high penetration of WPG resources in network without any need of its reinforcement [25].

Fig. 9.1 shows a schematic of the control and measure of the distribution network parameters based on AM. It is assumed that a control center is located in the primary substation. Local and remote measurement devices are used to record the system data. After the data analysis at the control center, the control commands are sent to generation units, transformers, storage systems, and the circuit breakers.

In this chapter, three AM strategies have been employed to keep the voltage within the permissible limits and increase the system reliability [26]:

1. Adjust of OLTC at secondary side of substation:

   In this scheme, AM focuses on keeping the voltage profiles in the allowable range. Because the network load is changing and that may cause voltage drop in the network, OLTCs are used to adjust voltage within allowable limits [23,26].

2. Control of RPCs:

   Application of RPC in distribution system can reduce power losses and by accurate management, the voltage profile can be corrected when voltage is not in allowable limits. The best control strategy is based on application of synchronized and coordinated OLTC and RPC control.

3. Use of battery storage systems (BSSs):

   BSSs are used in active distribution network to help solve the problem of intermittency of renewable energy sources, enhance system

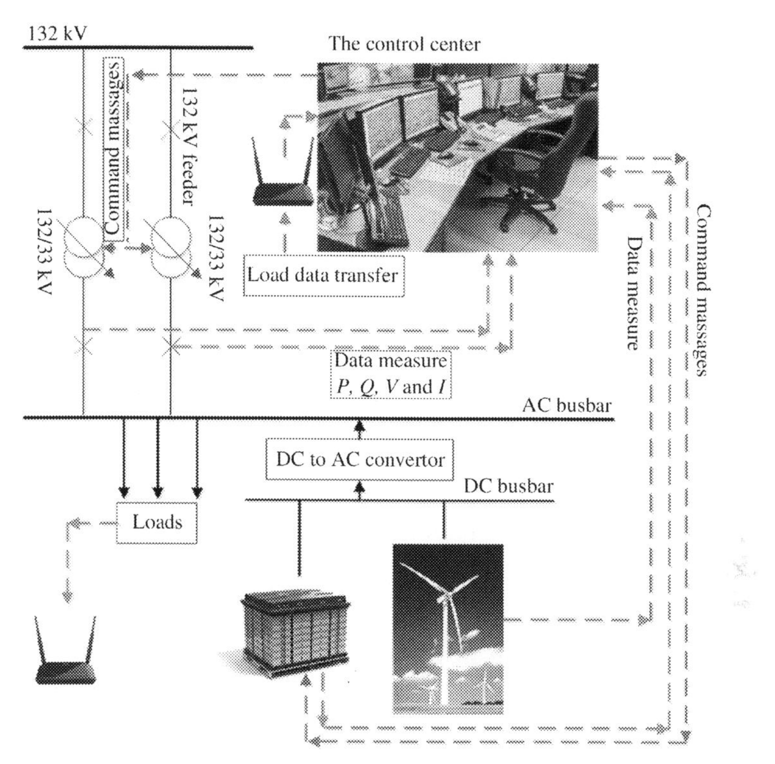

Figure 9.1 Schematic of AM in distribution network.

reliability, and improve flexibility of power system. This chapter presents management of BSSs as one of AM strategies. The objectives of DNOare focused on improving the energy efficiency and system reliability while keeping the charge and discharge constraints of BSSs.

This chapter introduces two operation cases of distribution network considered and compared with each other as follow:

- Case 1. Passive distribution networks:

  In these networks, both the renewable DG units and BSSs have been installed in the network to improve the technical features. When the penetration level of generation units is high, the unidirectional power flows may be changed to bidirectional. In this mode, the outputs of energy resources are not controlled by DNOs.

- Case 2. Active distribution networks:

  In active networks, measurement devices such as automatic meter reading (AMR) are located at load buses and important points. Moreover, OLTC in MV substation is controlled automatically with the aid of measuring primary system parameters. AM programs will be executed for optimal scheduling of distribution network.

9.2.2 The Hybrid System

A hybrid system which consists of WPG and BSS, as presented in Fig. 9.2, is proposed in this work as a measure to calculate the system reliability. In active distribution network, the communication links are also added as a parallel component to WPG, BSS, and load points. The integration of a number of WPG units into different sections of the power system can increase the reliability of their networks. In fact, the WTGs power output change with time as a result of the intermittent nature of the RERs. Therefore the BSS is used for more flexibility of the WPGs. BSS will also reduce the power outage time and increase the efficiency of the power system [27,28].

In the proposed reliability model for the distribution system, the WPG unit includes wind turbine (WT), AC/DC rectifier, and DC/AC inverter. Also a BSS includes battery bank, battery controller/ charger, and DC/AC inverter. These components will be integrated into a typical distribution network to improve the system reliability and the frequency and duration of sustained outages. In active distribution network, the communication links will be also considered as a parallel component to WPG, BSS, and load points (Fig. 9.2B). In the following, the reliability modeling of the WPG and ESS units is explained briefly.

9.3 RELIABILITY MODELS FOR THE WIND SYSTEM AND ESS

9.3.1 Reliability Model for the WT, AC/DC Rectifier, and DC/AC Converter System

The system reliability can be represented in three generic states such as: an up state that represents the full operation of each unit with several components, a derated state that represents the partial operation of the system, and a down state that represents down or nonoperational state of the system [29]. The reliability model for WPG included WT, AC/DC rectifier, and DC/AC converter system is illustrated in Fig. 9.3.

The major components of WPG system will be assumed as series configuration. Therefore the system will have two states: first state is a full operating state in a situation in which the wind resource is within the operating limits set and also the major components (WT, AC/DC rectifier, and DC/AC converter) are in the healthy states. Second state is

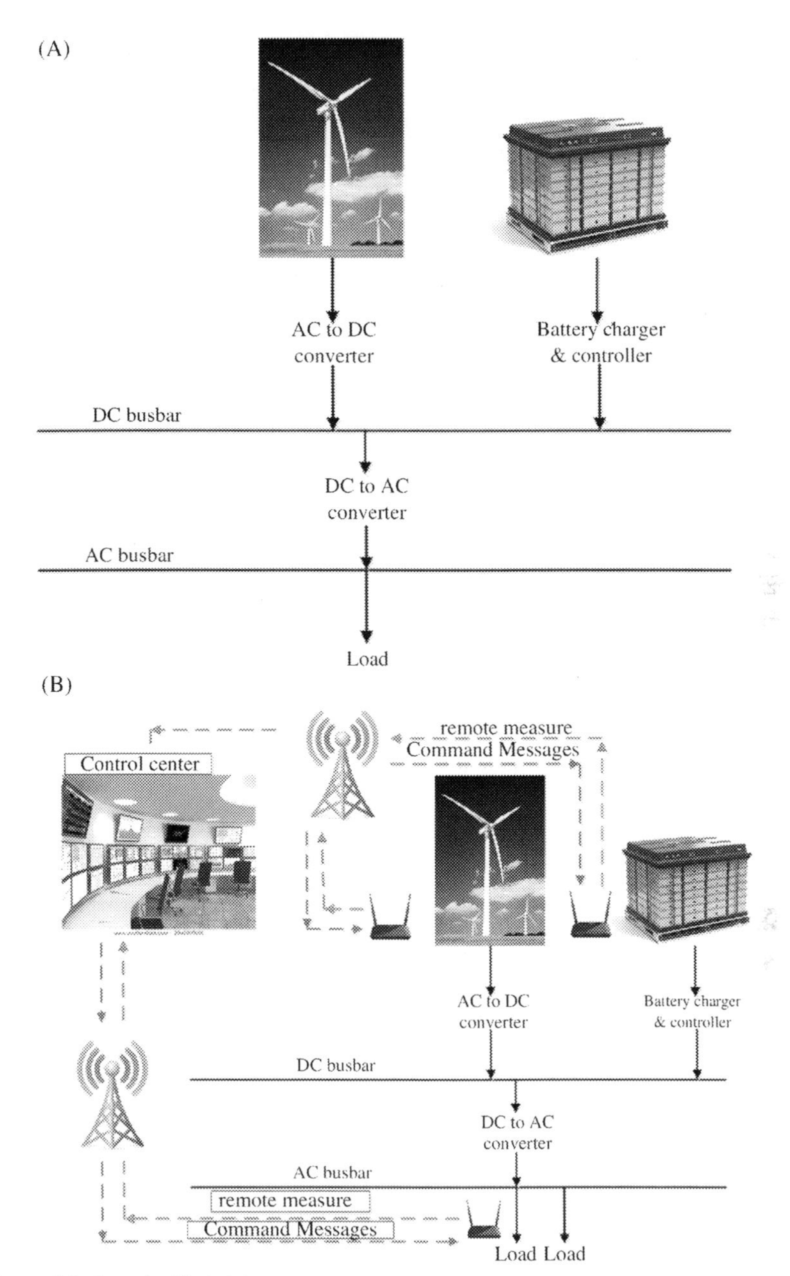

Figure 9.2 A typical hybrid system.

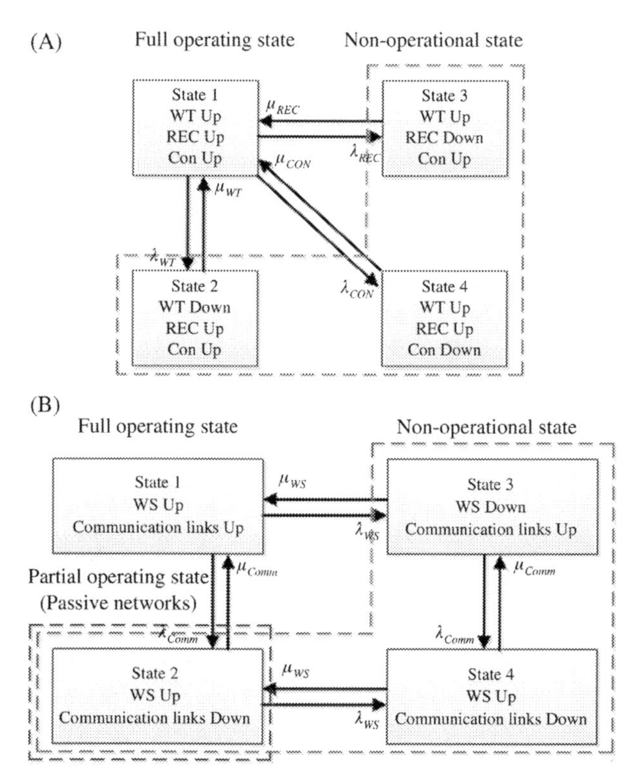

Figure 9.3 Markov model for wind system.

nonoperational state in which any components among the three major components of the system have failed. The stochastic transitional probability matrix of the system establishes this fact that the system will operate only if it is in state 1; otherwise the system will be in the down state. The states 2–4 have been observed to be in the failure states due to the series configuration of the system. In this study, the failure rate and repair rate for the WT, AC/DC rectifier, and DC/AC converter of the wind system are defined byλ_{WT}, $\mu_{WT,}\lambda_{REC}$, μ_{REC}, λ_{CON}, μ_{CON}, respectively. The transition matrix of the WPG system for Fig. 9.3A can be represented by Eq. (9.1).

$$P=\begin{bmatrix} 1-\lambda_{WT}-\lambda_{REC}-\lambda_{CON} & \lambda_{WT} & \lambda_{REC} & \lambda_{CON} \\ \mu_{WT} & 1-\mu_{WT} & 0 & 0 \\ \mu_{REC} & 0 & 1-\mu_{REC} & 0 \\ \mu_{CON} & 0 & 0 & 1-\mu_{CON} \end{bmatrix} \tag{9.1}$$

For this system a truncated matrix Q will be as follows [29]:

$$Q=[1-\lambda_{WT}-\lambda_{REC}-\lambda_{CON}] \tag{9.2}$$

The mean time to failure (MTTF) of the WPG system can be calculated by using the following equations [29,30]:

$$\begin{aligned} \text{MTTF} &= [[I]-[Q]]^{-1} = [\lambda_{WT}+\lambda_{REC}+\lambda_{CON}]^{-1} \\ &= \frac{1}{\lambda_{WT}+\lambda_{REC}+\lambda_{CON}} \end{aligned} \tag{9.3}$$

The total failure of the WPG system can be nearly obtained as follows:

$$\lambda_{WS}=\lambda_{WT}+\lambda_{REC}+\lambda_{CON} \tag{9.4}$$

The probability of the up and down states of each major component for the WPG system can be expressed as follows:

$$P_{Up}=\frac{\mu}{\lambda+\mu} \tag{9.5}$$

$$P_{Down}=\frac{\lambda}{\lambda+\mu} \tag{9.6}$$

The probability of the series components in the up state of the WPG system is denoted as follow:

$$P_{Up-WS}=\frac{\mu_{WT}}{\lambda_{WT}+\mu_{WT}}\times\frac{\mu_{REC}}{\lambda_{REC}+\mu_{REC}}\times\frac{\mu_{CON}}{\lambda_{CON}+\mu_{CON}} \tag{9.7}$$

which is the product of the probability of the major component for the WPG system.

We consider λ_{WS} and μ_{WS}as the WPG system failure rate and repair rate, respectively. Thenμ_{WS}obtains as follows:

$$P_{Up-WS}=\frac{\mu_{WS}}{\lambda_{WS}+\mu_{WS}} \tag{9.8}$$

$$\mu_{WS}=\frac{P_{Up-WS}\times\lambda_{WS}}{1-P_{Up-WS}}=\frac{\lambda_{WS}}{\left[(1+\frac{\lambda_{WT}}{\mu_{WT}})(1+\frac{\lambda_{REC}}{\mu_{REC}})(1+\frac{\lambda_{CON}}{\mu_{CON}})-1\right]} \tag{9.9}$$

$$\mu_{WS}=\frac{\lambda_{WT}+\lambda_{REC}+\lambda_{CON}}{\left[(1+\frac{\lambda_{WT}}{\mu_{WT}})(1+\frac{\lambda_{REC}}{\mu_{REC}})(1+\frac{\lambda_{CON}}{\mu_{CON}})-1\right]} \tag{9.10}$$

In active distribution network, the communication links will be also considered as a parallel component with WPG, BSS, and load points. It is assumed that a control center is located in the primary substation. After it is received and a the network data is analyzed, the control commands are sent to transformers, storage systems, and RPCs. In this case, the WPG system is assumed to be parallel to a communication link (Fig. 9.3B). The failure rate and repair rate for the WPG system and the communication link are defined byλ_{WS}, $\mu_{WS}\lambda_{Comm}$, μ_{Comm}, respectively. The transition matrix of the WPG system with considering AM can be represented by Eq. (9.11).

$$P=\begin{bmatrix} 1-\lambda_{WS}-\lambda_{Comm} & \lambda_{Comm} & \lambda_{WS} & 0 \\ \mu_{Comm} & 1-\mu_{Comm}-\lambda_{WS} & 0 & \lambda_{Comm} \\ \mu_{WS} & 0 & 1-\mu_{WS}-\lambda_{Comm} & \lambda_{WS} \\ 0 & \mu_{WS} & \mu_{Comm} & 1-\mu_{WS}-\mu_{Comm} \end{bmatrix} \tag{9.11}$$

In this case, the system will operate only if it is in state 1(or as AM mode); otherwise the system will be in the down state. Therefore, a truncated matrix Q will be as follows:

$$Q=[1-\lambda_{WS}-\lambda_{Comm}] \tag{9.12}$$

The MTTF of the WPG system can be calculated by using the following equations:

$$\begin{aligned} \text{MTTF} &= [[I]-[Q]]^{-1} = [\lambda_{WS}+\lambda_{Comm}]^{-1} \\ &= \frac{1}{\lambda_{WS}+\lambda_{Comm}} \end{aligned} \tag{9.13}$$

The total failure of the WPG system with AM can be nearly obtained as follows:

$$\lambda_{WS-AM}=\lambda_{WS}+\lambda_{Comm} \tag{9.14}$$

In this case, we have a 2-component parallel system which states 2, 3, and 4 are combined to form a cumulative down state. Therefore, the probability of the up state of the WPG system with AM is expressed as follow:

$$P_{Up-WS-AM}=\frac{\mu_{WS}}{\lambda_{WS}+\mu_{WS}}\times\frac{\mu_{Comm}}{\lambda_{Comm}+\mu_{Comm}} \tag{9.15}$$

The probability of cumulative down state will be summation of individual probabilities (mutually exclusive events).

$$P_{Down-WS-AM} = P_2 + P_3 + P_4 = \underbrace{\frac{\lambda_{Comm}}{\lambda_{Comm} + \mu_{Comm}} \times \frac{\mu_{WS}}{\lambda_{WS} + \mu_{WS}}}_{P_2} + \underbrace{\frac{\mu_{Comm}}{\lambda_{Comm} + \mu_{Comm}} \times \frac{\lambda_{WS}}{\lambda_{WS} + \mu_{WS}}}_{P_3} + \underbrace{\frac{\lambda_{Comm}}{\lambda_{Comm} + \mu_{Comm}} \times \frac{\lambda_{WS}}{\lambda_{WS} + \mu_{WS}}}_{P_4} \quad (9.16)$$

If we consider λ_{WS-AM} and μ_{WS-AM}as the WPG system failure rate and repair rate with considering AM, respectively, μ_{WS-AM}obtains as follows:

$$P_{Up-WS-AM} = \frac{\mu_{WS-AM}}{\lambda_{WS-AM} + \mu_{WS-AM}} \quad (9.17)$$

$$\mu_{WS-AM} = \frac{P_{Up-WS-AM} \times \lambda_{WS-AM}}{1 - P_{Up-WS-AM}} = \frac{\lambda_{WS} + \lambda_{Comm}}{\left[(1 + \frac{\lambda_{WS}}{\mu_{WS}})(1 + \frac{\lambda_{Comm}}{\mu_{Comm}}) - 1\right]} \quad (9.18)$$

9.3.2 Reliability Model for Battery, Battery Controller/Charger and Inverter System

The battery system has three major components that is battery, battery controller/charger, and inverter system. The reliability model for BSS included battery, battery controller/charger, and inverter system is illustrated in Fig. 9.4. The major components of battery system will be assumed as series configuration. Therefore system will have two states: first state is a full operating state in a situation in which the major components (battery, battery controller/charger and inverter) are in the healthy states. Second state is nonoperational state in which any components among the three major components of the system have failed. The system will operate only if it is in state 1; otherwise the system will be in the down state. The states 2–4 have been observed to be in the failure states due to the series configuration of the system. According to Fig. 9.4, the failure rate and repair rate for the battery, battery controller/charger, and inverter of the BSS system are defined by $\lambda_{Bat}, \mu_{BAT}, \lambda_{BC}, \mu_{BC}, \lambda_{CON}, \mu_{CON}$, respectively. The BSS has a similar

(A)

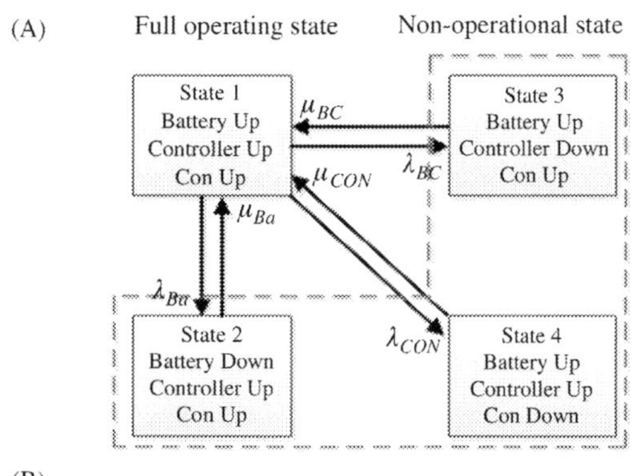

(B)

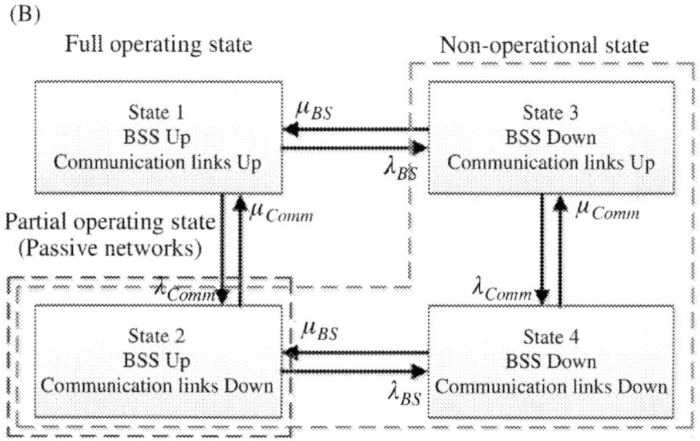

Figure 9.4 Markov model for electric storage system.

operation with the WPG system as discussed in Section 9.3.1. Therefore, the total failure rate and repair rate of the battery, battery controller/charger, and inverter system can be estimated by using the following equations:

$$\lambda_{BSS} = \lambda_{Bat} + \lambda_{BC} + \lambda_{CON} \tag{9.19}$$

$$\mu_{BSS} = \frac{\lambda_{Bat} + \lambda_{BC} + \lambda_{CON}}{\left[(1 + \frac{\lambda_{Bat}}{\mu_{Bat}})(1 + \frac{\lambda_{BC}}{\mu_{BC}})(1 + \frac{\lambda_{CON}}{\mu_{CON}}) - 1\right]} \tag{9.20}$$

The total failure rate and repair rate of the BSS with considering AM will be as follows:

$$\lambda_{BSS-AM} = \lambda_{BSS} + \lambda_{Comm} \tag{9.21}$$

$$\mu_{BSS-AM} = \frac{\lambda_{BSS} + \lambda_{Comm}}{\left[(1 + \frac{\lambda_{BSS}}{\mu_{BSS}})(1 + \frac{\lambda_{Comm}}{\mu_{Comm}}) - 1\right]} \tag{9.22}$$

$$P_{Up-BSS-AM} = \frac{\mu_{BSS-AM}}{\lambda_{BSS-AM} + \mu_{BSS-AM}} \quad (9.23)$$

9.4 MATHEMATICAL FORMULATION

The model is based on the optimal power flow and maximizes benefit of distribution company (DisCo) at daily short-term period. Meanwhile, the model is simulated based on the daily load curve. The problem formulation is provided in five subsections as follows.

9.4.1 Uncertainty Parameters Modeling

In this chapter, uncertainties of the load demand and the WPG output power are modeled by the scenario-based approach. Load and WPG power estimation errors are assumed as random variables with definite probability distribution functions (PDF) [31]. Fig. 9.5 shows a PDF for the load and WPG power forecast error in h^{th} demand level [32].

Each scenario corresponds with a single realization of the random variables in each time period. Also, each scenario has an associated probability of occurrence. for each scenario load demand and WPG output power forecast error can be expressed as follows:

$$P_{h,s}^{D} = P_{h,s}^{D_{forecasted}} + \Delta P_{h,s}^{D} \quad (9.24)$$

$$P_{h,s}^{WPG} = P_{h,s}^{WPG_{forecasted}} + \Delta P_{h,s}^{WPG} \quad (9.25)$$

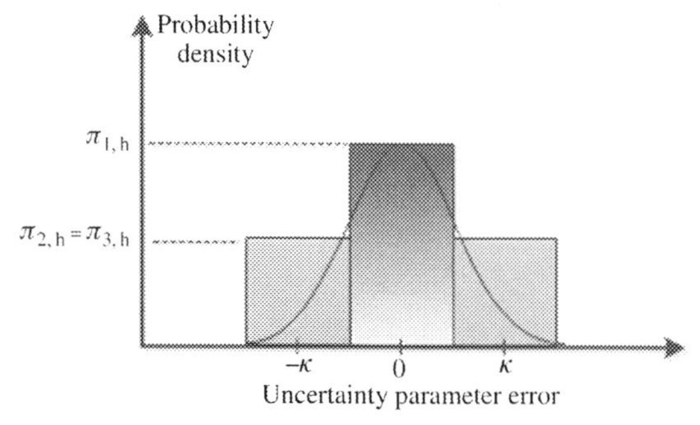

Figure 9.5 Typical discretization of PDF of uncertainty parameter error.

where $\Delta P_{h,s}^{D}$ and$\Delta P_{h,s}^{WPG}$ are load demand forecast error at time h in scenario s (MW).

In this chapter, it is assumed that the PDF of each random variable is discretized into three intervals centered on the zero mean. The width of each interval (κ) is the standard deviation of the forecast error for each parameter. The probability associated with each interval for the load demand and WPG output power are indicated by $\pi_l$andπ_{WS}, respectively. Uncertainty scenarios of these parameters are autonomous so that the total number of scenarios can be calculated by multiplying scenarios of each element as follows [33]:

$$\pi_{s,h} = \pi_{l,h} \times \pi_{WS,h} \tag{9.26}$$

The total number of scenarios, N_S, will be $l \times WS$ states.

9.4.2 Load and Electricity Price Modeling

The daily load variation is modeled by multiplication of two parameters. The first one is the base load ($P_{i,base}^{D}$, $Q_{i,base}^{D}$). Each hour of a day is defined as a demand level. Therefore, there will be 24 hours or demand levels which will be shown byN_h.

The second parameter is the demand level factor ($DLF_{h,s}$). This parameter defines the forecasted value of the "load to peak load ratio" in each demand level and varies between 0 and 1. A typical $DLF_{h,s}$ is shown in Fig. 9.6. Therefore, the demand of ith bus at h^{th} demand level and in s^{th} scenario is calculated as [34]:

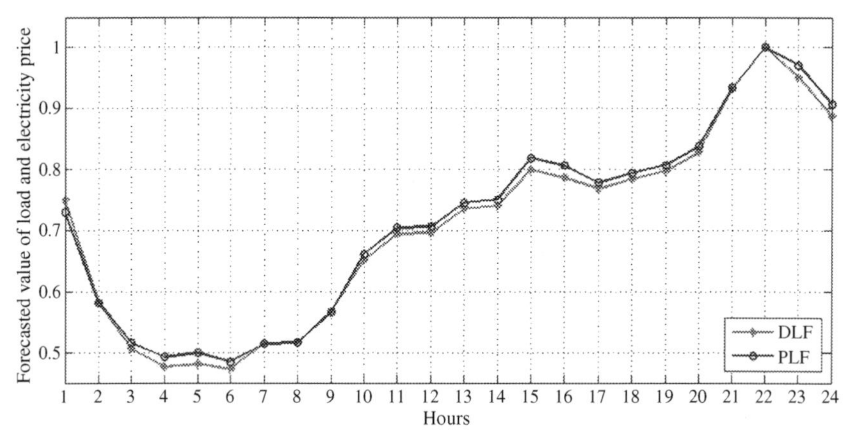

Figure 9.6 Typical demand and price level factors.

$$P_{i,h,s}^{D} = P_{i,base}^{D} \times DLF_{h,s} \tag{9.27}$$

$$Q_{i,h,s}^{D} = Q_{i,base}^{D} \times DLF_{h,s} \tag{9.28}$$

$$S_{i,h,s}^{D} = P_{i,h,s}^{D} + jQ_{i,h,s}^{D} \tag{9.29}$$

where $S_{i,h,s}^{D}$, $P_{i,h,s}^{D}$ and $Q_{i,h,s}^{D}$ are the apparent, active, and reactive powers of ith bus at h^{th} demand level and in s^{th} scenario, respectively.

The price of purchased energy from upstream grid is determined by the market operation. This value changes during each demand level. In this chapter, it is assumed that the price of electricity at h^{th} demand level can be determined as [34]:

$$\sigma_h = \rho_{base} \times PLF_h \tag{9.30}$$

where ρ_{base} is the base price and PLF_h is the hth price level factor which is assumed to be known. A typical PLF_h curve is illustrated in Fig. 9.6.

9.4.3 Wind System

Wind energy is undoubtedly one of the cleanest forms of producing power from a renewable source. Renewable energy facilities generally require less maintenance than dispatchable generators. It does not need the fuel and its availability reduces its operation costs. Even more importantly, renewable energy produces little or no waste generates such as carbon dioxide or other pollutants, and therefore has minimal impact on the environment. The output power of WTG depends on the characteristics of the wind regime, wind speed, air density, swept area of rotor, tower height, and aero-turbine performance, etc. The relationship between the power output of the WT and wind speed can be expressed as [30,35]:

$$P_{h,s}^{WPG}(\mathrm{v}(\mathrm{h,s})) = \begin{cases} 0 & \mathrm{v}(\mathrm{h,s}) \prec \mathrm{v}_{ci} \\ \dfrac{v(\mathrm{h,s}) - \mathrm{v}_{ci}}{v_r - \mathrm{v}_{ci}} P_r & \mathrm{v}_{ci} \leq \mathrm{v}(\mathrm{h,s}) \leq \mathrm{v}_r \\ P_r & \mathrm{v}_r \leq \mathrm{v}(\mathrm{h,s}) \leq \mathrm{v}_{co} \\ 0 & \mathrm{v}(\mathrm{h,s}) \succ \mathrm{v}_{co} \end{cases} \tag{9.31}$$

9.4.4 Objective Function

Operation model is a combination of the WPG units and other energy supply resources. DisCo has the ability to swap energy with the upstream network. The DisCo's objective function in short-term operations framework is maximizing its energy benefits for 1 day. Eq. (9.6) provides the mathematical formulation of a DisCo:

$$OF = \sum_{h=1}^{N_h} \sum_{s=1}^{N_s} \pi_{h,s} \left\{ \begin{array}{l} \sum_{i=1}^{N_{load}} \rho_{sell}^{P} \times P_{i,h,s}^{D} + \sum_{i=1}^{N_{load}} \rho_{sell}^{Q} \times Q_{i,h,s}^{D} - \sum_{g=1}^{N_g} \sigma_h \times P_{h,s}^{ss} - \sum_{g=1}^{N_g} \sigma_{Qfix} \times Q_{h,s}^{ss} \\ - \sigma_h \times P_{loss}^{tot} - \sum_{w=1}^{N_{wind}} \rho^{WPGo} \times P_{h,s}^{WPG} \times P_{Up-WS} \\ - C_k^{\deg} (\sum_{k=1}^{N_k} \frac{P_{k,h,s}^{disc}}{\eta_k^{disc}} + \eta_k^{C} \times P_{k,h,s}^{c}) \times P_{Up-BSS} \end{array} \right\} \tag{9.32}$$

The first and second term of (9.32) are revenues from the energy sold to customers by DNO. The third term of (9.32) is the cost of power purchased from the external network or energy market by DNO, and its value depends on the electricity market price (σ_h). The fourth term of (9.32) is the amount of payment for the reactive power purchased from the external network, and is described as a predetermined price and fix (λ_{Qfix}). The fifth term describes the benefits or costs due to the variation of network power losses, which are affected by the power generation by the wind and storage units. Sixth term shows the operation cost of the WPG system considered a few amounts or zero. The probability of full operation is multiplied in this term. The last term captures the degradation cost of batteries at the hth hours for the kth battery due to the charging/discharging activities. The probability of full operation is also multiplied in this term. DNO also has the ability to sell energy in the market. This condition occurs when $P_h^{ss}$is negative.

9.4.5 Constraints and Optimal Power Flow Equations

The power flow equations should be satisfied in i^{th} bus, h^{th} hours, and s^{th} scenario as follows:

$$P_{h,s}^{ss} + P_{i,h,s}^{WPG} - P_{i,h,s}^{D} + \sum_{k=1}^{N_k} (P_{k,h,s}^{disc} - P_{k,h,s}^{c}) = V_{i,h,s} \sum_{j} V_{j,h,s} (G_{ij} \cos\delta_{i,h,s} + B_{ij} \sin\delta_{j,h,s}) \tag{9.33}$$

$$Q_{h,s}^{ss} - Q_{i,h,s}^{D} = V_{i,h,s} \sum_{j} V_{j,h,s}(G_{ij}\cos\delta_{i,h,s} - B_{ij}\sin\delta_{j,h,s}) \tag{9.34}$$

where $P_{h,s}^{ss}$ and $Q_{h,s}^{ss}$ are active and reactive power generated (or absorbed) by the MV substation at h^{th} hour, and $V_{i,h,s}$ and $\delta_{i,h,s}$ show the magnitude and angle of voltage in i^{th} bus, h^{th} hour, and s^{th} scenario, respectively.

The voltage of each bus at h^{th} hour and s^{th} scenario should be kept within the safe operating limits:

$$V_i^{\min} \leq V_{i,h,s} \leq V_i^{\max} \tag{9.35}$$

The active and reactive powers of the substation are limited as follows:

$$P_{ss}^{\min} \leq P_{h,s}^{ss} \leq P_{ss}^{\max} \tag{9.36}$$

$$Q_{ss}^{\min} \leq Q_{h,s}^{ss} \leq Q_{ss}^{\max} \tag{9.37}$$

The power flow through the line connected to nodes i, j is limited by its thermal limit for all demand levels:

$$S_{ij,h,s} \leq S_{ij}^{\max} \tag{9.38}$$

The tap setting of tap-changer l at h^{th} demand level and s^{th} scenario is assumed as follows:

$$T_l^{\min} \leq T_{l,h,s} \leq T_l^{\max} \tag{9.39}$$

Due to the abundant nature, the wind power resources are mostly used along with energy storage. Eqs. (9.40–9.43) express the ESS constraints [36]. Eq. (9.40) shows the limits of charge and discharge power of battery k in demand level h so that if binary variable $b_{k,h}^{c}$ is 1, it is in charging mode. Also if binary variable $b_{k,h}^{disc}$ is 1, it is in discharging mode. Otherwise it is neither charging nor discharging mode. Eq. (9.41) shows the capacity of battery k in each demand level h is limited to a minimum and maximum value. Eq. (9.42) indicates at the same time the battery cannot be in both charge and discharge mode. Eq. (9.43) indicates changes of the battery capacity at any demand level (hour) $h+1$ compared to demand level (hour) h.

$$0 \leq P_{k,h,s}^{c} \leq b_{k,h}^{c} P_{k,h}^{c,\max},\ 0 \leq P_{k,h,s}^{disc} \leq b_{k,h}^{disc} P_{k,h}^{disc,\max} \tag{9.40}$$

$$E_k^{\min} \le E_{k,h,s} \le E_k^{\max} \tag{9.41}$$

$$b_{k,h,s}^{c} + b_{k,h,s}^{disc} \le 1;\ b_{k,h,s}^{c},\ b_{k,h,s}^{disc} \in \{1, 0\} \tag{9.42}$$

$$E_{k,h+1,s} = E_{k,h,s} + (\eta_k^C \times P_{k,h,s}^{c} - \frac{P_{k,h,s}^{disc}}{\eta_k^{disc}}) \tag{9.43}$$

Objective is maximization OF with considering constraints (9.9–9.27) and for both active and passive management of network.

$$\begin{array}{l} Max\ \text{OF} \\ \text{subject to} \\ \text{Eqs. (33) to (43)} \end{array} \tag{9.44}$$

9.5 TEST SYSTEM DATA AND ASSUMPTIONS

In this study, simulations have been carried out on the nine-bus test system. The single-line diagram of this test system is represented in Fig. 9.7. The data of this test system and loads are given in Table 9.1 [37]. Distribution test system includes high voltage distribution substation 132–33 kV which feeds eight load points and each branch has been separated from network by an isolator switch. Maximum capacity of each branch of the network is 15 MVA. Power factor of all points is assumed to be 0.95 lag. WPG_1 to WPG_4 are installed at the nodes 6–9,

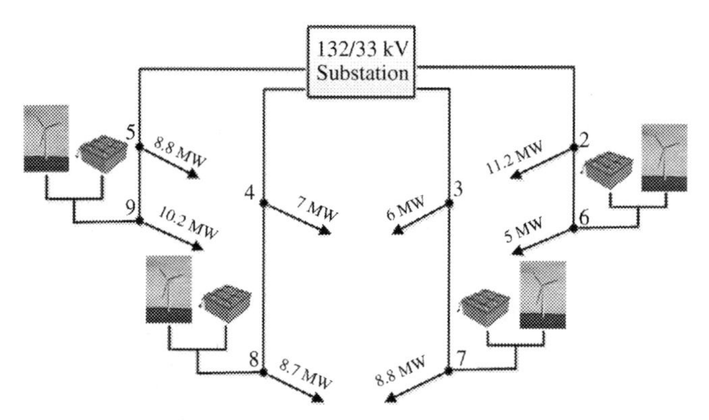

Figure 9.7 Test system schematic.

Table 9.1 Network data of 9-bus test system

From	To	R (Ω)	X (Ω)	Length of branch (km)	Load value (MW)
1	2	2	4	4	11.2
2	6	2.8	5.5	5.5	5
1	3	1.4	1.5	1.5	6
3	7	2.78	5.5	5.5	8.8
1	4	2.26	4.5	4.5	7
4	8	2.4	5	5	8.7
1	5	1.7	1.7	1.7	8.8
5	9	2.1	4	4	10.2

Table 9.2 Economical and technical data

Parameters	Values	Parameters	Values
$V_i^{\max}$(p.u.)	1.05	ρ_{sell}^{P}($/MWh)	80
$V_i^{\min}$(p.u.)	0.9	ρ_{sell}^{Q}($/Mvarh)	40
$\mathrm{PF}_{\mathrm{DG}}$	0.95	ρ^{WPGo}($/MWh)	67
ρ_{base}	72	σ_{Qfix}($/MWh)	36

Table 9.3 The BSS and WPG Units data

The BSS Unit's data				The WPG Unit's data	
E_k^{cap}(MWh)	3	$P_{k,h}^{c,\max}$(MW)	0.7	v_{ci}(m/s)	2
$E_k^{\min}$(MWh)	0.2	$P_{k,h}^{disc,\max}$(MW)	0.7	v_r(m/s)	14
$E_k^{\max}$(MWh)	3	η_k^C	0.95	v_{co}(m/s)	22
$C_k^{\deg}$($/MWh)	2.7	η_k^{disc}	0.95	P_r(MW)	4

respectively. The WPG units have a capacity of 4 MW. Four numbers of ESS are installed as parallel to the WPG units at the nodes 6–9. For this purpose, battery units with the capacity of 2 MWh are considered with charging/discharging power ratings of 500 kW. The minimum and maximum energy stored in the batteries are 200 kWh and 3000 kWh, respectively. RPC1 and RPC2 are installed in bus 8 and 9, respectively. Their capacities are 4 MVar and to reactive power compensation. Duration of each load level is one hour. Total operating period is equal to 24 hours. The base price of energy purchased from the grid is assumed to be 72 $/MWh. The remainder of parameters is listed in Tables 9.2–9.4. The detailed information about the system reliability indices and the characteristics of the components that constitute the WPG and BSS system are presented in Table 9.5. Active distribution networks are smarter compared

Table 9.4 Load demand and WPG output power scenarios and their probabilities

	Load demand (%)	$\pi_{l,h}$	WPG output power (%)	$\pi_{WS,h}$	$\pi_{s,h}$
s_1	90	0.2	20	0.1	0.02
s_2	100	0.6	20	0.1	0.06
s_3	110	0.2	20	0.1	0.02
s_4	90	0.2	60	0.8	0.16
s_5	100	0.6	60	0.8	0.48
s_6	110	0.2	60	0.8	0.16
s_7	90	0.2	100	0.1	0.02
s_8	100	0.6	100	0.1	0.06
s_9	110	0.2	100	0.1	0.02

Table 9.5 Reliability indices of different components of the WTG, ESS, and PV units and their combinations based on the Markov model [30,38]

Parameters	Failure rate (f/year)	Repair rate (repair/year) in PM mode	Repair rate (repair/year) in AM mode
WT	0.05	15	20
AC/DC converter	0.152	41.25	55.23
DC/AC converter	0.143	39.10	52.14
Battery	0.0312	39	51.95
Controller/charger	0.125	33.9	45.21
Communication link	0.1	-	35
WPG system (WT + AC/DC converter + DC/AC converter)	0.345	32.55	43.4
BSS (battery + controller + DC/AC converter)	0.2992	36.6	48.9
WPG system + Communication link	0.445	-	41.32
BSS + Communication link	0.3992	-	44.55

to passive network. Hence, in this chapter have assumed repair rate in AM mode is more than PM mode listed in Table 9.5. The reliability evaluation is performed on the distribution system based on the following two cases: passive distribution network, active distribution network.

9.6 SIMULATION RESULTS

In order to illustrate the AM effect on the proposed model, two cases (AM and PM) have been evaluated. The result has been obtained with

Table 9.6 The results of network operation for nine bus system

	In passive management	In active management
Without Markov model		
Benefit of DNO ($/day)	79979	82279
P_{loss} (MWh)	31.32	25.12
Q_{loss} (MVarh)	48.86	38.60
With Markov model		
Benefit of DNO ($/day)	79889	82161
P_{loss} (MWh)	31.40	25.21
Q_{loss} (MVarh)	49.00	38.78

considering Markov model for WPG system and BSS compared to without reliability index. The short-term scheduling model is studied with considering nine scenarios to model the uncertainty of wind speed and load forecasting error. Results are obtained for 9-bus test system. Table 9.6 summarizes the obtained results of benefit function. In order to compare and demonstrate the effectiveness of AM method in the short-term scheduling of distribution system, these results for AM method are also calculated and listed in Table 9.6. It is assumed that the location of the installed WPGs and ESSs in both PM and AM methods are identical. The expected benefit of DNO without Markov model is equal to 79,979 and 82,279 dollars per day in PM and AM, respectively. Also the expected benefit of DNO with considering Markov model will be equal to 79,889 and 82,161 dollars per day in PM and AM, respectively. In AM, the expected value of power loss reduced. The expected active and reactive power loss without Markov model for PM are 31.32 and 48.86 while their values in AM are 25.12 and 38.60, respectively.

Fig. 9.8 indicates the received active power values of upstream grid. In the first scenario the load value and WPG are less than the forecasted values. The third scenario is the worst condition so that WPG is less than the forecasted value while the load value is more than the forecasted value. Since the peak load occurs at 22 hours , the received active power values of grid in PM and AM state are 71.63 and 70.88 MW in the third scenario, respectively. In all scenarios the received active power values of grid in AM state is a bit less than PM. Fig. 9.8B indicates the received active power values of grid for sixth and ninth scenarios.

Fig. 9.9 indicates the received reactive power values of upstream grid. With installing the RPCs at bus 8 and 9, in all scenarios the received

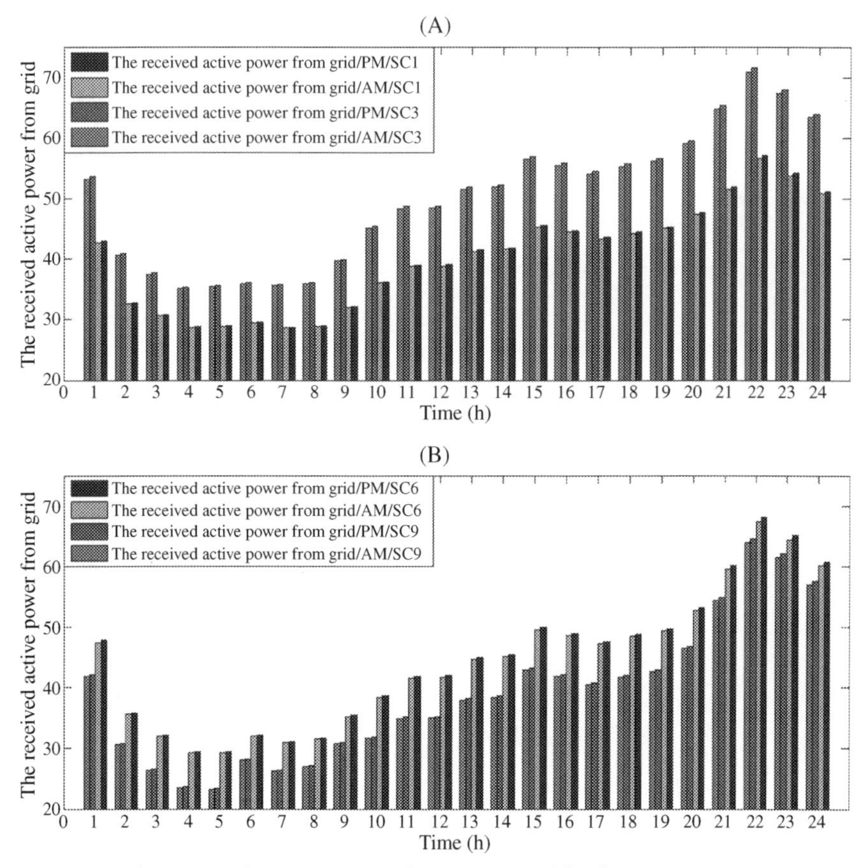

Figure 9.8 The received active power of upstream grid for four scenarios.

reactive power values of grid in AM state will be less than PM. The received reactive power values are shown for four scenarios as samples. Since the peak load occurs at 22 hours, the received reactive power values of grid in PM and AM state are 37.73 and 28.46 Mvar in the third scenario, respectively.

Fig. 9.10 indicates the output power of WTs for daily scheduling period. These powers will be affected by wind speed. These values will change different scenarios. This figure shows the output power of WTs for 20%, 60%, and100% capacity. The wind power spillway is zero because its operation cost is very less than electricity market price.

Fig. 9.11 shows the voltage profiles for passive and active networks. The voltage curve is shown for the peak load which occurs at the 22nd

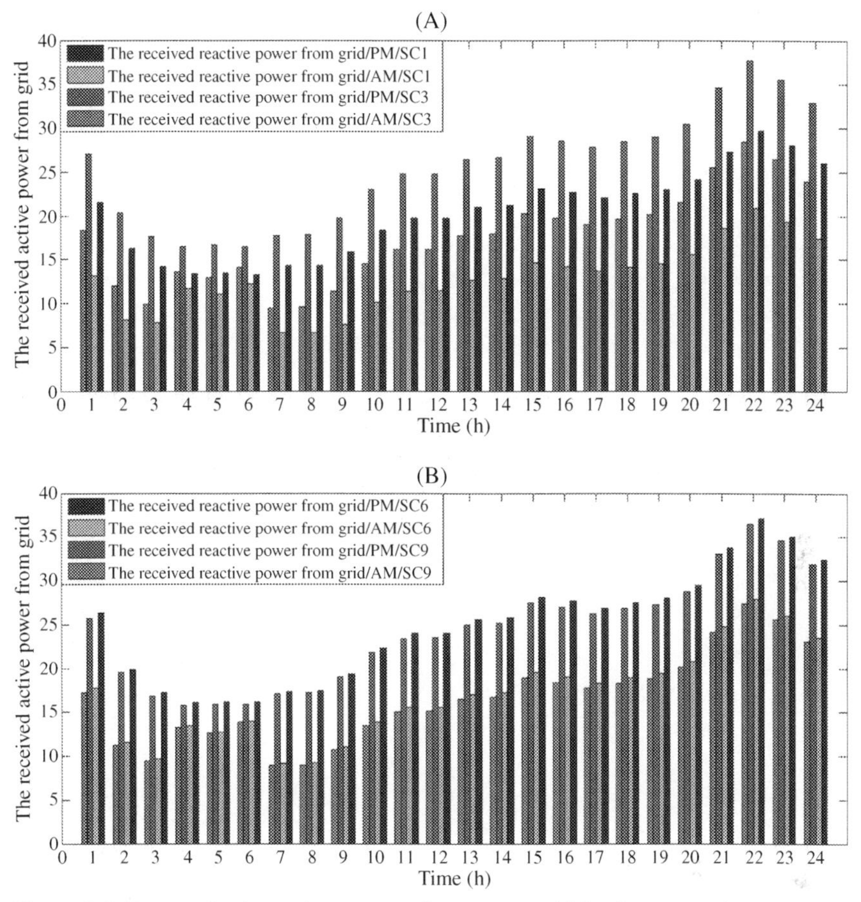

Figure 9.9 The received reactive power of upstream grid for four scenarios.

hour. According to Fig. 9.11, the voltage drop increases as the distance from substations increases in the passive networks. The proper integration of renewable DG units in the network can improve the voltage profile. The voltage profiles for the AM and PM mode with WTG in peak hour have been maintained within the permissible limits. The voltage curves are drawn for four scenarios and have been compared together.

In this study it is assumed that DNO is responsible for making the decision to the batteries' charge/discharge programs. Fig. 9.12A shows the hourly battery power output in passive and active network. These values are subject to the battery charger limit (i.e., [− 700 KW, + 700 KW]) at any time step. In normal conditions, because the grid

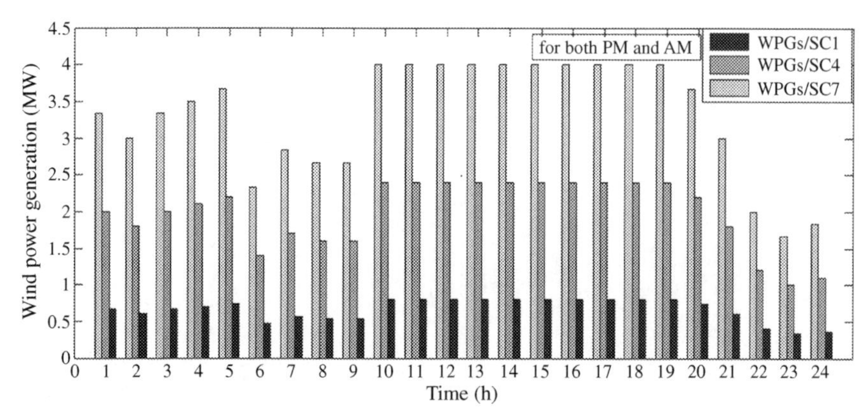

Figure 9.10 Wind turbines output power for daily scheduling period.

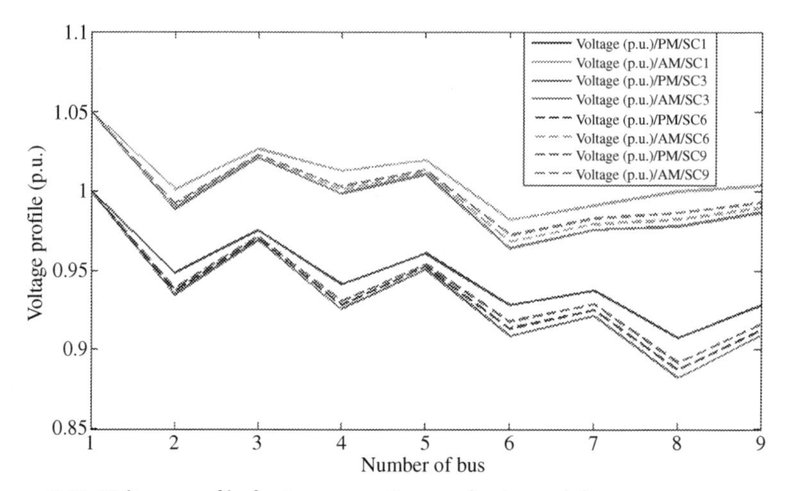

Figure 9.11 Voltage profile for two operation modes in peak hour.

electricity price is cheaper in the early morning, the battery bank starts to store as much energy as possible. Accordingly, the battery banks start to return the stored energy to the distribution network at the peak load time. Battery output power for all scenarios is the same. Fig. 9.12B indicated the hourly RPCs power output for the first and third scenarios for samples. Because of the low operation cost of RPCs, their power spillway isn't high. In the off peak load hours, the operation cost of RPCs will be more than market price.

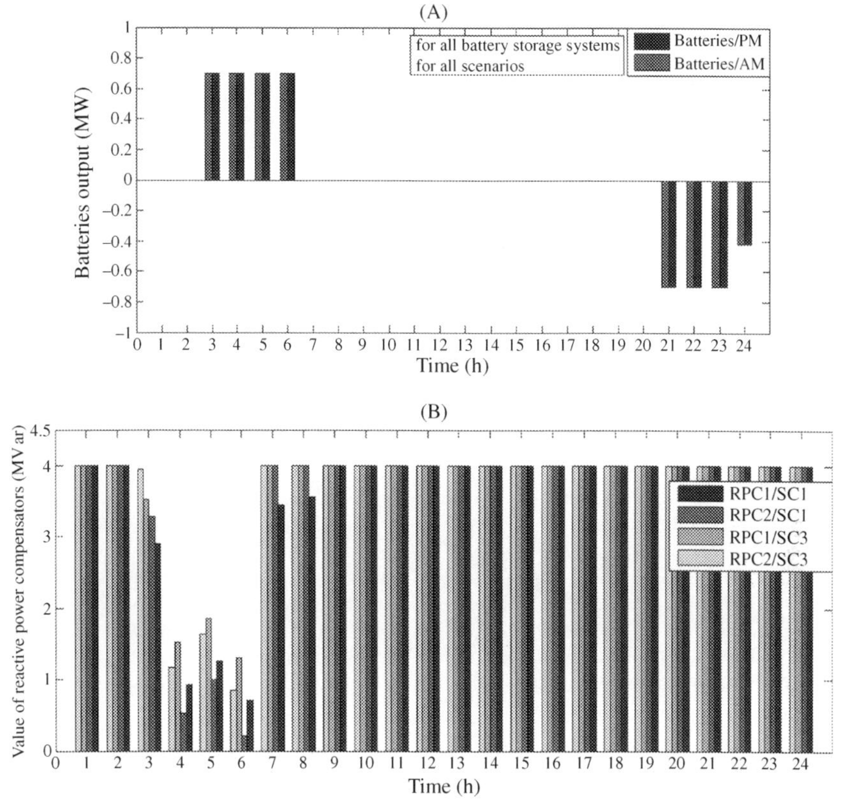

Figure 9.12 Optimal scheduling of storage systems and reactive power compensators.

9.7 CONCLUSION

In this chapter, active network management has been presented to the short-term scheduling of distribution system with the integration of WPG units. The result of AM is increasing DNO benefit. The BSS, which is assumed in parallel with the WPG units can decrease the DNO costs and power losses. A Markov model is proposed to a reliability modeling of WPG and BSS. A Markov model provides the characteristics of the major components of the WPG and BSS used to modeling reliability. From studied results, it has been derived that voltage profile and losses of the system intensely depend on management method of system. It has been shown that appropriate condition of operation has great effects on the DNO benefit. Using AM has several benefits in comparison to passive management which cannot be neglected. The AM method addition to power losses also reduces the received power from the upstream network.

REFERENCES

[1] P. Zhang, W. Li, S. Li, Y. Wang, W. Xiao, Reliability assessment of photovoltaic power systems: Review of current status and future perspectives, Appl. Energy 104 (2013) 822–833.

[2] Lund, P. Large-scale urban renewable electricity schemes—integration and interfacing aspects. Energy Convers. Manage. 2012;63: 162–172.

[3] S. Gill, I. Kockar, G.W. Ault, Dynamic optimal power flow for active distribution networks, IEEE Trans. Power Syst. 29 (1) (2014) 121–131.

[4] J. Zhang, H. Cheng, C. Wang, Technical and economic impacts of active management on distribution network, Int. J. Electr. Power Energy Syst. 31 (1) (2008) 130–138.

[5] S. Abapour, K. Zare, B. Mohammadi-Ivatloo, Evaluation of technical risks in distribution network along with distributed generation based on active management, IET Gener. Transm. Distribut. 8 (4) (2014) 609–618.

[6] S. Carr, G.C. Premier, A.J. Guwy, R.M. Dinsdale, J. Maddy, Energy storage for active network management on electricity distribution networks with wind power, IET Renewa. Power Gener. 8 (3) (2014) 249–259.

[7] S. Gill, M. Dolan, A. Emhemed, I. Kockar, M. Barnacle, G. Ault, et al., Increasing renewable penetration on islanded networks through active network management: A case study from Shetland, IET Renew. Power Gener. 9 (5) (2015) 453–465.

[8] Kane, L., Ault, G., Gill, S. An assessment of principles of access for wind generation curtailment in active network management schemes, in: Proc. 22nd International Conference and Exhibition on Electricity Distribution (CIRED 2013): 237–240.

[9] F. Capitanescu, I. Bilibin, E.R. Ramos, A comprehensive centralized approach for voltage constraints management in active distribution grid, IEEE Trans. Power Systems 29 (2) (2014) 933–942.

[10] F. Pilo, G. Pisano, G.G. Soma, Optimal coordination of energy resources with a two-stage online active management, IEEE Trans. Industrial Electronics 58 (10) (2011) 4526–4537.

[11] P. Siano, P. Chen, Z. Chen, A. Piccolo, Evaluating maximum wind energy exploitation in active distribution networks, IET Gener. Transm. Distribut. 4 (5) (2010) 598–608.

[12] P. Vasant (Ed.), Innovation in Power, Control, and Optimization: Emerging Energy Technologies: Emerging Energy Technologies, IGI Global, Malaysia, 2011.

[13] J. de Souza, M.J. Rider, J.R.S. Mantovani, Planning of distribution systems using mixed-integer linear programming models considering network reliability, J. Control Automat. Electr. Systems 26 (2) (2015) 170–179.

[14] A.R. Abbasi, A.R. Seifi, Considering cost and reliability in electrical and thermal distribution networks reinforcement planning, Energy 84 (2015) 25–35.

[15] G. Muñoz-Delgado, J. Contreras, J.M. Arroyo, Multistage generation and network expansion planning in distribution systems considering uncertainty and reliability, IEEE Trans. Power Systems 31 (5) (2016) 3715–3728.

[16] L.F. Rocha, C.L.T. Borges, G.N. Taranto, Reliability evaluation of active distribution networks including islanding dynamics, IEEE Trans. Power Systems 32 (2) (2016) 1545–1552.

[17] A. Safdarian, Z.D. Merkebu, M. Lehtonen, M. Fotuhi-Firuzabad, Distribution network reliability improvements in presence of demand response, IET Gener. Transm. Distribut. 8 (12) (2014) 2027–2035.
[18] M. Moeini-Aghtaie, H. Farzin, M. Fotuhi-Firuzabad, R. Amrollahi, Generalized analytical approach to assess reliability of renewable-based energy hubs, IEEE Trans. Power Systems 32 (1) (2017) 368–377.
[19] M. Rahmani-andebili, Reliability and economic-driven switchable capacitor placement in distribution network, IET Gener. Transm. Distribut. 9 (13) (2015) 1572–1579.
[20] M. Al-Muhaini, T.H. Gerald, Evaluating future power distribution system reliability including distributed generation, IEEE Trans. Power Delivery 28 (4) (2013) 2264–2272.
[21] R. Billinton, Y. Gao, Multistate wind energy conversion system models for adequacy assessment of generating systems incorporating wind energy, IEEE Trans. Energy Conver. 23 (1) (2008) 163–170.
[22] R.N. Allan, M.G. Da Silva, Evaluation of reliability indices and outage costs in distribution systems, IEEE Trans. Power Systems 10 (1) (1995) 413–419.
[23] S. Abapour, K. Zare, B. Mohammadi-Ivatloo, Dynamic planning of distributed generation units in active distribution network, IET Gener. Transm. Distrib 9 (12) (2015) 1455–1463.
[24] S. Abapour, E.K. Alireza, A. Frakhor, M. Abapour, Optimal integration of wind power resources in distribution networks considering demand response programs, Electrical and Electronics Engineering (ELECO), 2015 9th International Conference on, IEEE, 2015.
[25] Fu, Y., Ch, W., Zh, L., et al., Optimal DG Integration in Active Distribution Network Based on S-OPF. In Springer Intelligent Computing in Smart Grid and Electrical Vehicles, 2014, pp. 378–387.
[26] Abapour, S., Zare, K., Mohammadi-Ivatloo, B., Maximizing penetration level of distributed generations in active distribution networks. In IEEE Smart Grid Conference (SGC), 2013, pp. 113–118.
[27] Z. Zhou, M. Benbouzid, J.F. Charpentier, F. Scuiller, T. Tang, A review of energy storage technologies for marine current energy systems, Renew. Sustain. Energy Rev. 18 (2013) 390–400.
[28] M.N. Kabir, Y. Mishra, G. Ledwich, Z. Xu, R.C. Bansal, Improving voltage profile of residential distribution systems using rooftop PVs and Battery Energy Storage systems, Appl. Energy 134 (2014) 290–300.
[29] R. Billinton, R.N. Allan, Reliability Evaluation of Engineering Systems, Plenum Press, New York, 1992.
[30] T. Adefarati, R.C. Bansal, Reliability assessment of distribution system with the integration of renewable distributed generation, Appl. Energy 185 (2017) 158–171.
[31] A.J. Conejo, M. Carrión, J.M. Morales, Decision-making Under Uncertainty in Electricity Markets, Springer, New York, 2010.
[32] T. Niknam, M. Zare, J. Aghaei, Scenario-based multiobjective volt/var control in distribution networks including renewable energy sources, IEEE Trans. Power Delivery 27 (4) (2012) 2004–2019.
[33] A. Rabiee, A. Soroudi, B.M. Ivatloo, M. Parniani, Corrective voltage control scheme considering demand response and stochastic wind power, IEEE Trans. Power Syst. 29 (6) (2014) 2965–2973.
[34] A. Soroudi, M. Ehsan, A possibilistic–probabilistic tool for evaluating the impact of stochastic renewable and controllable power generation on energy losses in distribution networks—a case study, Renew. Sustain. Energy Rev. 15 (1) (2011) 794–800.

[35] S. Nojavan, H.A. Aalami, Stochastic energy procurement of large electricity consumer considering photovoltaic, wind-turbine, micro-turbines, energy storage system in the presence of demand response program, Energy Conver. Manage. 103 (2015) 1008–1018.
[36] D.T. Nguyen, B.L. Le, Optimal bidding strategy for micro-grids considering renewable energy and building thermal dynamics, IEEE Trans. Smart Grid 5 (4) (2014) 1608–1620.
[37] K. Khalkhali, S. Abapour, S.M. Moghaddas-Tafreshi, M. Abapour, Application of data envelopment analysis theorem in plug-in hybrid electric vehicle charging station planning, IET Gener. Transm. Distribut. 9 (7) (2015) 666–676.
[38] C. Li, X. Ge, Y. Zheng, C. Xu, Y. Ren, C. Song, et al., Techno-economic feasibility study of autonomous hybrid wind/PV/battery power system for a household in Urumqi, China, Energy 55 (2013) 263–272.

CHAPTER 10

Calculation of the Participants' Loss Share in the Advanced Distribution Network

Sina Ghaemi[1] and Kazem Zare[2]
[1]Azarbaijan Shahid Madani University, Tabriz, Iran
[2]University of Tabriz, Tabriz, Iran

10.1 INTRODUCTION

Today, distributed generations' (DGs) important role in the advanced distribution networks is undoubted, because their optimum performance in the network leads to obtain more benefits. These benefits can be technical, economic, and environment like loss reduction, power quality improvement, cost reduction, and reduction of CO_2 emissions [1,2]. Moreover, penetration of the other new technologies in the network like energy storage system (ESS), and price respective loads causes to change the performance of the network from passive form to active one. Therefore, due to increasing the new technologies penetration in the distribution networks, it is needed to adopt the new policy that causes to encourage them to participate in the network actively.

As known, distribution network losses are the major concern of the distribution companies (DISCOs) that should deal with them, because in that level, since, the voltage magnitude is low and the ratio of lines resistance to inductance is high, the amount of the losses is considerable. Hence, network's participants' performance should be controlled in a way that leads to decrease the network losses. Therefore, it is needed to determine each participant's loss share in the network to control their performance in the optimum manner.

Several studies on loss allocation (LA) approach have been done in the transmission level previously. In Ref. [3], regardless of considering the network topology, network loss has been divided among participants according to their active power consumption. Marginal loss coefficient and direct loss coefficient [4–6] are known as the other popular

Operation of Distributed Energy Resources in Smart Distribution Networks
DOI: https://doi.org/10.1016/B978-0-12-814891-4.00010-2

approaches, which are utilized in transmission and distribution networks. Whereas, implementation of them in the distribution network cannot lead to better results because of their dependency on Newton- Raphson power flow approach. Authors in Ref. [7] have proposed the Z-bus method for LA in the transmission network but it cannot be suitable for distribution network according to the [8]. Likewise, authors in Ref. [9], have introduced a succinct approach for LA; but, it is seen that the obtained results in some situations, which are mentioned in Ref. [8] cannot be reliable.

The LA issue is noticeable in the distribution network, because of the amount of the active loss in that level and its impact on the obtained benefits of each participant. Several attempts have been made to determine the suitable approach for LA in the distribution network recently. Branch current decomposition method has been introduced in Ref. [8] in order to compute the loss share of each load by verifying each node's injection current's share from each branch current. Authors in Ref. [10] have presented power summation method for computing LA by considering the DGs in the distribution network. Likewise, energy summation approach has been presented in Ref. [11]. Authors in Ref. [12] have utilized the graph theory for energy LA, whereas, it cannot be suitable in a way that DGs are integrated. Savier et al. [13] compute the loss share of each load according to the node's injected current, which is named exact method. Authors in Ref. [14] performed the energy LA using the different load curves. Ghofrani et al. [15], have determined the participants' loss share in three steps. Firstly, the loss share of the generation units has been determined in the nodes, that the amount of the generation is more than consumption, then the loads; contribution from the loss have been determined in nodes which the amount of the consumption is more than the generation. Finally, no allocated losses have been distributed among participants based on the normalization approach. Voltage based approach has been introduced for computing the loss share in Ref. [16]. The main concern of this method is to reduce the assumption which is used in LA methods. Jagtap et al. [17], have done LA in the distribution networks using the different loads and DGs models. Shapley value approach has been utilized in Ref. [18] in order to perform the LA procedure. Authors in Ref. [19] have proposed the LA according to the contractual power in the radial restructured distribution network. Authors in Ref. [20] have minimized the distribution network losses according to the ability of reconfiguration in the network and DGs presence. A bibliographical survey has been applied in Ref. [21] in order to decrease the amount of the distribution network loss. Shapely value approach has been utilized for the weakly meshed distribution network in Ref. [22] so as to allocate the network loss among participants.

The main problem in the LA approach is the nonlinear bond between the loss and passing current in each line and all the mentioned references have tried to deal with this problem and reduce the amount of the assumption in their works. LA approaches can be compared according to different parameters such as economic factors, amount of assumption, exactness, and data usage [7] in order to clarify each one's valuation.

In this chapter, network's participants are categorized in two different groups namely producer and consumer of the electricity power. For instance, ESS owner depends on its performance during the day, and can be a producer or consumer of the power but DG owner is always known as a producer. Afterwards, each participant's effect on each network branch losses is determined and then according to its generated or consumed power, loss share of the particular participant is calculated. It is tried to propose fairly LA among different participants and finally according to the various factors, which are utilized for comparison among different LA methods, the effectiveness of the proposed LA method has been manifested and it is concluded that it can be easily implemented to the advanced distribution network, where there exists various types of the participants.

This chapter includes four sections. In Section 10.2, which is made of two subsections, the main formulation of the recommended LA. Simulation and result of the chapter are demonstrated in Section 10.3 and finally, Section 10.4 provides the conclusion of this chapter.

10.2 THE PROPOSED LOSS ALLOCATION APPROACH

In this section, the new method for LA is presented and it can be implemented for both producers and consumers of the network. Using the branch oriented methods, in which each participant's impact on the each line's loss is determined, leads to the acceptable results. Therefore, in this chapter, a new branch oriented approach is introduced as well. This section includes two subsections. In the first subsection, it is clarified that each participant influences which line's branch and then the recommended method is clarified in the next subsection.

10.2.1 Determination of the Participant's Effect on Each Network's Branch Loss

According to the radial operation of the distribution network, it is seen that the current passing through each branch is equal to sum of the injected current to the nodes, which are located beyond that particular branch. In order to ease the problem, Fig. 10.1 shows the radial distribution network that is composed of the 5 node.

From this figure, it can be concluded that, the current of the branch 2, is equal to the sum of the current which is injected to node 4 and 5. Therefore, before determining the loss share of each node, it is needed to clarify that each bus in which branch has been caused to the loss. So as to obtain this target participant matrix is utilized which is mentioned in Ref. [19]. The participant matrix can be written for the particular network as follows:

In this matrix, each row demonstrates branch numbers and each column refers to the network's nodes.

$$\boldsymbol{P}_{iden} = \begin{bmatrix} 0 & 1 & 1 & 1 & 1 \\ 0 & 0 & 1 & 1 & 1 \\ 0 & 0 & 0 & 1 & 0 \\ 0 & 0 & 0 & 0 & 1 \end{bmatrix}$$

As shown, node 3 and 4 have caused the loss in the branch 2. In addition it can be seen that for the particular bus like 4, it has caused the loss in branches 3, 2, and 1 and it has no impact on the loss of the branch 4.

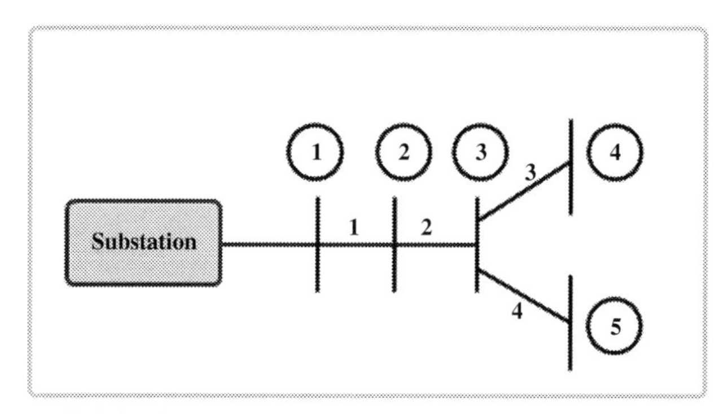

Figure 10.1 Radial distribution network.

10.2.2 Calculation of the Participants' Loss Share

In this subsection, the procedure of the LA among participants is discussed. In order to explain the recommend method more understandingly, the allocation of the particular branch's loss between two participants (e.g., LA of the branch 2 in Fig. 10.1 between load 4 and 5) has been considered, then it is extended for all of the participants.

The branch loss has the relation with square of its current as follows:

$$S_{loss}^{branch} = Z_{line} \cdot I_{line} \cdot I_{line}^{*} \tag{10.1}$$

As mentioned, current of the branch is composed of the injected current of the two participants. Therefore, the branch loss can be rewritten according to Eq. (10.2):

$$S_{loss}^{branch} = Z_{line} \cdot (I_1 + I_2) \cdot (I_1^* + I_2^*) \tag{10.2}$$

In this approach, the loss share of each participant relates to its voltage node and active and reactive power consumption. So, injected current of each node can be changed using voltage node and power consumption. Hence, the branch loss can be modified as following:

$$S_{loss}^{branch} = Z_{line}\left(\frac{p_1^2 + q_1^2}{V_1^2} + \frac{p_2^2 + q_2^2}{V_2^2}\right) + Z_{line}\left(\frac{p_1 - jq_1}{V_1^*} \times \frac{p_2 + jq_2}{V_2}\right) + Z_{line}\left(\frac{p2 - jq_2}{V_2^*} \times \frac{p_1 + jq_1}{V_1}\right) \tag{10.3}$$

It is assumed that,

$$V_1^* V_2 = A + jB \tag{10.4}$$

Then, the loss of the branch is expressed as follows:

$$S_{loss}^{branch} = Z_{line}\left(\frac{p_1^2 + q_1^2}{V_1^2} + \frac{p_2^2 + q_2^2}{V_2^2}\right) + Z_{line}\left(\frac{2A(p_1 p_2)}{A^2 + B^2} + \frac{2A(q_1 q_2)}{A^2 + B^2}\right) + Z_{line}\left(\frac{2Bp_1 q_2}{A^2 + B^2} - \frac{2Bp_2 q_1}{A^2 + B^2}\right) \tag{10.5}$$

As shown from the Eq. (10.5), loss of the particular branch is associated with voltage and power consumptions of the nodes, which caused the loss in that. As seen, branch loss includes six terms. First and second terms are easily related to the each node. Whereas, the decomposition of the next four terms between them can be a complex problem. In this approach a novel coefficients has been defined based on the each node's power consumption and then, with the help of theses coefficients, the loss share of each one from the particular node is calculated. Hence, the loss share of load 1 and 2 from the particular branch can be computed as follows:

$$S_{loss}^{1} = Z_{line}\left(\frac{p_1^2+q_1^2}{V_1^2}\right) + Z_{line}\left(\alpha_{1,2}\frac{2A(p_1p_2)}{A^2+B^2} + \beta_{1,2}\frac{2A(q_1q_2)}{A^2+B^2}\right) + Z_{line}\left(\omega_{1,2}\frac{2Bp_1q_2}{A^2+B^2} - \zeta_{1,2}\frac{2Bp_2q_1}{A^2+B^2}\right) \quad (10.6)$$

$$S_{loss}^{2} = Z_{line}\left(\frac{p_2^2+q_2^2}{V_2^2}\right) + Z_{line}\left(\alpha_{2,1}\frac{2A(p_1p_2)}{A^2+B^2} + \beta_{2,1}\frac{2A(q_1q_2)}{A^2+B^2}\right) + Z_{line}\left(\omega_{2,1}\frac{2Bp_1q_2}{A^2+B^2} - \zeta_{2,1}\frac{2Bp_2q_1}{A^2+B^2}\right) \quad (10.7)$$

Where the new coefficients are defined as follows:

$$\alpha_{1,2} = p_1^2/p_1^2 + p_2^2, \qquad \alpha_{2,1} = p_2^2/p_1^2 + p_2^2 \quad (10.8)$$

$$\beta_{1,2} = q_1^2/q_1^2 + q_2^2, \qquad \beta_{2,1} = q_2^2/q_1^2 + q_2^2 \quad (10.9)$$

$$\omega_{1,2} = p_1^2/p_1^2 + q_2^2, \qquad \omega_{2,1} = q_2^2/p_1^2 + q_2^2 \quad (10.10)$$

$$\zeta_{1,2} = q_1^2/q_1^2 + p_2^2, \qquad \zeta_{2,1} = p_2^2/q_1^2 + p_2^2 \quad (10.11)$$

According to the defined coefficients in Eqs. (10.8)–(10.11), the sum of the loss share of load 1 and 2 is equal to the loss share of the branch.

$$S_{loss}^{branch} = S_{loss}^{1} + S_{loss}^{2} \quad (10.12)$$

Therefore, the loss share of each node is equal to the sum of its portion from the loss of the branches located beyond it. Finally, loss share of the *i*th node, which can be the location of the producer or consumer is computed using Eqs, (10.13)–(10.15):

$$S_{loss}^{i} = \sum_{l=1}^{N_{branch}} P_{iden}(l,i)\left(Z_{line,l}\left(\frac{p_i^2+q_i^2}{V_i^2}\right) + Z_{line,l}\left(\begin{array}{l}\sum_{j=1/i}^{N_{bus}} P_{iden}(l,j)(\alpha_{i,j}\frac{2A(p_ip_j)}{A^2+B^2}\\ +\beta_{i,j}\frac{2A(q_iq_j)}{A^2+B^2}\Big)\end{array}\right) + Z_{line,l}\left(\sum_{j=1/i}^{N_{bus}} P_{iden}(l,j)\left(\omega_{i,j}\frac{2Bp_iq_j}{A^2+B^2} - \zeta_{i,j}\frac{2Bp_jq_i}{A^2+B^2}\right)\right)\right) \quad (10.13)$$

$$S_{loss}^{active,i} = real\{S_{loss}^{i}\} \quad (10.14)$$

$$S_{loss}^{reactive,i} = \mathrm{Im}\{S_{loss}^{i}\} \quad (10.15)$$

In generally, participants of the radial distribution network are sorted in two groups: producer or consumer. In this method, the sign of the consumed power of each consumer is positive and for each producer, this sign is negative.

This approach is compatible for the nodes in which consumption of the power or production of that occur separately. However, it is possible that, in some nodes, consumption and production happen simultaneously. So, for these nodes, like *i*, which is shown in Fig. 10.2, the net injected power is formulated as follows:

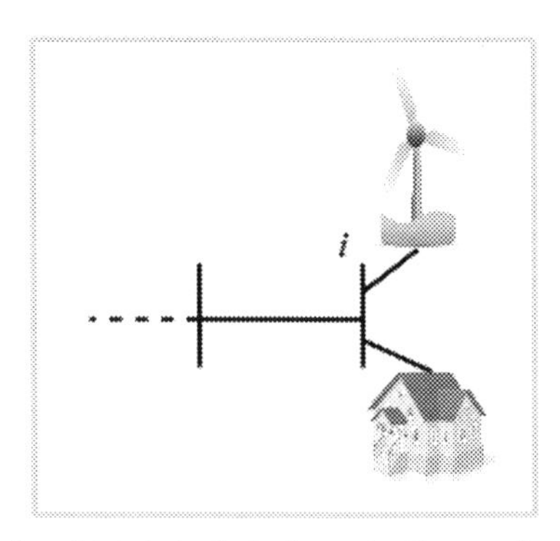

Figure 10.2 Simple node which is included production and consumption.

$$I_i = I_i^{cons.} + I_i^{prod.} \tag{10.16}$$

For these nodes, cross terms have been produced between each other and are needed to decompose them between them fairly. Hence, the new coefficients should be defined between each other as well. The loss share of the each producer and consumer in the same node like i, are formulated using Eqs. (10.17) and (10.18):

$$S_{loss}^{i,cons.} = \sum_{l=1}^{N_{branch}} P_{iden}(l,i)(Z_{line,l}\left(\begin{array}{c}\dfrac{p_{i,cons.}^2 + q_{i,cons.}^2}{V_i^2} + \alpha_{c,p}\dfrac{p_i^{cons.}p_i^{prod.}}{V_i^2} \\ + \beta_{c,p}\dfrac{q_i^{cons.}q_i^{prod.}}{V_i^2}\end{array}\right) + Z_{line,l}\left(\begin{array}{c}\sum_{j=1/i}^{N_{bus}} P_{iden}(l,j)(\alpha_{i,j}\dfrac{2A(p_ip_j)}{A^2+B^2} \\ + \beta_{i,j}\dfrac{2A(q_iq_j)}{A^2+B^2})\end{array}\right) + Z_{line,l}\left(\sum_{j=1/i}^{N_{bus}} P_{iden}(l,j)\left(\omega_{i,j}\frac{2Bp_iq_j}{A^2+B^2} - \zeta_{i,j}\frac{2Bp_jq_i}{A^2+B^2}\right)\right)) \tag{10.17}$$

$$S_{loss}^{i,prod.} = \sum_{l=1}^{N_{branch}} P_{iden}(l,i)(Z_{line,l}\left(\begin{array}{c}\dfrac{p_{i,prod.}^2 + q_{i,prod.}^2}{V_i^2} + \alpha_{p,c}\dfrac{p_i^{cons.}p_i^{prod.}}{V_i^2} \\ + \beta_{p,c}\dfrac{q_i^{cons.}q_i^{prod.}}{V_i^2}\end{array}\right) + Z_{line,l}\left(\begin{array}{c}\sum_{j=1/i}^{N_{bus}} P_{iden}(l,j)(\alpha_{i,j}\dfrac{2A(p_ip_j)}{A^2+B^2} \\ + \beta_{i,j}\dfrac{2A(q_iq_j)}{A^2+B^2})\end{array}\right) + Z_{line,l}\left(\sum_{j=1/i}^{N_{bus}} P_{iden}(l,j)(\omega_{i,j}\frac{2Bp_iq_j}{A^2+B^2} - \zeta_{i,j}\frac{2Bp_jq_i}{A^2+B^2})\right)) \tag{10.18}$$

Where,

$$\alpha_{c,p} = p_{cons.}^2 / p_{cons.}^2 + p_{prod.}^2, \qquad \alpha_{p,c} = p_{prod.}^2 / p_{cons.}^2 + p_{prod.}^2 \tag{10.19}$$

$$\beta_{c,p} = q^2_{cons.}/q^2_{cons.} + q^2_{prod.}, \qquad \beta_{2,1} = q^2_{prod.}/q^2_{cons.} + q^2_{prod.} \tag{10.20}$$

The process of the LA for among the network participants has been shown in Fig. 10.3.

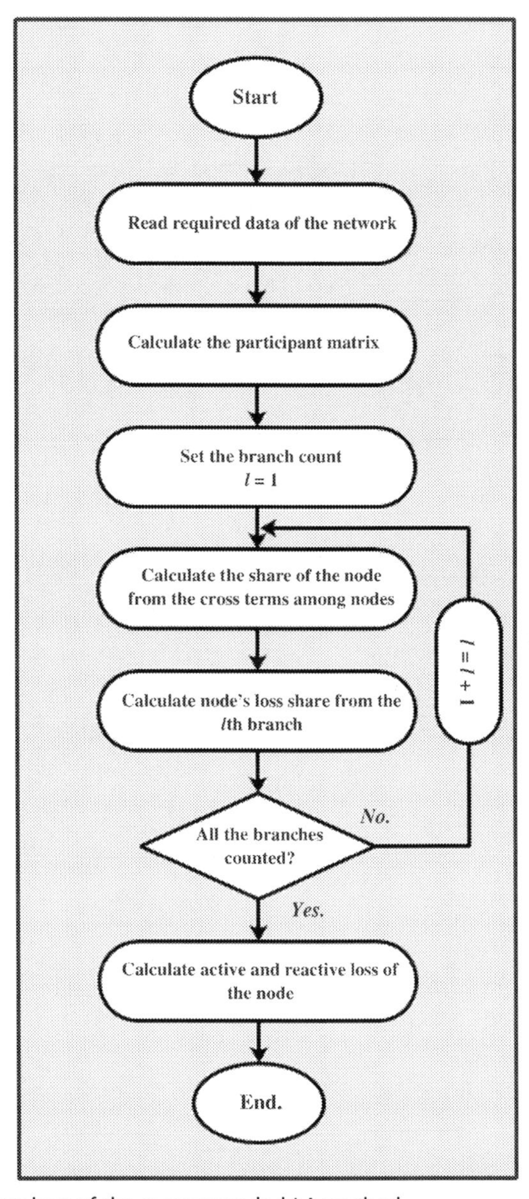

Figure 10.3 Flowchart of the recommended LA method.

10.3 SIMULATION AND RESULTS

In this part, the recommended approach for LA has been utilized for calculating the loss portion of each participant in the test distribution network. 33-bus distribution network has been selected as a test system, which is modified by three different DGs. Network data has been provided in appendix. The Fig. 10.4 shows the test system.

Each DG's generated active and reactive powers and the location of them have been shown in Table 10.1.

Fig. 10.5 shows the active loss share of each node before and after the presence of the DGs in the network.

In addition, Fig. 10.6 provides the reactive loss share of each node before and after the penetration of the DGs which are known as producers of the network.

As shown from Figs. 10.5 and 10.6, the loss share of node 30 is bigger than the others before installation of the DGs. Because the particular node has utilized more branches for providing its required power.

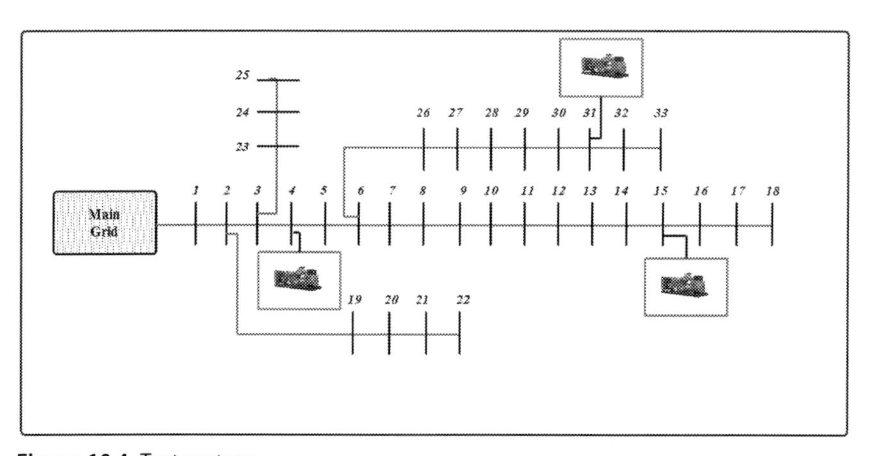

Figure 10.4 Test system.

Table 10.1 Information of the distributed generations

DG number	Active power (kW)	Reactive power (kVAr)	Location
1	2000	800	4
2	1000	400	15
3	500	0	31

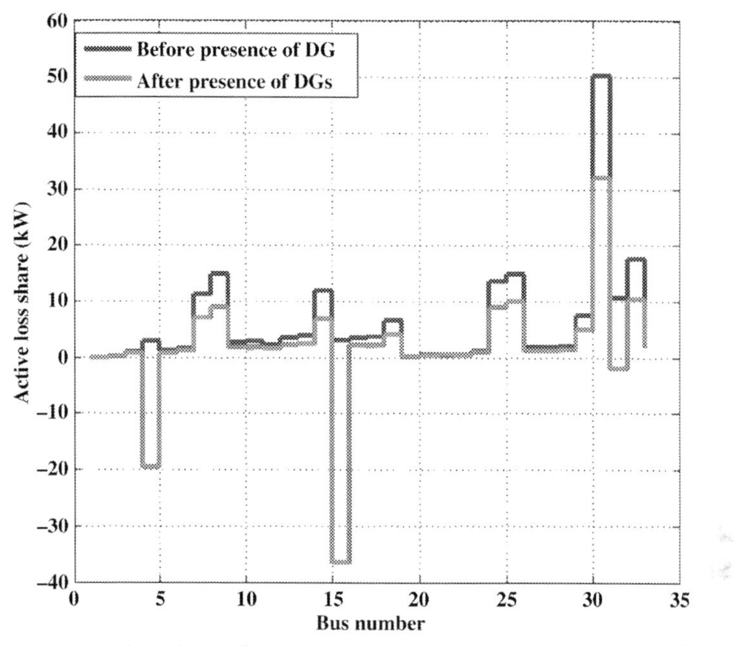

Figure 10.5 Active loss share of each node before and after the presence of DGs.

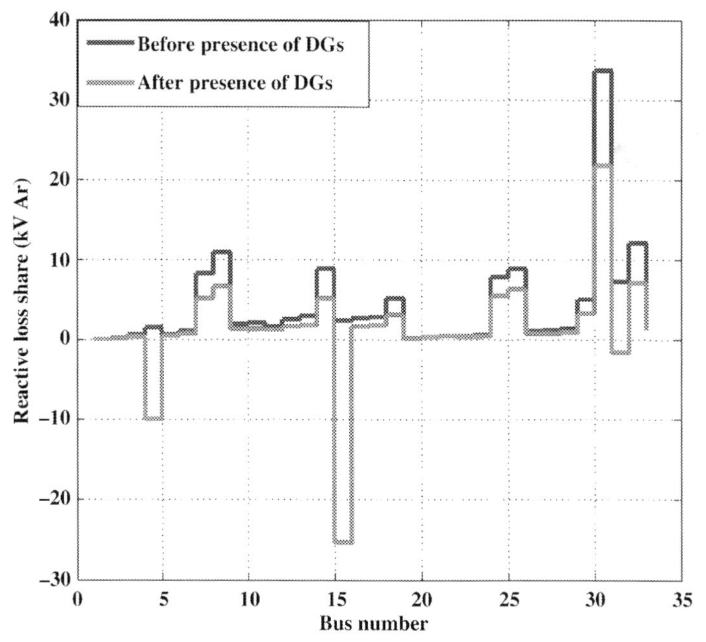

Figure 10.6 Reactive loss share of each node before and after the presence of DGs.

Furthermore, as can be seen from Figs. 10.5 and 10.6, the loss share of the nodes, which production of the power occurs besides the consumption, is negative. Since, in these nodes, the amount of the production is more than the consumption, obtaining of these results is logical.

Table 10.2 presents active and reactive loss share of the producer and consumer, which are located at the same node.

It can be seen from this table that, the active and reactive loss portion of the producer are negative. It means that, the presence of that leads to reduce the network loss and from this point of view, its presence is beneficial for DN. On the other hand, it is also shows the reason why loss share of those nodes becomes negative.

As discussed, the defined method for LA in the radial distribution network is known as a branch oriented approach. So, the loss share of each node (producer and consumer) from loss of each branch is calculated. Table 10.3 provides the active loss share of each node from loss of five different branches.

It is shown that, all of the participants have effect on the loss share of branch #1. Whereas, this cannot be true for other branches. The reason of this results is that, first branch is located beyond all of the nodes and is known as a main branch and each node has utilized this branch for providing its required power. However, for the other branches, when the loss share of the nodes on them is zero, it means that they cause no loss in those branches.

LA method should be able to satisfy its user when it has suitable and logical results. It is expected that for two different participants, which are located near each other and have the same power consumption, the difference between loss shares of these nodes is not considerable. In addition, for the two participants who are located far from each other but they have the same consumption or production, it is expected that there exists differences between their loss share as a reason of the distance. In this method, e.g., for the node #9 and #10 which are located near each other

Table 10.2 Loss share of each producer and consumer in the same node

Node	Consumer		Producer	
	Active (kW)	Reactive (kVAr)	Active (kW)	Reactive (kVAr)
4	1.8522	0.9453	− 21.4641	− 10.9634
15	1.9673	1.4888	− 38.5897	− 26.8036
31	5.9256	4.0177	− 7.9310	− 5.5818

Table 10.3 Active loss share of each node from loss of five different branches

	Active loss (kW)				
	Branch #1	**Branch #2**	**Branch #5**	**Branch #8**	**Branch #26**
1	0	0	0	0	0
2	0.2417	0	0	0	0
3	0.1889	0.8593	0	0	0
4	0.3385	1.5474	0	0	0
5	0.1042	0.4733	0	0	0
6	0.0952	0.4319	0.5400	0	0
7	0.6568	3.0157	3.6160	0	0
8	0.6661	3.0583	3.6673	0	0
9	0.0977	0.4430	0.5538	0	0
10	0.0983	0.4457	0.5573	0.3041	0
11	0.0763	0.3463	0.4351	0.3060	0
12	0.1160	0.5269	0.6576	0.2381	0
13	0.1168	0.5306	0.6622	0.3540	0
14	0.3618	1.6543	2.0188	0.3564	0
15	0.0902	0.4085	0.5110	1.0073	0
16	0.0998	0.4527	0.5660	0.2841	0
17	0.1000	0.4537	0.5673	0.3107	0
18	0.1864	0.8485	1.0476	0.3114	0
19	0.1864	0	0	0.5483	0
20	0.1871	0	0	0	0
21	0.1872	0	0	0	0
22	0.1873	0	0	0	0
23	0.2035	0.9266	0	0	0
24	1.6926	7.8494	0	0	0
25	1.6984	7.8760	0	0	0
26	0.1007	0.4570	0.5710	0	0
27	0.1010	0.4581	0.5725	0	0.0711
28	0.0968	0.4388	0.5486	0	0.0675
29	0.3383	1.5450	1.8865	0	0.2515
30	1.9749	9.2942	11.0072	0	1.8759
31	0.4403	2.0135	2.4408	0	0.3262
32	0.7142	3.2803	3.9257	0	0.5396
33	0.1230	0.5591	0.6972	0	0.0890

with the same power consumption, active loss share of the node #9 is 2.5881 kW and for node #10 this amount is 2.8837 kW. It is seen that the difference is not high. On the other hand, for the nodes #6 and #17 which are located far from each other, the portion of node #6 from the active loss is 1.1106 kW and for the node #17 this amount is 2.2869 kW.

So, it is seen that, the difference between them is acceptable. Therefore, this method can achieve reasonable results.

It should be considered that, the reason of reduction in the network loss after presence of the producers is related to them not to the consumers. Because their consumption after the presence of the producer has not been changed. Hence, it is expected that the portion of the producers from loss reduction will be significant. In this method, total active loss has reduced to 60.5323 kW. From this amount, share of the producers is −67.9848 kW and the portion of the consumers is 128.5171 kW. Active loss of the network before penetration of the DGs is 202.15 kW. Therefore, it is seen that, although the amount of the loss share of the consumers has reduced after DGs presence, the main reason of the reduction is related to the DGs, and the spatial cross subsidies fail to happen in the recommended approach for LA.

Finally, it is valuable to mentioned that, determination of the optimum amount of powers produced by DGs are important and have a direct effect on the network loss. Because, if the amount of the generated powers by DGs is not relevant with the amount of the network demands, it may lead to increase the network loss instead of reducing that. Fig. 10.7 shows the amount of the network loss and loss share of each DG in two different scenarios of power generation. (At the first scenario amount of the generated power of each DG is equal to the mentioned data in Table 10.1 and in the second one these amounts reduce to their half).

It can be seen that, the amount of the powers, which are generated by DGs, change the amount of the network loss significantly. Therefore

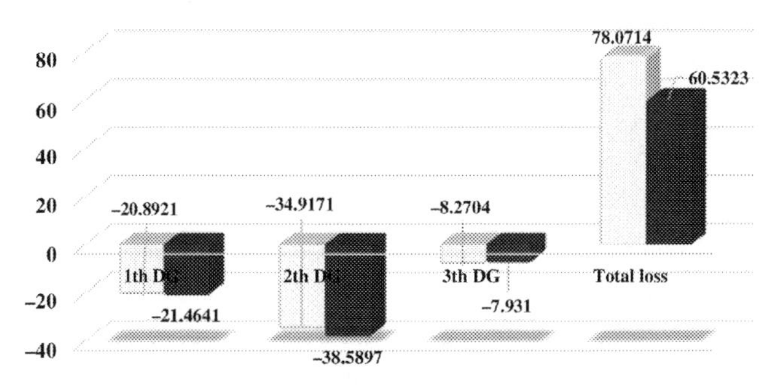

Figure 10.7 Active loss share of each DG in two different scenarios and network loss.

determination of the optimum generated powers of the DGs is essential but it is not the major concern of this paper although, it has enough potential to be studied in future works.

10.4 CONCLUSION

This chapter provided to propose the new method for LA in the advanced distribution network in which the importance of the loss is measured from both technical and economic perspectives. In this chapter, the branch oriented based approach was suggested for LA. In this method, the loss shares of each producer and consumer are related to their power consumption and their node's voltage magnitude. The defined coefficients in this approach were implemented in order to decompose the cross terms among participants fairly. Finally, the results of this study illustrate that the proposed method for LA can be acceptable because the obtained loss shares from this method are reasonable. In addition, spatial cross subsidies were not produced in this approach.

APPENDIX

Table 10.4 demonstrates the required data of the 33-bus distribution network.

Table 10.4 33-Bus distribution network data

Line no.	*R* (Ω)	*X* (Ω)	*P* (kW)	*Q* (kVAr)
1	0.0922	0.0477	100	60
2	0.4930	0.2511	90	40
3	0.3660	0.1864	120	80
4	0.3811	0.1941	60	30
5	0.8190	0.7070	60	20
6	0.1872	0.6188	200	100
7	1.7114	1.2351	200	100
8	1.0300	0.7400	60	20
9	1.0400	0.7400	60	20
10	0.1966	0.0650	40	30
11	0.3744	0.1238	60	35
12	1.4680	1.1550	60	35
13	0.5416	0.7129	120	80
14	0.5910	0.5260	60	10

(Continued)

Table 10.4 (Continued)

Line no.	*R* (Ω)	*X* (Ω)	*P* (kW)	*Q* (kVAr)
15	0.7463	0.5450	60	20
16	1.2890	1.7210	60	20
17	0.7320	0.5740	90	40
18	0.1640	0.1565	90	40
19	1.5042	1.3554	90	40
20	0.4095	0.4784	90	40
21	0.7089	0.9373	90	40
22	0.4512	0.3083	90	40
23	0.8980	0.7091	420	200
24	0.8960	0.7011	420	200
25	0.2030	0.1034	60	25
26	0.2842	0.1447	60	25
27	1.0590	0.9337	60	20
28	0.8042	0.7006	120	70
29	0.5075	0.2585	200	600
30	0.9744	0.9630	150	70
31	0.3105	0.3619	210	100
32	0.3410	0.5302	60	40

REFERENCES

[1] N. Khalesi, N. Rezaei, M.R. Haghifam, DG allocation with application of dynamic programming for loss reduction and reliability improvement, Int. J. Electr. Power Energy Systems 33 (2) (2011) 288–295.

[2] U. Sultana, A.B. Khairuddin, M.M. Aman, A.S. Mokhtar, N. Zareen, A review of optimum DG placement based on minimization of power losses and voltage stability enhancement of distribution system, Renew. Sustain. Energy Rev. 63 (2016) 363–378.

[3] A.J. Conejo, J.M. Arroyo, N. Alguacil, A.L. Guijarro, Transmission loss allocation: A comparison of different practical algorithms, IEEE Trans. Power Systems 17 (3) (2002) 571–576.

[4] J. Mutale, G. Strbac, S. Curcic, N. Jenkins, Allocation of losses in distribution systems with embedded generation, IEE Pro. Gener. Transm. Distribut. 147 (1) (2000) 7–14.

[5] E. Carpaneto, G. Chicco, J.S. Akilimali, Loss partitioning and loss allocation in three-phase radial distribution systems with distributed generation, IEEE Trans. Power Systems 23 (3) (2008) 1039–1049.

[6] Carpaneto, E., Chicco, G., Akilimali, J.S., Computational aspects of the marginal loss allocation methods for distribution systems with distributed generation, in: Electrotechnical Conference, 2006. MELECON 2006. IEEE Mediterranean (pp. 1028–1031). IEEE, 2006, May.

[7] A.J. Conejo, F.D. Galiana, I. Kockar, Z-bus loss allocation, IEEE Trans. Power Systems 16 (1) (2001) 105–110.

[8] E. Carpaneto, G. Chicco, J.S. Akilimali, Branch current decomposition method for loss allocation in radial distribution systems with distributed generation, IEEE Trans. Power Systems 21 (3) (2006) 1170–1179.

[9] W.L. Fang, H.W. Ngan, Succinct method for allocation of network losses, IEE Proc. Gener. Transm. Distribut. 149 (2) (2002) 171–174.
[10] M. Atanasovski, R. Taleski, Power summation method for loss allocation in radial distribution networks with DG, IEEE Trans. Power Systems 26 (4) (2011) 2491–2499.
[11] M. Atanasovski, R. Taleski, Energy summation method for loss allocation in radial distribution networks with DG, IEEE Trans. Power Systems 27 (3) (2012) 1433–1440.
[12] Macqueen, C.N., Irving, M.R., An algorithm for the allocation of distribution system demand and energy losses, in: Power Industry Computer Application Conference, 1995. Conference Proceedings., 1995 IEEE (pp. 234–239). IEEE, 1995, May.
[13] J.S. Savier, D. Das, An exact method for loss allocation in radial distribution systems, Int. J. Electr. Power Energy Systems 36 (1) (2012) 100–106.
[14] J.S. Savier, D. Das, Energy loss allocation in radial distribution systems: A comparison of practical algorithms, IEEE Trans. Power Deliv. 24 (1) (2009) 260–267.
[15] Z. Ghofrani-Jahromi, Z. Mahmoodzadeh, M. Ehsan, Distribution loss allocation for radial systems including DGs, IEEE Trans. Power Deliv. 29 (1) (2014) 72–80.
[16] K.M. Jagtap, D.K. Khatod, Loss allocation in distribution network with distributed generations, IET Gener. Transm. Distribut. 9 (13) (2015) 1628–1641.
[17] K.M. Jagtap, D.K. Khatod, Loss allocation in radial distribution networks with various distributed generation and load models, Int. J. Electr. Power Energy Systems 75 (2016) 173–186.
[18] S. Sharma, A.R. Abhyankar, Loss allocation of radial distribution system using Shapley value: A sequential approach, Int. J. Electr. Power Energy Systems 88 (2017) 33–41.
[19] S. Ghaemi, K. Zare, Loss allocation in restructured radial distribution networks considering the contractual power, IET Gener. Transm. Distribut. 11 (6) (2016) 1389–1397.
[20] R.S. Rao, K. Ravindra, K. Satish, S.V.L. Narasimham, Power loss minimization in distribution system using network reconfiguration in the presence of distributed generation, IEEE Trans. Power Systems 28 (1) (2013) 317–325.
[21] S. Kalambe, G. Agnihotri, Loss minimization techniques used in distribution network: Bibliographical survey, Renew. Sustain. Energy Rev. 29 (2014) 184–200.
[22] S. Sharma, A.R. Abhyankar, Loss allocation for weakly meshed distribution system using analytical formulation of Shapley value, IEEE Trans. Power Systems 32 (2) (2017) 1369–1377.

CHAPTER 11

Multi-objective Modeling and Optimization for DG-Owner and Distribution Network Operator in Smart Distribution Networks

Gianni Celli, Emilio Ghiani, Susanna Mocci, Fabrizio Pilo and Gian Giuseppe Soma
University of Cagliari, Cagliari, Italy

11.1 INTRODUCTION

The liberalization of the electricity market has broken the natural monopoly in power systems. Vertically integrated electric utilities disappeared in many countries, and new electric companies, committed to network operation and expansion, to produce and to sell energy, had been constituted. Such new entities can play in the markets or act as regulated bodies committed to serve communities under the regulation framework imposed by national regulators in the name of the society. Looking at the players in new liberalized environment, one can recognize energy producers, one or more transmission system operators (TSOs), several distribution system operators (DSOs), and consumers. Producers are no longer confined in the transmission level since, in order to harvest renewable energy, the paradigm of distributed generation (DG) has been fully implemented in many countries and new hybrid figures such as prosumers (customers that produce and consume energy) are common.

TSOs are responsible for the operation and the expansion of transmission grids and they care about adequacy and security of the system by purchasing energy and services in the relevant markets. DSOs are responsible for the operation and the expansion of local distribution systems with the goal to keep the quality of service at the levels imposed by regulators. All the operators are strongly committed to increase efficiency and cut system costs. In the EU Winter Package 2016 the future development of the distribution system is clearly depicted in such a way that DSOs

Operation of Distributed Energy Resources in Smart Distribution Networks
DOI: https://doi.org/10.1016/B978-0-12-814891-4.00011-4

must change their traditional "fit and forget" planning approach and the consequent passive management, by including in their operation and planning plans the services offered in market. As an example, the society is eager to make energy production less polluting by increasing the use of renewable energy sources (RES); DSOs are committed to increase efficiency in the usage of public money without jeopardizing the quality of service obtained so far, and imposing economical or technical barriers to RES; energy producers (particularly RES owners) want to be free of producing and selling energy and placing offers in well-functioning markets; consumers are interested in high quality energy at the lowest prices and also to bid on energy and service markets, provided that current technology and market would make this possible.

This new challenging scenario raises new models and methodologies for technical and economic studies. In particular, new planning approaches for smart distribution networks should have the following characteristics [1]:

- to deal with uncertainty and explicitly with risks;
- to use a probabilistic approach in load and generation modeling;
- to consider time-varying load and generation;
- to be based on multi-objective or multi-criteria optimization frameworks;
- to be able to select multiple, diverse, trade-off solutions;
- to be decision focused and use mathematical decision techniques;
- to use appropriate time schedules and planning horizons;
- to integrate the operation within the planning process;
- to enable reuse of model, data, and solutions;
- to make use of massive data produced by modern digital society.

For these reasons, it is fundamental to develop software-planning tools able to assess the several planning options available, so that distributors can find those that better accomplish their planning purposes.

Typically, the optimal solution has to achieve a comfortable level of performance on several conflicting criteria by minimizing the associated cost. A systematic approach helps the decision maker finding the preferred solution. Normally, a structured decision-making process involves five steps:

1. identify the objectives,
2. identify the options for achieving the objectives,
3. define the evaluation criteria to compare the options,
4. analyzing the options, and
5. making the choices.

In this challenging context, the need to find compromising solutions for the conflicting goals of system stakeholders, and the difficulty of defining a unique objective function, leads to multi-objective (MO) approaches. The overall optimal power system optimization is then achieved when the stakeholders simultaneously accomplish their own objectives. In some cases, these objectives contradict each other and cannot be handled by conventional single optimization techniques. In particular, in a multi-objective optimization problem no single solution exists that simultaneously optimizes each ibjective function (OF). Moreover, the use of MO methodologies gives information on the consequences of the decision with respect to all objective functions assumed in planning.

11.2 MULTI-OBJECTIVE PROGRAMMING

11.2.1 General Formulation of Multi-Objective Problem

The nature of real-world problems is intrinsically multi-objective. Thus, multi-objective optimization hass become very popular and important for scientists and engineers. It can be defined as the problem of finding [2] "a vector of decision variables which satisfies constraints and optimizes a vector function whose elements represent the objective functions. These functions form a mathematical description of performance criteria, which are usually in conflict with each other. Hence, the term 'optimize' means finding such a solution which would give the values of all the objective functions acceptable to the decision maker."

In a mathematical form, a MO optimization problem can be expressed as in (11.1).

$$\begin{cases} \min f(x) = & [f_1(x), f_2(x) \ldots f_m(x)]^T \\ x \in \Omega & \\ c_j(x) = 0 & j = 1 \ldots n \\ h_k(x) \leq 0 & k = 1 \ldots p \end{cases} \tag{11.1}$$

where x represents a decision vector, Ω is the solution domain, $f_i(x)$ denotes the objective functions, and $c_j(x)$ and $h_k(x)$ are the equality end inequality constraints respectively.

Differently from single-objective optimization problems that may have a unique optimal solution, MO problems (as a rule) present a possibly uncountable set of solutions. This set is found through the use of the Pareto optimality theory. A solution belongs to the Pareto set, or it is said

Pareto optimal, non-dominated, Pareto efficient or non-inferior, if no improvement is possible in one objective without worsening in any other objective. Thus, identifying the Pareto set is a key point for the decision maker's selection of a compromising solution that satisfies all the objectives as better as possible. The Pareto set identification is illustrated in Fig. 11.1.

It is crucial to observe that in the absence of preference information all non-dominated solutions are considered equivalent.

Several MO optimization algorithms have been proposed in the iterature and applied to various problems [3–5]. In principle, the presence of several goals in a problem involves a set of optimal solutions (largely known as Pareto-optimal solutions), instead of a single optimal solution.

In the literature, the MO methods adopted for smart distribution system optimization are divided into two main groups [4].

The first group makes use of single-objective technique and a priori information. By changing the master objective function,several solutions of the Pareto set are identified. The use of single-objective optimization methods is known as the "classical approach" to MO optimization. The classical approach asks the user to perform an a priori decision making, by assigning preferences to the objectives under consideration, such that the final product is the solution that matches with those specifications. The weighted-sum and the ε-constrained methods are the ones that are most widely used in this category and provide a single least-cost solution [4]. Classical optimization methods suggest converting the MO optimization problem to a single-objective optimization problem by emphasizing one particular Pareto-optimal solution at a time. When such a method is

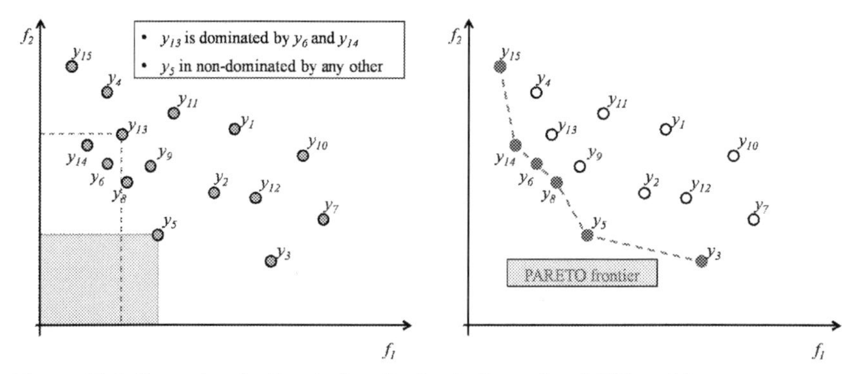

Figure 11.1 Example of a Pareto frontier for 2-dimensional MO problem.

to be used for finding multiple solutions, it has to be applied many times, hopefully finding a different solution at each simulation run.

On the contrary, MO evolutionary algorithms (MOEAs) are characterized by their ability to find multiple Pareto-optimal solutions in one single simulation run. In fact, since evolutionary algorithms (EAs) work with a population of solutions, they can be extended to maintain a diverse set of solutions during the optimization process.

11.3 OBJECTIVE FUNCTIONS IN THE SMART DISTRIBUTION NETWORK PLANNING

In traditional planning of distribution systems, the most significant optimization problems that have to be solved can be recapped in the following list:

- losses minimization problem,
- optimal placement of capacitors,
- optimal voltage regulation,
- optimal network expansion,
- service continuity and reliability optimization,
- optimal number, type and location of automation devices.

None of these optimization problems regards a single OF, but several ones are involved. For instance, when capacitor banks are added to distribution systems they can have an effect on power factor correction, loss reduction, voltage profile improvement; or if the number, type and location of automation devices is considered the objectives that need to be studied should be the cost of the devices, the cost of energy not supplied, the reliability indexes. However, the traditional approach to the distribution network planning is to convert the optimization into a single-objective formulation by summing up all these terms with suitable weights.

In smart distribution systems all previous optimization problems are still actual, but new ones arise mainly focused on the optimal exploitation of distributed energy resources (as distributed generation, distributed energy storage, demand side response, etc.). The network calculations are further complicated due to the stochastic behavior of most of these resources and the additional uncertainties that they introduce. Moreover, new actors are involved in the smart distribution network planning (DG owners, prosumers, energy retailers, aggregators, etc.), all of them with their own needs and targets to achieve. Therefore, a true MO approach becomes essential to take account of the different goals and identify the best planning compromising solutions.

In the next section, the required probabilistic network calculation is described; in addition, the different OFs that can be frequently encountered in optimization problems of smart distribution networks are defined.

11.3.1 Probabilistic Network Calculation

By considering that DG introduces new uncertainties in distribution studies, probabilistic models should be used in distribution modeling and planning.

If loads and generators are modeled with an average annual power, some contingencies cannot be evaluated. In order to overcome this problem, loads and generators have been considered in the power flow calculation via their daily load (generation) curves, assumed valid for all the days of the year. Generators able to follow the daily curve of the local load can allow a strong reduction of branch currents in a specific portion of the network, while generators with an opposite behavior determine a less important power flow decrease or, in some cases, the worsening of the existing situation (e.g., a possible overload current and/or overvoltage that could appear on weak lateral at the time of minimum load due to the simultaneous high generation). Instead, generators characterized by an almost constant production during the day may be able to better compensate the power flows in the main feeders, where the compensation effect of the several loads supplied reduces the variability of the equivalent daily curve. To perform these evaluations, the daily load (generation) curves have been discretized into 24 intervals (each one of the duration of 1 h). The implemented probabilistic load flow (PLF) takes into account the probability density function (*pdf*) of loads and generators. Moreover, in order to guarantee the correctness of the results, it is important to consider the existing correlation between DG units, between generators and loads and also between loads. Once calculated the current flowing in each branch and the voltage of each node with their *pdf*, it is possible to choose the correct size of each conductor and to verify all the technical constraints, taking into account the uncertainties related to these electrical variables. In particular, the voltage magnitude at each node must lie within the upper and lower permissible voltage limits; the current magnitude at each branch must lie within the upper acceptable level. Additional constraints can be added to consider also the admissible short circuit current in the networks as well as the limitations related to maximum and minimum DG production.

11.3.2 Network Upgrading

The OF is related to the capital expenditures (CAPEX) in distribution systems is necessary to take into account investments that can be needed to upgrade a distribution network to face the natural growth of energy demand, and to satisfy the connection requests from new loads or power producers according to a smart distribution scenario.

The costs for network upgrading, C_U, can be easily assessed by using (11.2):

$$C_U = \sum_{j=1}^{N_b} C_{0j} = \sum_{j=1}^{N_b} \left(B_{0j} + M_{0j} - R_{0j}\right) \tag{11.2}$$

where N_b is the number of network branches, C_{0j} is the present cost of the jth branch, and B_{0j}, M_{0j}, and R_{0j} are respectively its building, management, and residual costs transferred to the cash value at the beginning of the planning period by using economical expressions based on the inflation rate, the interest rate, and the load growth rate (all of them constant).

11.3.3 Energy Losses

The OF described in this section is the main term of the operational expenditures (OPEX) in distribution systems and attempts to minimize the energy losses arising from the N_b branches.

The customer's demand evolution has been modeled as a piecewise linear curve. Due to this statement, it is acceptable for planning studies to assume that also the branch current grows linearly, making the assessment of the Joule energy losses (E_L) easy through the expression in (11.3) (in kWh/year).

$$E_L = \frac{8760}{1000} \sum_{j=1}^{N_b} 3 \cdot r_j \cdot L_j \left(\int_0^N I_{jk}^2 dy \right) = 26,28 \cdot N \sum_{j=1}^{N_b} r_j \cdot L_j \left(I_{fj}^2 + I_{0fj}^2 + I_{fj} \cdot I_{0fj} \right) \tag{11.3}$$

where I_{0fj} and I_{fj} are respectively the currents of the jth branch at the beginning and at the end of the period, N is the period duration in years,

r_j and L_j are respectively the conductor's resistance per kilometer and the length (in km) of the jth branch.

The cost of power losses can be obtained considering a unitary cost of losses ($/kWh); then, this cost has to be actualized in order to obtain its net present value (NPV).

11.3.4 Continuity of Supply

The continuity of supply is related to the duration of a branch fault that is usually divided into two phases: fault location and fault repair [6]. Reclosers can restrict the area of influence of a fault, reducing the number of customers affected by long-term interruptions during the fault location phase. In this stage, intentional islanding may be used to supply unfaulted portions of the network automatically separated from the faulted section. The repair stage consists of the time required to isolate the faulted branch, connect any emergency ties, and repair the fault. DG, enabling power to be restored to the nodes downstream the sectionalized branch, can lead to significant reliability improvements [7]. Load flow studies should be performed to check that voltages and currents are within their operative ranges and that DG units have a sufficient probability to pick up the loads in the islanded network.

With a reference to the different continuity of supply OFs, the following terms can be adopted:

- wnergy not supplied,
- cost of energy not supplied,
- SAIDI reliability index,
- SAIFI reliability index.

The energy not supplied E_{NS} can be evaluated by means of (11.4).

$$E_{NS} = \sum_{j=1}^{N_b} \lambda_j \cdot L_j \cdot \left(\sum_{i=1}^{N_{loc,j}} P_{0i} \cdot t_{loc} + \sum_{i=1}^{N_{rep,j}} P_{0i} \cdot t_{rep} \right) \tag{11.4}$$

where λ_j is the branch fault rate (number of faults per year and km of feeder), L_j is the branch length (km), N_{loc} and N_{rep} are the number of nodes isolated during the fault location and repair stages respectively, P_{0i} is the node power (kW) at the beginning of the period, and t_{loc} and t_{rep} are the durations of the fault location and repair stages (h) respectively. N_{rep} depends on the presence of emergency ties, whereas N_{loc} and t_{loc} decreases with the number of reclosers, because the fault location becomes easier and faster. The optimal number and position of reclosers is determined with the algorithm in [7].

The cost of energy not supplied can be obtained considering a unitary cost ($/kWh); then, this cost has to be actualized in order to obtain its net present value (NPV).

An alternative way to represent the continuity of supply OF is the use of some reliability indexes.

The reliability indexes for distribution networks are frequently quantified by means of two different objectives:

- the system average interruption frequency index (SAIFI);
- the system average interruption duration index (SAIDI).

The indexes can be calculated with the expressions (11.5) and (11.6), respectively. In the same equations, fr_i is the bus failure rate, NC_i is the number of customers in the ith bus, U_i is the annual outage for customers in the ith bus, whereas n is the overall number of busses of the network.

$$SAIFI = \frac{\sum_{i=1}^{n} fr_i \cdot NC_i}{\sum_{i=1}^{n} NC_i} \tag{11.5}$$

$$SAIDI = \frac{\sum_{i=1}^{n} U_i \cdot NC_i}{\sum_{i=1}^{n} NC_i} \tag{11.6}$$

11.3.5 Cost of Energy

In a given distribution network, the cost for feeding the loads of the network can be subdivided considering the purchasing of energy from the transmission grid, $(CkWh)_G$, and the one produced by DG, $(CkWh)_{DG}$.

In a competitive electricity market, different retail sale rates of the energy produced by a DG unit, that depends on the technology adopted (mini gas turbine, CHP, wind turbine, etc.), and of the energy fed by transmission system have to be considered. By assuming a constant power demand growth rate and by calculating the amount of energy generated per year by each generator on the basis of its power production probability density function, it is possible to assess the value of the energy that is necessary to buy from both the transmission system and for paying the energy of the DG installed in the network during the study period. By resorting to an average value of the energy rate in the planning period, it is easy to calculate the terms $(CkWh)_G$ and $(CkWh)_{DG}$, opportunely transferred to the cash value at the beginning of the planning period, so that they can be comparable with the other OF costs. The authors are aware that calculating the cost of energy from transmission and DG is not an easy task, and that many cost terms cannot be exactly known. It should be noticed that more complex cost models can be easily introduced without modifying the general approach proposed.

11.3.6 Voltage Profile

Voltage profile regulation is one of the major concerns related to the use of DG in smart distribution networks, given that in numerous situations it constitutes a barrier to the development of new renewable generation in distribution systems. It is well known that, to completely overcome all the problems related to the voltage regulation in distribution systems, every generator must participate to the regulation but, even maintaining the actual situation (e.g., does not contribute to voltage regulation), significant improvements may be obtained by placing generation units of suitable size in the right locations.

A possible index that can be considered as representative of the voltage profile, OF_{vp}, can be considered as in (11.7), where V_i represents the voltage phasor in the ith node, V_r the rated voltage, and n is the overall number of busses of the network.

$$OF_{vp} = \frac{\sum_{i=1}^{n} ||V_i| - V_r|}{n} \tag{11.7}$$

11.3.7 Environmental Impact of DG

Currently, there is a strong effort of the European governments to cut greenhouse gas emissions, particularly those caused by the electrical energy generation processes. It is known that DG technologies are often characterized by a high-energy efficiency (e.g., cogeneration plants) or by a low environmental impact (renewable sources). Thus, DG can be used to produce power in a cleaner way in respect to the traditional generation plants, without forgetting that the DG presence itself can improve the energy efficiency of the distribution networks by reducing Joule losses. For this reason, it is useful to introduce in the planning process new goals that take into account the different environmental impact of the DG units connected to the distribution system.

In the proposed approach, a first step in this direction has been taken by distinguishing the generation typologies depending on their specific

CO_2 emissions per kWh produced. Obviously, this approach advantages the renewable sources, like wind and photovoltaic, which could be assumed theoretically with zero CO_2 emissions. Actually, the emission due to the building and disposal processes of those plants should be considered; however, also in this case the greenhouse gas emissions are particularly lower compared with those typical of the other technologies. On the contrary, generally the renewable sources may introduce minor benefits for the network planning and operation. In fact, on equal sizes, a wind turbine shows a lower number of working hours per year (and consequently an annual average energy generated); therefore, its influence on the power flow reduction will be lower. Moreover, the presence of land constraints (due to land availability, host country laws and regulations, social issues, etc.) can reduce the number of available sites where it is possible to allocate specific DG technologies, limiting their benefits to the network.

11.4 CLASSICAL MO OPTIMIZATION METHODS: ε-CONSTRAINED METHOD AND WEIGHTED-SUM APPROACH

Generally speaking, classical MO optimization methods converting the MO optimization problem to a single-objective optimization problem by emphasizing one particular Pareto-optimal solution at a time. When such a method is to be used for finding multiple solutions, it has to be applied many times, hopefully finding a different solution at each simulation run. In order to solve an MO optimization problem, it is fundamental to find the non-inferior solutions (Pareto-optimal solutions or Pareto frontier) and finally choose a solution from this set.

Generating the Pareto set can be computationally expensive and often infeasible, because the complexity of the underlying application prevents exact methods from being easily applicable. For this reason, a number of stochastic search strategies have been developed. They usually do not guarantee to identify optimal trade-offs but try to find a good approximation, i.e., a set of solutions with objective vectors that are (hopefully) not too far away from the optimal objective vectors.

Thus, a MO analysis provides very useful information not only by finding single particular solutions that are non-dominated, but also by deciphering the shape, extension, and correlation of the trade-offs between objectives.

A typical approach for MO constrained optimization is the ε-constrained method [3], that is an algorithm transformation technique, which can convert algorithms for unconstrained problems to algorithms for constrained problems, using the ε level evaluation, which compares search points based on the pair of objective values and constraint violation of them. In particular the ε-constrained technique selects a particular Nth objective as the primary objective function $f_N(x)$ and makes the other objective functions $f_i(x)$ become constraints, with the preference index ε_i, of a single optimization problem. The resulting problem can be stated as in (11.8).

$$\begin{cases} \min f_N(x) & \\ x \in \Omega & \\ c_j(x) = 0 & j = 1 \ldots n \\ h_k(x) \leq 0 & k = 1 \ldots p \\ f_i(x) \leq \varepsilon_i & i = 1 \ldots m \text{ and } i \neq N \end{cases} \tag{11.8}$$

where ε_i represents the maximum permissible levels of the ith objective.

In general, ε_i can be defined as follows: $\varepsilon_i = \varepsilon_i^* + \Delta\varepsilon_i$, where ε_i^* is a global non-inferior value of the ith objective and $\Delta\varepsilon_i$ denotes the trade-off preference assigned by the decision maker. The trade-off preferences can be seen as the compromised values between the conflicting objectives. The decision maker on the basis of experience and/or system operation policies can establish them.

The block diagram of the described MO optimization algorithm is depicted in Fig. 11.2.

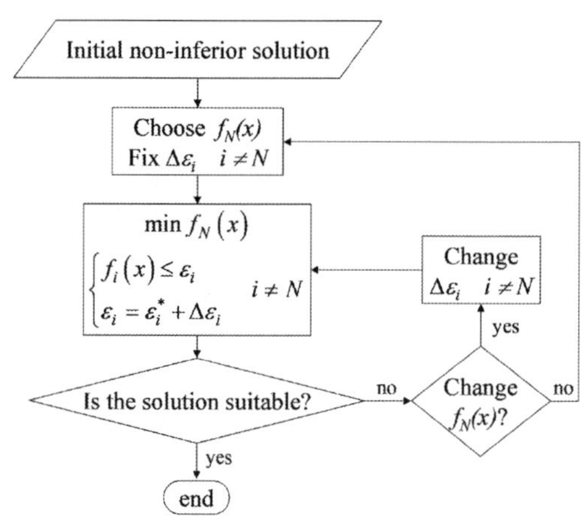

Figure 11.2 Block diagram of the ε-constrained optimization algorithm.

The starting configuration is achieved by minimizing the weighted sum of all the objective functions.

Afterwards, the iterative application of the ε-constrained technique allows generating the Pareto set, which satisfies the decision maker's requirements. At the beginning of the iterative MO procedure, the planner defines the master objective function, f_N, whereas the other slave functions, f_i, are regarded as constraints. Furthermore, the decision maker also has to choose acceptable trade-off variation levels, $\Delta\varepsilon_i$, for each single objective function f_i. Once the setting of these parameters is completed, the procedure can automatically generate a non-dominated solution.

Deep knowledge of the problem is required to define adequate master objectives and constraint levels or aggregation method and weights, respectively. These procedures can be very useful to find single solutions when information is known as a priori. On the other hand, several solutions of the Pareto set can be found by changing the aggregation function or the master objective iteratively.

The weighting method is an alternative way for searching the optimum of an OF as the sum of the different OFs considered in the MO problem, and the set of solutions can easily be obtained by applying different values assigned to the weights [3]. The mathematical formulation of the optimization with the weighing method can be expressed as in (11.9).

$$\begin{cases} \min \sum_{i=1}^{N} w_i \cdot \quad f_i(x) \\ x \in \Omega \\ c_j(x) = 0 \qquad j = 1 \ldots n \\ h_k(x) \leq 0 \qquad k = 1 \ldots p \end{cases} \tag{11.9}$$

where w_i are the weights associated with each of the f_i (x) $(I = 1, 2, \ldots, N)$ OFs.

Since frequently the OFs can be expressed with different measurement units, weights can be used to make comparable quantities, as well as, of course, to define the relative importance of each OF. In the ideal case, each global minimum of scalar OF defined by the weighted sum with its particular set of weights represents a point of the Pareto front. An easy way to determine Pareto front of a problem is to minimize various combinations of sub-objectives, systematically varying the weight, and solving the optimization problem several times.

These methods have their limitations: the weighted-sum method may require a long operating time with a large number of objectives and the solutions found strongly depend on the shape of the Pareto frontier. Similarly, the ε-constrained method requires strong a priori knowledge of the problem and it is not suitable for a large number of objectives. However, this can prove to be very time-consuming and the solutions depend on the shape of the Pareto frontier and the aggregation method.

11.5 MO EVOLUTIONARY ALGORITHMS: NSGA-II

The complexity of the future distribution system, with multiple players sharing the responsibility for the proper operation, suggests the use of "true" MO algorithms that produce a set of Pareto optimal solutions without the use of subjective weights.

These algorithms fall in the group of multi-objective optimization methods based on MOEA [8,9]. MOEA manage sets of possible solutions simultaneously, and permit identification of several solutions of the Pareto front at once.

During the past twenty years a large number of multi-objective evolutionary algorithms (MOEA) has been developed. The main classification of these algorithms is in first generation and second-generation MOEA. The second generation of MOEA is characterized by the use of elitism. At present, two of the most recognized algorithms of the second generation are the NSGA-II [8,9], and the strength Pareto evolutionary algorithm 2 (SPEA2) [10].

These algorithms allow finding an accurate, diverse, and well-spread Pareto front and they guarantee to produce useful information for the subsequent decision-making process. Even though with specific formulation and modifications, many authors have proposed MO approaches for the smart distribution networks optimization problem. The existing approaches have been analyzed in detail and their limitations from both a theoretical and empirical standpoint have been described. As a conclusion, the recent NSGA-II demonstrates the capacity to generate a rich set of trade-offs between the examined objectives, does not require a priori preference articulation and develops concave portions of the Pareto approximate front [11]. Recent MO frameworks presented the integration of stochastic and controllable DER in the distribution grid [12], the inherent time-varying behavior of demand and distributed generation (particularly when renewable sources are used), the fact that load models can significantly affect the optimal location and sizing of DER in distribution systems [13], and strategies to achieve an integration of DG units in LV and MV distribution grids while optimizing several relevant objectives [14].

11.5.1 NSGA-II description

The key point of the NSGA-II algorithm is represented by the classification procedure of the individuals of a generic population. This process is based on the concept of Pareto dominance. If coordinates of a vector $x \in \mathfrak{R}_n$ measure negative attributes (loss, cost, quantities of "bad ones," etc.), x Pareto dominates a vector $y \in \mathfrak{R}_n$ ($x \prec y$) if $x_i \leq y_i$ for all coordinates i, with strict inequality for at least one coordinate. If an alternative x is not Pareto dominated in a given set of alternatives, it is Pareto optimal. Therefore, the Pareto optimal set (front) of individuals in a given population is represented by those solutions that are non-dominated by any other solution in the same population or, in different words, that cannot be improved in any OF without deteriorating some of the other OFs considered.

The NSGA-II algorithm sorts a population into different non-domination levels (fronts). Initially, it finds the Pareto optimal set of the current population ($FRONT = 1$), then it discounts temporarily these solutions and searches again the Pareto optimal set among the remaining individuals of the population ($FRONT = 2$). This process is repeated until all fronts are identified and associated to all individuals. This attribute (the non-domination rank) is one of the two elements that characterizes the fitness of the solutions.

The second attribute is related to the need to preserve diversity into the Pareto optimal front. The original NSGA used the well-known sharing function approach, based on a sharing parameter, which sets the extent of sharing desired in a problem. This parameter is related to the distance metric chosen to calculate the proximity measure between two population members. The parameter denotes the largest value of that distance metric within which any two solutions share each other's fitness. The main drawback of this methodology is that its performance depends on the value of this parameter, usually sets by the user.

To overcome this problem, the c that does not require any user-defined parameter for maintaining diversity among population members. It estimates the density of solutions surrounding a particular solution in each non-dominated set, by calculating for each OF the average normalized distance of the two nearest neighbors (crowding distance) and summing all these single contributions, as indicated in (11.10).

$$CD(i) = \sum_{j=1}^{N_{OF}} \frac{OF_j(i+1) - OF_j(i-1)}{OF_j^{\max} - OF_j^{\min}} \tag{11.10}$$

where N_{OF} is the number of OF considered, and $OF_j^{\min}$ and $OF_j^{\max}$ are the minimum and maximum values of the jth objective function. Obviously, the crowding distance computation requires sorting every non-dominated set according to each objective function value in ascending order of magnitude. For each objective function, the boundary solutions (solutions with smallest and largest function values) are assigned an infinite distance value. By doing so, it is now possible to compare two solutions in a specific non-dominated front, and to recognize the more crowded solution as the one that has the smaller crowding distance value. Having characterized each individual of the population with these two attributes, it is possible to define a crowded-comparison operator ($\prec_n$) that guides the selection process of the NSGA-II, as in (11.11).

$$(i \prec j) \text{ if} \begin{cases} (\text{RANK}_i < \text{RANK}_j) \\ \text{or} \\ (\text{RANK}_i = \text{RANK}_j) e (CD_i > CD_j) \end{cases} \tag{11.11}$$

That is, between two solutions with differing non-domination ranks the solution with the lower (better) rank is preferable. Otherwise, if both solutions belong to the same front, then the solution that is located in a lesser-crowded region is preferable (Fig. 11.3).

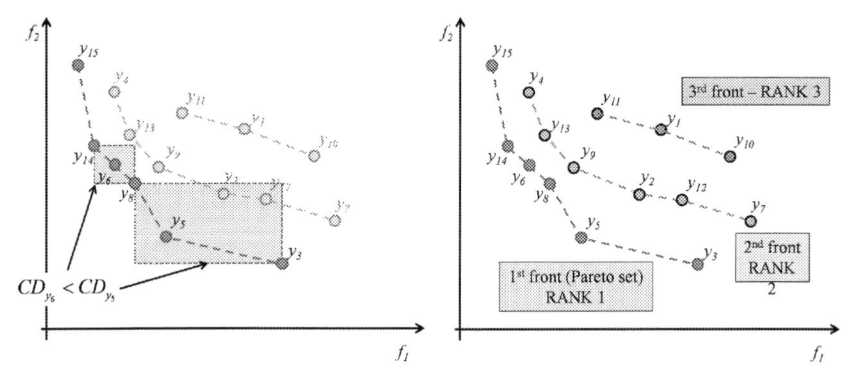

Figure 11.3 NSGA-II crowded-comparison approach.

Fig. 11.4 showsthe flow chart of the MO evolutionary algorithm adopted. It starts by randomly generating an initial parent population, $\mathbf{G_0}$, of N_I individuals and evaluating the OFs for each of them. The population is sorted based on the non-domination. After that, binary tournament selection, crossover, and mutation operators are applied to create the first offspring population, $\mathbf{O_1}$, of size N_I.

After this initial phase, the optimization procedure continues in a slightly different manner, in order to introduce elitism. First, a combined population of parents and offspring is formed ($\boldsymbol{C_{gen}} = \boldsymbol{G_{gen-1}} \cup \boldsymbol{O_{gen}}$), having size $2N_I$. Then, the population $\boldsymbol{C_{gen}}$ is sorted according to non-domination. Since all previous and current population members are included in $\boldsymbol{C_{gen}}$, elitism is ensured. The new population of parents, $\boldsymbol{G_{gen}}$, is formed by choosing primarily the best non-dominated front, P_1, of the

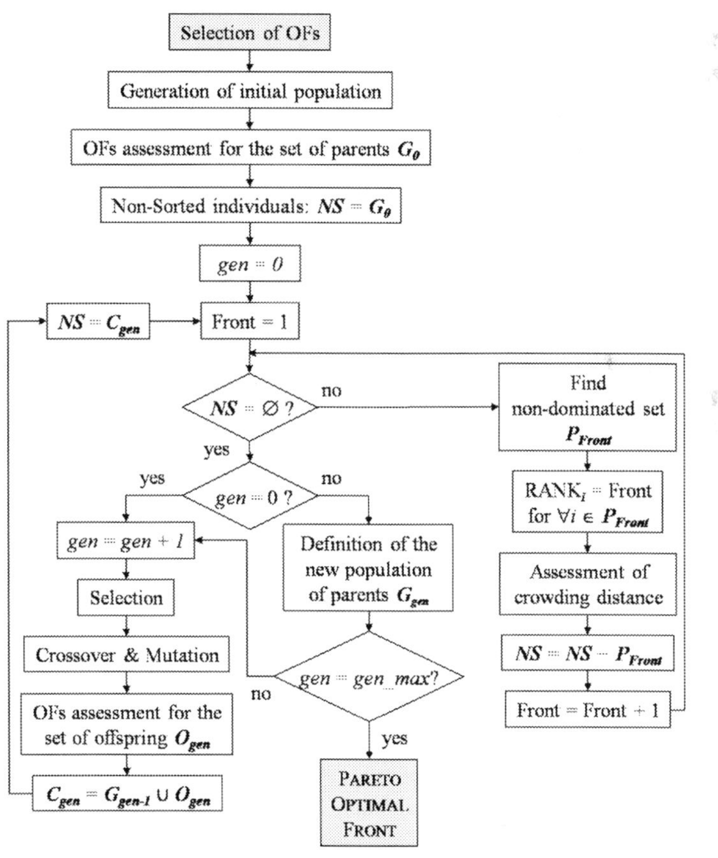

Figure 11.4 Flow chart of the MO evolutionary algorithm implemented.

combined population, and then by adding the subsequent fronts in ascending order (P_2, P_3, etc.) until no more sets can be accommodated. Usually, the last front chosen cannot be included integrally in the new population. Therefore, to exactly choose N_I population members, the solutions of the last front are sorted using the crowded-comparison operator in descending order and the best solutions needed to fill all population slots are chosen. The new population $\boldsymbol{G_{gen}}$ of size N_I is now used for selection, crossover, and mutation to create a new population $\boldsymbol{O_{gen+1}}$ of size N_I. It is important to notice that a binary tournament selection operator has been implemented but the selection criterion is based on the crowded-comparison operator.

Any power system optimization problem is highly constrained. Many technical, economic, and regulatory constraints exist that limit the region of feasible solutions. Among them it can be cited: thermal capacity of conductors and transformers, maximum overvoltage and voltage drop in the network nodes different for normal and emergency operation conditions, maximum short circuit current, maximum number and duration of long interruptions suffered by customers, land constraints for generation allocation, maximum amount of DG that can be accepted on a specific distribution network, etc.

For constrained problems, the efficiency of any EAs is linked with a good management of the infeasible solutions. In fact, simply disregard those solutions that can limit the ability of the algorithm to explore the search region and to find optimal solutions.

The constraint-handling method adopted in this chapter makes use of the binary tournament selection, in which two solutions are picked from the population, and the better one is chosen. In the presence of constraints, each solution can be either feasible or infeasible. Thus, there may be at most three situations:

1. both solutions are feasible,
2. one is feasible and other is not,
3. both are infeasible.

In case 1, the crowded-comparison operator is applied. In case 2, the feasible solution is selected. Finally, in case 3, it is arbitrarily assigned to both solutions, an equal probability to be selected. The effective implementation of this procedure is obtained by including all the infeasible solutions in an additional last front (the worst). Consequently, any feasible solution always has a better non-domination rank than any infeasible solution, while all feasible solutions are ranked according to their non-domination level and their crowding distance values.

11.5.2 Advanced Multi-objective Methods and Decision-making Techniques

MO programming is a powerful tool, but one of its strengths could also be interpreted as a weakness: providing more than one solution, it leaves the final choice open to the subjectivity of the planner, rather than to objectivity and transparency.

For this reason, the MO technique has to be preferably combined with the application of decision making techniques that assign a fitness value to each planning alternative in the Pareto set by assessing the overall expected goodness or the risk of its implementation. These tools can be based on the probability choice method that minimizes the expected costs of each alternative in all envisaged scenarios, or on the risk analysis [15,16] that identifies the preferred solution as the one that minimizes the regret felt by a decision maker after verifying that the decision made was not optimal, given the futures that in fact have occurred. A third approach has also been proposed in the literature that, resorting to the stability area's concept, combines the two aforementioned ones and helps to recognize the best solution when it is difficult to assign a probability of occurrence to each scenario [17].

11.6 SMART DISTRIBUTION SYSTEM OPTIMIZATION EXAMPLE: THE OPTIMAL DG SIZING AND SITING PROBLEM

11.6.1 Problem Description

The optimal planning of DG on smart distribution networks is a classical MO problem in smart distribution optimization.

The DG placement during an evolution of an existing distribution network has simultaneously the potentiality of achieving great savings (e.g., deferment of investments and reduction of power losses) or causing technical problems (e.g., overvoltages and/or overloads), depending on the DG size and location.

In the current power distribution scenario DSOs need to improve the distribution planning process so that the potential benefits of DG can be exalted, maintaining the risks within an acceptable level. The DSOs cannot completely decide where DG will appear and the real situation will be surely quite different from the optimal one, by vanishing many of the efforts made by the planner. The application of MO programming allows choosing in a set of non-inferior solutions the one that better fits planner's subjective point of view.

The authors have proposed a MO version of an optimization tool, based on the ε-constrained method, but they had to recognize that the need to make some a priori decisions was a drawback of the procedure since it was in some cases too much dependent on the planning engineering subjective point of view [5]. To overcome these problems, a MO approach, based on the NSGA-II, has been adopted to solve the optimal allocation problem of DG by the same authors [18,19].

The presented optimization procedure permits the definition of the optimal size and allocation of DG in a given network as a compromise among different non-inferior solutions produced by the application of the MO methodology. This way of proceeding may be very useful because it allows the planner to be aided by a software tool, able to take into account quickly and precisely all possible combinations in a real size distribution network scenario, leaving the decision maker the control on the process to make the final decision.

In the proposed approaches, candidate solutions to the optimization problem play the role of individuals in a population, and the fitness function determines the quality of the solutions. Evolution of the population then takes place after the repeated application of the above operators. In particular, a genetic algorithm (GA) is implemented. GA is one of the most popular types of evolutionary algorithm for seeking the solution of a problem in the form of strings of binary numbers, or more complex coding, that reflect something about the problem being solved, in this case size and position on a network of DGs, by applying operators such as recombination and mutation.

The implemented evolutionary algorithm starts by randomly generating an initial population of possible solutions. For each solution a value of DG penetration is chosen between zero and a maximum limit, fixed by the planner on the ground of economic and technical justifications. A number of DG units of different sizes are then randomly chosen until the total amount of power installed reaches the DG penetration level assigned. At this point, the DG units are randomly located among the network nodes and the objective function is evaluated verifying all the technical constraints; if one of them is violated, the solution is penalized. Regarding the population size, the best results have been found assuming it is equal to the number of network nodes.

Once the initial population is formed, the genetic operators are repeatedly applied in order to produce the new solutions. In particular, a classical "remainder stochastic sampling without replacement" scheme has been adopted for the selection operator [20], and a "uniform crossover" has been chosen by which each vector's element is swapped with probability 0.5. For the mutation operator, all the vector elements are mutated, with a small mutation probability, choosing a different value in the defined alphabet. If one technical constraint is violated or the total amount of DG exceeds the maximum level of DG penetration, the new solution is penalized.

Finally, according to the GA "Steady State" typology, the new population is formed comparing old and new solutions and choosing the best among them. The algorithm stops when the maximum number of generations is reached or when the difference between the objective function value of the best and the worst individuals becomes smaller than a specified value.

In the next section, after the definition of solution coding in the GA algorithm IS adopted, two application examples of the approaches proposed by the authors will be presented.

11.6.2 Solution Coding

The first important aspect of a correct GA implementation is the coding of the potential solution.

Referring to the DG allocation problem, an alternative is a particular set of generators, of different type and size, allocated in some nodes of a given distribution network, while the coordinates are the value of the OFs assessed for this DG allocation.

If the network structure is fixed, all the branches between nodes are known, and the evaluation of the objective functions depends only on size, type, and location of DG units. For this reason each solution can be coded by using a vector, whose size is equal to the number of MV/LV nodes, in which each element contains the information on the presence of a DG unit. A binary coding would be sufficient to solve the DG siting problem (one for presence and zeroe for absence of DG unit), but not to choose the DG size and type. Therefore, it is necessary to define a wider alphabet of integers: fixed the number T of generator types and the number S of generator sizes, different for each DG typology considered, each element of the vector solution can assume any integer included between

0 and $\sum_{i=1}^{T} S_i$. By arbitrarily ordering the generator types, the following definition of the alphabet can be assumed:

0	no DG located in the node;
$1\ldots S_1$	size indexes of the 1st DG type
$S_1 + 1\ldots S_2$	size indexes of the 2nd DG type
$\vdots$	$\vdots$
$S_{T-1} + 1\ldots S_T$	size indexes of the Tth DG type

11.6.3 ε-Constrained Method—Application Example

In order to show the capability of the MO methodology to solve the problem of the optimal DG allocation, a ε-constrained method adopted by the authors to permit the planner to decide the best compromise between cost of network upgrading, cost of energy losses, cost of energy not supplied, and cost of energy required by the served customers is presented [5]. The definition of these OFs has been discussed in Section 3. Such objectives should be met subject to the network power flow equations as well as to the limits on the bus voltages, steady state current, and short circuit currents.

The methodology has been applied to a small portion of a distribution network, constituted by 142 MV/LV nodes and 2 primary substations has been considered as illustrated in Fig. 11.5.

The period taken into consideration for the planning study is twenty-years long, with all nodes existing at the beginning of the period. In the proposed case test, the network will be developed during the considered time horizon. This assumption will not result in the loss of generality, and shorter planning horizon can be considered in the strategic planning procedure.

The annual average active power delivered to the MV nodes is, at the beginning of this period, about 5 MW. For each MV/LV node a constant power demand growth rate of 3% per year has been assumed (this assumption has been made for the sake of clarity, but there are no restrictions to define a power demand growth rate differentiated for each node and for each sub-period). The majority of the branches are of the overhead type, but some buried cables exist.

The optimization algorithm can choose different sizes of DG generators within a discrete number of prefixed sizes. In the proposed application, generator sizes of 200-400-600 kW have been adopted. The price

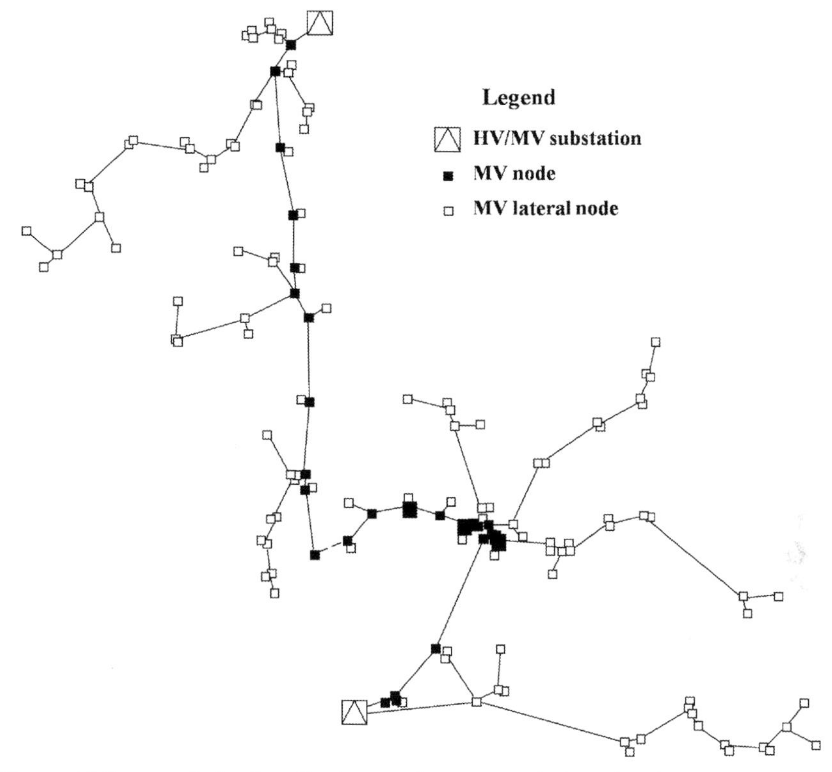

Figure 11.5 ε-constrained method example ––case study network.

of the energy purchased from the wholesale electricity market has been assumed equal to 40 $/MWh, whereas the price of the energy supplied by DG has been considered equal to 45 $/MWh; this hypothesis is valid under the consideration the DG is expected to reach the grid parity condition. In the presence of a liberalized electricity market, different retail sales rate of the energy produced by a DG unit should be considered depending on the technology adopted (mini gas turbine, CHP, wind turbine, etc.), the regulatory actions and the willingness to harness renewables.

In this application, the master function is optimized with the same GA used for finding the starting solution. Of course, the GA objective function must be modified according to the planner's choice: in this phase, the GA looks for those solutions that improve the master function, and simultaneously complies with constraints remaining within the allowed variation margin of deterioration in one or more objective functions. By doing so, a set of potential non-dominated solutions can be

Table 11.1 Costs progression in MO iterative procedure (case study 1 – master FO: energy losses cost)

	Starting configuration	MO iteration #1	MO iteration #2
Network upgrading cost [M$]	0.9	0.8	0.4
Energy losses cost [M$]	1.1	0.8	0.7
Cost of energy not supplied [M$]	1.2	1.3	1.1
Cost of energy required by the customers [M$]	23.8	24.2	24.9
Total cost [M$]	27.0	27.1	27.1
DG level [%]	0.00	13.64	44.34

achieved with a new GA run. The optimization process can be stopped when the solutions are acceptable or it can be restarted with different values of $\Delta\varepsilon_i$ until a satisfactory result is obtained. In particular, the planner can decide that a network characterized by a lower building and maintenance cost to the detriment of energy losses or not supplied energy costs is more suited to his needs (provided that such increase remains within the prefixed threshold). Obviously, the final decision, about which solution from the Pareto set can be advantageously adopted, relies on the decision maker. This feature of the procedure can be very useful in the present scenario, which requires the planner to consider several alternatives for different uncertain futures.

The application of the GA to minimize the generalized cost of the network has allowed finding the starting configuration. The global cost of the network during the assigned study period is equal to 27.0 M$ (Table 11.1) without any operating DG. Such a high network cost is due to the significant growth rate of the demand, which requires the enforcement of a large number of branches.

The attempt to minimize the global cost has led to a solution with many lines close to their maximum capacity and for this reason the cost of the energy losses counts for a significant percentage of the generalized cost of the network. The use of DG as an electric supply option can reduce both costs.

Very often the planner needs more alternatives to evaluate, and sometimes he can prefer to reduce the cost of losses instead of improving service quality, depending on strategic decisions, regulatory directives regarding the electric service, and budget restrictions.

As showed in the following examples, the proposed MO optimization process permits finding out alternative configurations, characterized by different costs for each single function constituent of the global cost. In each optimization stage the MO algorithm looks for alternative solutions, which improve the master function to the detriment of the slave functions.

Two different cases have been investigated. In the first case study, the energy losses cost has been regarded as the master objective function.

The value assumed by the cost of the energy losses in the initial network configuration is equal to 1.1 M$ (Table 11.1). Two consecutive steps of iteration have been run. In the first iteration, a deterioration margin of the slave functions, necessary to allow the GA a more exhaustively exploration in the space of solutions, has been admitted. By doing so, the new optimized configuration permits reducing the energy losses cost in percentage of the 23% against an increase in the global cost. This losses reduction is obtained with the integration of a DG penetration level of 13.64%. In the second iteration, by admitting a larger variation in the other slave functions, a new DG arrangement has been found by the GA: the total network cost is increased in relation to both the cost of the starting network and the cost after the first iteration. The planner could tolerate this worsening if the main objective is to reduce the energy losses. In the new optimal solution the cost of energy losses decreases from the value of 0.8 M$ to the value of 0.7 M$. The penetration level of DG increases from the value of the 13.64% to the value of 44.34% and consequently the cost of the purchased energy is higher. That is to say that in this example the losses can be reduced by spending more money to buy energy from the DG resources. The optimal position of DG is indicated in Fig. 11.6.

In the second case (Table 11.2), the cost of energy not supplied has been regarded as the master objective function.

In this case the planner aims at reducing the number and the duration of service interruptions by positioning DG in suited locations. In Table 11.3 the data used for reliability calculations are presented: in particular, the unitary cost for the energy not supplied, C_{kWhns}, the ASSD cost and the branches fault rates has been reported.

Even though many standards and almost all the distributors do not generally allow resorting to "intentional islanding" operation, in order to emphasize the effect of DG it has been hypothesized that this practice can take place [7]. Indeed, among the many new features of new active

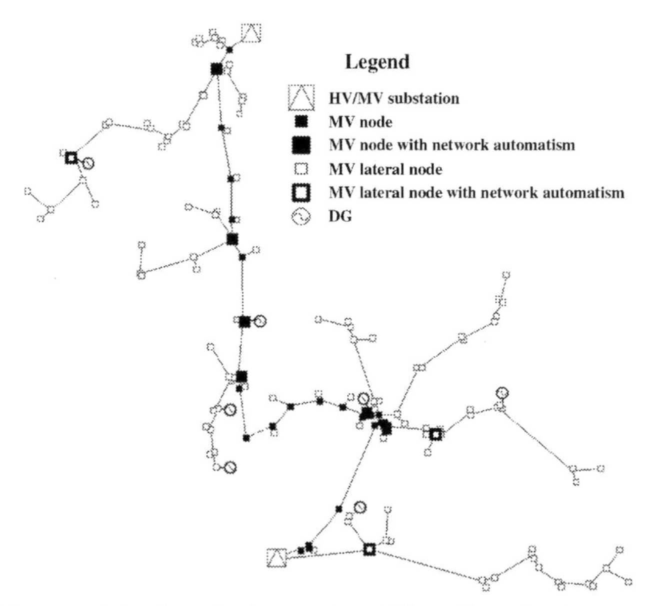

Figure 11.6 ε-constrained method example——DG configuration (case study 1, MO iteration #2).

Table 11.2 Costs progression in MO iterative procedure (dase wtudy 2 – master FO: energy not supplied cost)

	Starting configuration	MO iteration #1	MO iteration #2
Network upgrading cost [M$]	0.9	0.5	0.5
Energy losses cost [M$]	1.1	1.1	1.0
Cost of energy not supplied [M$]	1.2	0.9	0.8
Cost of energy required by the customers [M$]	23.8	24.6	24.9
Total cost [M$]	27.0	27.0	27.2
DG level [%]	0.00	30.70	44.34

Table 11.3 Data used for reliability calculation

C_{kWhns}	**6 $/kWh**	
C_{ASSD}	**20,000 $**	
	Overhead lines	Buried cables
λ	0.15 (year km)$^{-1}$	0.10 (year km)$^{-1}$
t_{loc}	1 h	1 h
t_{rep}	5 h	8 h

networks, the possibility of using DG to supply loads in self-sustaining islands seems to have a real interest for the local distribution companies. In [7] Celli et al. have pointed out that the practice of intentional islanding can be really useful for those nodes that suffer for low service continuity (e.g., nodes that cannot use alternative energy routes during faults in the main feeder), provided that DG could feed the island and network automation is optimally located. In the proposed example, the capability of optimizing the location of DG has been advantageously used to find a network arrangement able to give the customers a much more reliable service avoiding the construction of new emergency ties.

The starting network configuration is equal to the previous case study, in which the energy not supplied cost has the value of 1.2 M$. With the first optimization step this value is reduced to 0.9 M$ (Table 11.2). By relaxing the constraints on the slave functions, a further optimization permits reducing energy not supplied cost up to 0.8 M$ thanks to a new allocation of DG (Fig. 11.3). It is worth noticing that in this case, generators are located at the end of long and heavy loaded lateral edges to serve as back up energy sources during upstream faults. Each area is accepted if the probability that generation exceeds if the load demand is greater than 30%.

Global benefits on energy not supplied energy are clearly recognizable, but benefits are much more significant for those customers that suffer for poor quality due to their position in the network. For example, nodes 1 and 2 in Fig. 11.7, due to the coordination between DG and network automation, drastically reduce their number and duration of long interruptions (Table 11.4).

It can be easily noticed that system indexes cannot capture the importance of intentional islanding for customers seeking premium power contracts. Finally, it is prominent to observe that this amelioration can only be achieved by accepting a major generalized cost of the network and this fact justifies the adoption of MO approaches, that allows the planner to stress some terms of the objective function in order to get some specific results (e.g., reliability improvements under regulatory actions or market pressure).

11.6.4 NSGA-II Approach——Application Example

The NSGA-II has been applied to a small portion of a real distribution network constituted by 37 MV/LV nodes and two primary substations

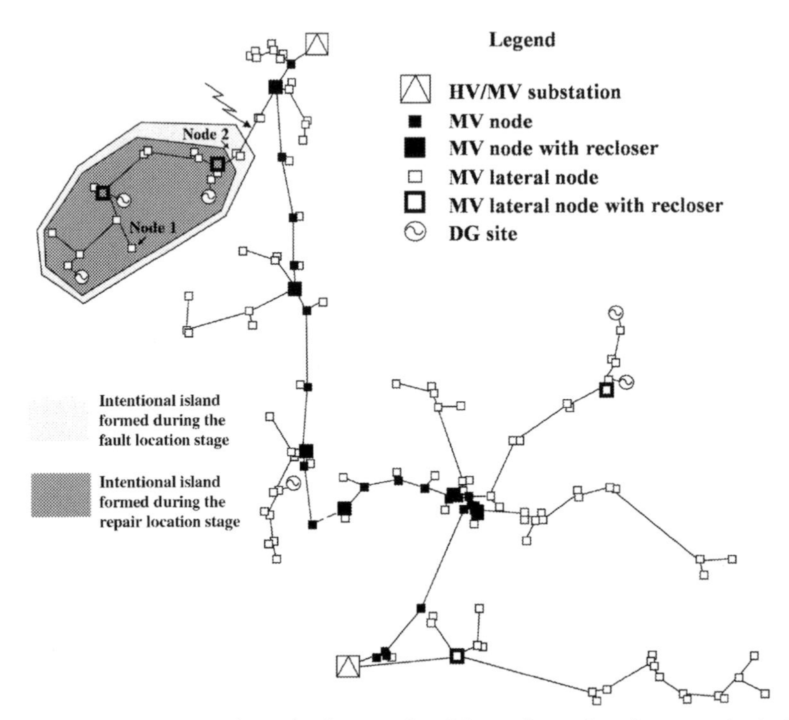

Figure 11.7 ε-constrained method example—DG configuration (case study 2, MO iteration #2).

Table 11.4 Reliability indexes in the network depicted in Fig. 11.7

	With DG and Islanding		With DG and Islanding	
SAIFI	0.66		0.91	
SAIDI	5 h 48'		6 h 32'	
	Node 1	Node 2	Node 1	Node 2
Number of interrup.	0.65	0.37	1.67	1.67
Duration of interrup.	0 h 54'	0 h 21'	6 h 35'	2 h 35'

(Fig. 11.8). An existing overhead open loop feeder between the two substations with four overhead laterals characterizes the network topology.

The period taken into consideration for the planning study is twenty years long. In that case, a test for the network will be developed during the considered time horizon., This assumption will not result in the loss of generality, and shorter planning horizon can be considered in the strategic planning procedure.

Four typologies of loads have been considered (residential, industrial, tertiary, and agricultural), modeled with the daily load curves depicted in

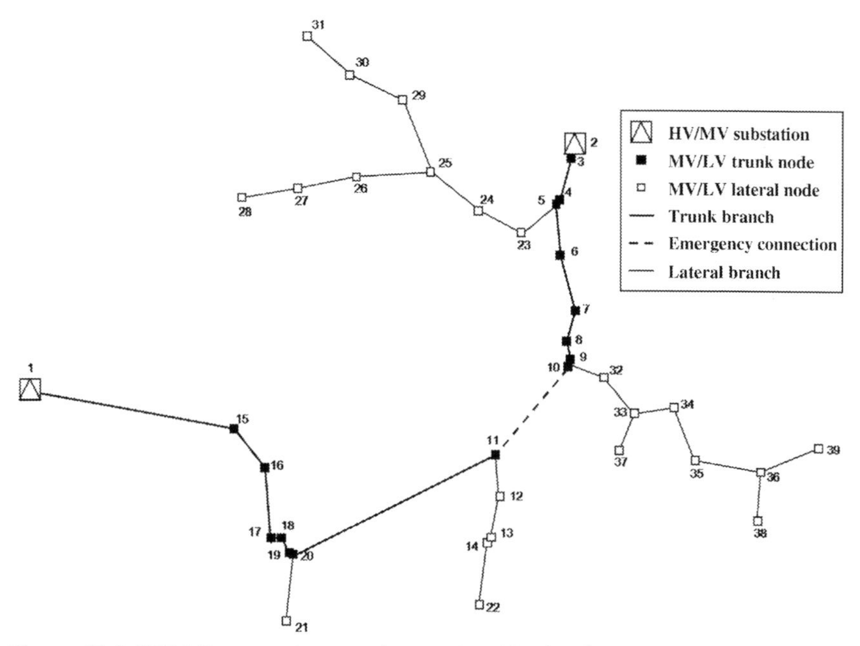

Figure 11.8 NSGA-II approach example—existing MV distribution network.

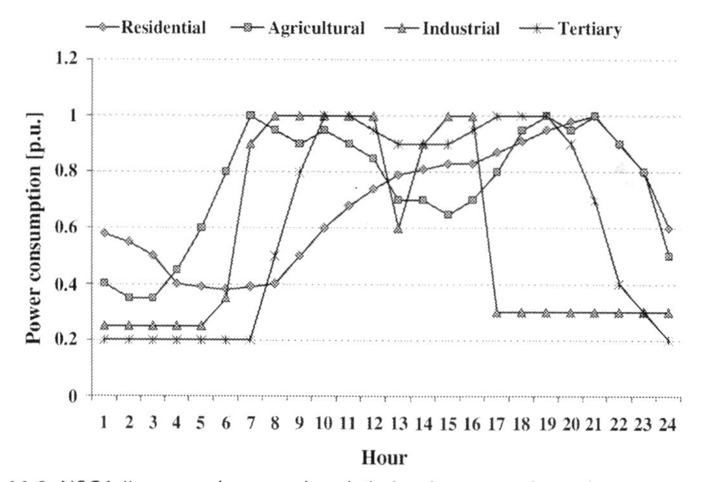

Figure 11.9 NSGA-II approach example—daily load curves adopted.

Fig. 11.9. For each MV/LV node a constant power demand growth rate of 3% per year has been assumed. Also three typologies of generators have been taken into account for a possible allocation in the network: wind turbine (WT), CHP, and gas turbine (GT). Their daily curves are shown in Fig. 11.10.

The generator sizes chosen for the optimization and the alphabet of the solution coding used by the MO evolutionary algorithm is illustrated in Table 11.5.

The specific CO_2 emissions for wind turbine have been supposedly negligible, whereas for CHP and gas turbine they have been assumed respectively equal to 0.35 and 0.60 kg/kWh. In order to simplify the optimization procedure, the specific CO_2 emission has been assumed constant for each generator type independently from its size and its working set point. The value related to the energy absorbed from the transmission grid, 0.53 kg/kWh, is estimated on the basis of the Italian mix of power plants. The optimizations have been carried out selecting two OFs: the Joule energy losses and the CO_2 emissions. Two simulations have been presented: in the first study, no environmental constraints have been considered, while in the second simulation it has been assumed that the allocation of WT is forbidden in the nodes 32÷39 (the lateral supplied from the trunk node 10). For both cases, it has been considered a

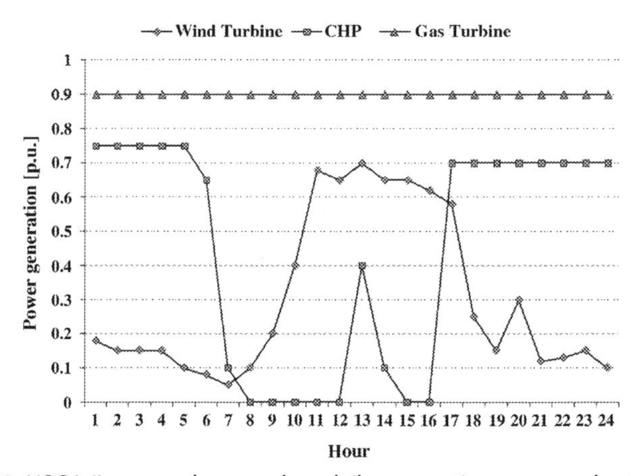

Figure 11.10 NSGA-II approach example— daily generation curves adopted.

Table 11.5 Alphabet used for coding the solution in the NSGA-II approach

code	DG Type	DG Size [kW]
0	No DG allocated	–
1–2	Wind turbine	1000/3000
3–4	CHP	800/1500
5–6	Gas turbine	500/1000

maximum amount of DG allocated equal to 60% of the total network load at the beginning of the study period. The Pareto optimal fronts for both cases are shown in Fig. 11.11; numerical results and DG set for the individuals of Pareto optimal front are reported in Tables 11.6–11.7, in ascending order with respect to losses.

It can be seen that in all solutions the level of WT generation is very high, in order to limit the CO_2 emissions (in fact, the CO_2 emission for WT is assumed equal to zero), while the other types of generation (CHP and GT) allow reducing the network losses; this reduction mainly depends on the chosen generator size (Table 11.5) and the daily generation curves (Fig. 11.10) related to these typologies, in respect to WT. By the generator size point of view, the sizes for no renewable DG are closer to load: this fact reduces the level of currents in the network and, consequently, the Joule losses [21]. Instead, the different daily generation curves allow, in particular hours of the day (e.g., in the first 7 h of day of Fig. 11.10), a good balance between generation and load; moreover, for CHPs, the energy production in the peak hours is equal to zero while the great majority of load have the maximum of demand. In other words, the algorithm is able to find a set of configurations that includes the best solutions with reference to each OF, but also the best compromises among

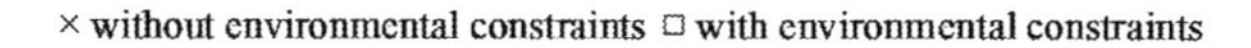

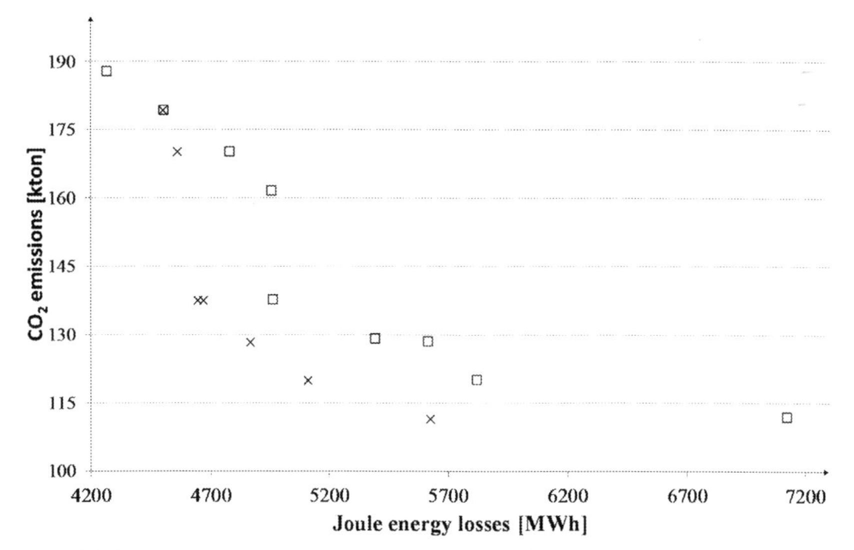

Figure 11.11 NSGA-II approach example—Pareto optimal fronts.

Table 11.6 Pareto optimal front and DG set (without environmental constraint—NSGA-II approach)

N.	Losses [MWh]	CO_2 [kton]	DG Type		
			WT	CHP	GT
1)	4505	179	2 × 1000	–	3 × 500
2)	4564	170	2 × 1000	1 × 800	2 × 500
3)	4646	138	4 × 1000	1 × 800	1 × 500
4)	4669	138	4 × 1000	–	3 × 500
5)	4867	128	4 × 1000	1 × 800	–
6)	5110	119	5 × 1000	1 × 800	–
7)	5626	111	6 × 1000	–	–

Table 11.7 Pareto optimal front and DG set (with environmental constraint—NSGA-II approach)

N.	Losses [MWh]	CO_2 [kton]	DG Type		
			WT	CHP	GT
1)	4269	188	1 × 1000	1 × 800	3 × 500
2)	4507	179	2 × 1000	–	3 × 500
3)	4780	170	2 × 1000	1 × 800	1 × 1000
4)	4959	162	4 × 1000	–	–
5)	4964	138	4 × 1000	1 × 800	1 × 500
6)	5395	129	5 × 1000	–	1 × 500
7)	5617	128	4 × 1000	2 × 800	–
8)	5821	120	5 × 1000	1 × 800	–
9)	7122	112	6 × 1000	–	–

DG technologies with a different amount of CO_2 emissions and different impact on the network energy loss reduction. That is, the middle solutions are the more robust ones in respect to the considered OFs [22].

However, it is significant to notice that every optimal configuration identified reduces both Joule losses and CO_2 emissions in comparison to the no DG allocation case, respectively by 34% and 29% on the average; moreover, the network upgrading cost is about 2,400 k€ for all the configurations, against the 4,950 k€ of the no DG case.

The previous remarks are valid also for the results of the optimization with the land constraints. However, the constrained allocation (is not possible allocate WT in the nodes 32÷39) leads to a set of DG allocations with a high level of losses in comparison to the non-constrained one. In particular, let us consider the last individuals in the two optimizations: the correspondent DG configurations are showed in Figs. 11.12 and 11.13.

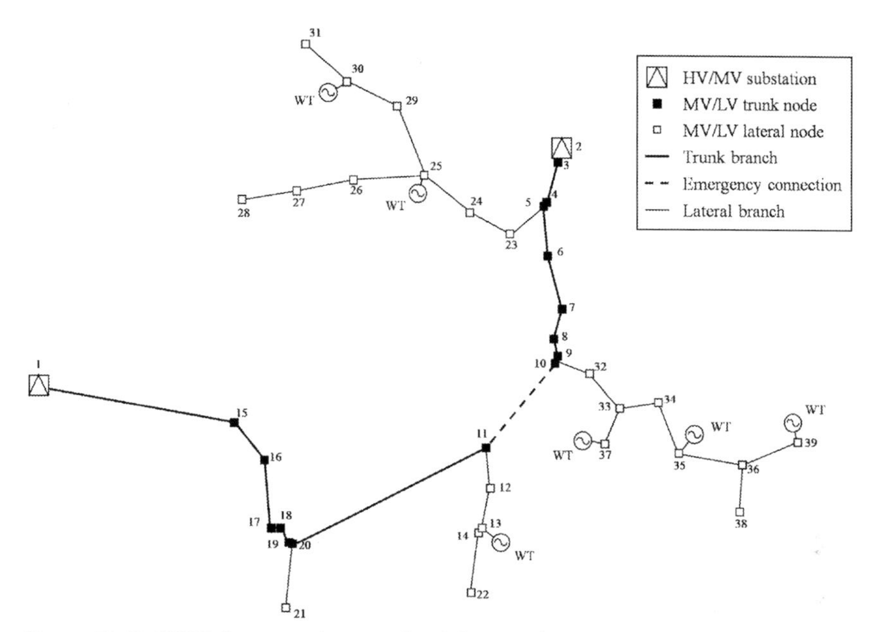

Figure 11.12 NSGA-II approach example—DG set without environmental constraint.

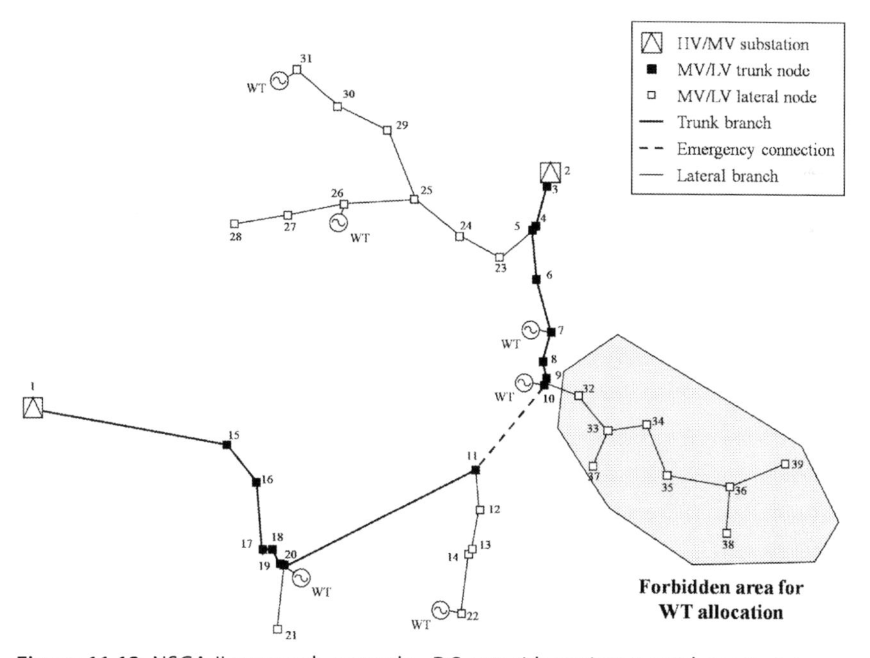

Figure 11.13 NSGA-II approach example—DG set with environmental constraint.

The environmental constraints do not allow the allocation of the WTs in the best point in order to reduce the losses, whereas the emissions are almost equal; the little difference in the CO_2 level depends on the different losses distribution in the network.

11.7 CONCLUSION

Multi-objective programming is recognized to be one of the most effective ways for optimizing transparently and objectively the distribution system evolution toward smart distribution systems, taking into account the multiple needs of different stakeholders. In order to drive the search of the optimal DG allocation it is possible to adopt many objective functions toward a solution not only economically and technically convenient, but also robust in respect to the other goals. It is worthy to note that independently from the MO algorithm used, a post optimization analysis has to be performed with typical decision theory methodologies (as cost-benefit analysis), in order to help the planner in the choice of the preferable alternative among all the solutions provided by the MO optimization. In order to show the potential and usefulness of MO approaches this chapter has explored the main characteristics of this optimization methodology. Two case studies with different optimization objectives have been presented. Suggestions for further investigation on available literature are also provided.

REFERENCES

[1] G. Celli, E. Ghiani, F. Pilo, G.G. Soma, New electricity distribution network planning approaches for integrating renewable, WIREs Ener Environ 2 (2) (2013) 140–157.

[2] A. Osyczka, Multicriteria optimization for engineering design, in: John S. Gero (Ed.), Design optimization, Academic Press, 1985, pp. 193–227.

[3] C.A. Coello Coello, G.B. Lamont, D.A. Van Veldhuizen, Evolutionary Algorithms for Solving Multi-Objective Problems, Springer, US, 2007.

[4] A. Alarcon-Rodriguez, G. Ault, S. Galloway, Multi-objective planning of distributed energy resources: A review of the state-of-the-art, Renew Sust Ener Rev (2010).

[5] G. Celli, E. Ghiani, S. Mocci, F. Pilo, A multiobjective evolutionary algorithm for the sizing and siting of distributed generation, IEEE Trans Power Syst 20 (2) (May 2005) 750–757.

[6] G. Celli, F. Pilo, Optimal sectionalizing switches allocation in distribution networks, IEEE Trans. on Power Del 14 (3) (July 1999) 1167–1172.

[7] F. Pilo, G. Celli, S. Mocci, "Improvement of reliability in active networks with intentional islanding", 2004 IEEE international conference on electric utility deregulation, restructuring and power technologies. Proceedings, 2004, pp. 474–479 Vol. 2.

[8] K. Deb, Multi-objective optimization using evolutionary algorithms, John Wiley and Sons, 2001.

[9] K. Deb, A. Paratap, S. Agarwal, T. Meyarivan, A fast and elitist multi-objective genetic algorithm: NSGA-II, IEEE Trans Evol Comp (2002).

[10] E. Zitzler, M. Laumanns, S. Bleuler, A Tutorial on evolutionary multiobjective optimization, in: X. Gandibleux, M. Sevaux, K. Sörensen, V. T'kindt (Eds.), Metaheuristics for multiobjective optimisation. Lecture Notes in Economics and Mathematical Systems, vol 535, Springer, Berlin, Heidelberg, 2004.

[11] M. Berry, D.J. Cornforth, G. Platt, An introduction to multiobjective optimisation methods for decentralised power planning, Pow Ener Soc Gen Meet (2009).

[12] A. Alarcon-Rodriguez, G.W. Ault, "Multi-objective planning framework for stochastic and controllable distributed Energy resources", IET Renewable Power Generation.

[13] L.F. Ochoa, A. Padilha-Feltrin, G.P. Harrison, Evaluating distributed time-varying generation through a multiobjective index, IEEE Trans Pow Del (2008).

[14] Singh D.D. Singh, K.S. Verma, Multiobjective optimization for DG planning with load models, IEEE Trans Pow Syst (2009).

[15] S. Kannan, S. Baskar, J.D. McCalley, P. Murugan, Application of NSGA-II algorithm to generation expansion planning, IEEE Trans Power Syst (2009).

[16] R. Fang, D.J. Hill, A new strategy for transmission expansion in competitive electricity markets, IEEE Trans Power Syst (2003).

[17] P. Caramia, G. Carpinelli, A. Russo, P. Verde, Decision theory criteria and cable sizing in non sinusoidal conditions, Int J Elec Pow Ener Sys (2001).

[18] G. Celli, S. Mocci, F. Pilo, G.G. Soma, "A Multi-objective Approach for the optimal distributed generation allocation with environmental constraints", in ProcEedings of 10th international conference on probabilistic methods applied to power systems, PMAPS 2008.

[19] G. Celli, F. Pilo, "Optimal distributed generation allocation in MV distribution networks". in Proc. IEEE PICA Conference, Sydney, May 20–24, 2001.

[20] D.E. Goldberg, Genetic algorithms in Search, optimization & machine learning, Addison Wesley, 1989.

[21] G.P. Harrison, A. Piccolo, P. Siano, A.R. Wallace, Exploring the tradeoffs between incentives for distributed generation developers and DNOs, IEEE Trans Power Syst (2007).

[22] G. Carpinelli, G. Celli, S. Mocci, F. Pilo, A. Russo, "Optimization of embedded generation sizing and siting by using a double trade-off method", IEE Proceedings on generation, transmission & distribution, 2005.

CHAPTER 12

Demand Response and Line Limit Effects on Mesh Distribution Network's Pricing

Sayyad Nojavan, Sadjad Sarkhani and Kazem Zare
University of Tabriz, Tabriz, Iran

12.1 INTRODUCTION

The advent of distributed generation resources (DGs) and demand response (Dr) programs in deregulated electricity markets has changed the traditional role of distribution companies from being just an energy buyer/seller to a situation in which they are not restricted only to buying energy from the wholesale market and selling it to their customers, but rather they can participate in the market by optimal utilization of their DGs and Dr programs. Therefore, the benefits of optimal planning and operation of DG units are studied in [1–4].

12.1.1 Literature Review

Since initial development of electricity markets is started from the supply side of the power system, relatively few studies have been made about Discos in comparison with Gencos. Reference [5] proposes an energy acquisition model for Disco in which Disco owns enough DG to meet its customers' energy requirements and has no need to purchase energy from the wholesale market. Interruptible load (IL) option and supplying energy from both DG and wholesale market is not considered in this model. A day-ahead energy acquisition model is proposed for a Disco with DG and IL options in [6]. A method to determine the optimal bidding strategy of a retailer in a short-term electricity market is presented in [7]. The aim is to minimize the cost of energy supply in the sequence of trading opportunities that provide the day ahead and intraday markets. It does not take into account any DG or IL. An hourly-ahead profit model for a Disco are presented in [8,9] considering the high penetration of distributed generation. In this model, Disco is active in both energy and reserve markets

Operation of Distributed Energy Resources in Smart Distribution Networks
DOI: https://doi.org/10.1016/B978-0-12-814891-4.00012-6

for selling the surplus production. A bi-level energy acquisition model for Disco is introduced in [10] in which the upper sub-problem maximizes Discos' revenue and the lower sub-problem minimizes operating costs. To solve the proposed models in [10], they have formulated it as nonlinear complementarity constraints. While, in general, nonlinear constraints is not appropriate for time computing.

In previous works, the bidding problem of Gencos and Discos is formulated as a bi-level optimization problem [e.g. 11–16] and several methods have been utilized to solve them. For example, methods such as penalty interior point algorithm [11], primal-dual interior point algorithm [12], non-interior point algorithm [13], Karush-Kuhn-Tucker (KKT) optimality conditions [14,15], and nonlinear complementary method [16] have been applied on generation side. Also, a method for determining the price bidding strategies of market participants consisting of Gencos and Discos in a day-ahead electricity market is proposed in [17], while taking into consideration the load forecast uncertainty and demand response programs. Fuzzy genetic algorithm is used for finding the best bidding strategies of market players. So, an operational decision making a model for Disco in a competitive environment with DG units and IL options is developed in [18]. This model considers a full ac model of the distribution network and its interconnection with the transmission grid. In the developed model DG units and interruptible loads are located at the distribution level and their impacts on feeder losses, voltage profiles, and power flow of the distribution system are accurately accounted for. Finally, a generic operations framework for a distribution company is proposed in [19] that is a two-stage hierarchical model in which the first deals with Disco's activities in the day-ahead stage and the second deals with Disco's activities in real-time.

12.1.2 Procedure and Contribution

In this paper, a bi-level model is proposed for Disco to obtain optimal energy acquisition strategies for maximizing its profit while supplying its end consumers' energy requirements. It is assumed that Disco has access to three resources to supply its consumers, namely, it has DG; and can interrupt a part of its end consumers' load as a demand response program (interruptible load), and can purchase energy from the wholesale market. In this bi-level model, Disco's profit is maximized in the upper level while system operating cost is minimized and system constraints are satisfied in the lower level. It should be mentioned that linearized constraints have several advantages compared to non-linear ones, such as finding their global optimal solutions is easier, they have shorter run time (especially in large scale problems), and they have more simple implementation. So, strong duality theory was used to solve the proposed bi-level model by converting it into a mixed-integer linear problem. Therefore, the main contribution of this paper is that to use duality theory to convert the proposed bi-level energy acquisition model of Disco into a mixed-integer linear problem, which is solvable using conventional optimization solvers.

With considering the above discuss, the contribution of this paper are as below:

1. to propose a bi-level model for a Disco to obtain optimal energy acquisition strategies;
2. to formulate a mathematical program with equilibrium constraints (MPEC) to derive the optimal energy acquisition strategies in (1) using Karush–Kuhn–Tucker (KKT) conditions;
3. to convert the MPEC in (2) into an equivalent mixed-integer linear constraints using the strong duality theorem.

12.1.3 Paper Organization

The remainder of this paper is organized as follows. Section 2 provides the mathematical formulation of Disco which derives the optimal energy acquisition strategies for maximizing its profit. Section 3 comprises and discusses the obtained results from a case study. Finally, Section 4 provides the relevant conclusions.

12.2 PROBLEM FORMULATION

As mentioned before, we have proposed a bi-level model for Disco to obtain optimal energy acquisition strategies for maximizing its profit in which Disco's profit is maximized in the upper level while system operating costs are minimized and system constraints are satisfied in the lower level.

12.2.1 Upper Level

As mentioned above, Discos' profit is maximized in the upper level considering its own DGs operation constraints. Discos profit is defined as the revenue it collects from energy selling to end customers minus the cost it pays for the same amount of energy.

Assuming that a Disco sells energy to its consumers with a fixed price and does not take advantage from the IL compensation, revenue of Disco from energy selling is calculated as follows:

$$Rd(m) = \sum_{i \in Nd(m)} \sum_{t \in T} \overline{\lambda}(P^0_{d,it} - P_{IL,it}) \tag{12.1}$$

The costs of Disco are related to its DG cost and cost of buying energy from the wholesale market, which can be formulated as follows:

$$Cd(m) = \sum_{i \in Nd(m)} \sum_{t \in T} \left\{ C_{dg,it}(P_{dg,it}) + \lambda_{it}(P^0_{d,it} - P_{IL,it} - P_{dg,it}) \right\} \tag{12.2}$$

where, $C_{dg,it}(P_{dg,it})$ is generation cost of DG and is formulated as follows:

$$C_{dg,it}(P_{dg,it}) = a_{dg,it}P^2_{dg,it} + b_{dg,it}P_{dg,it}, \; i \in Ndg, t \in T \tag{12.3}$$

Disco's profit in the upper level will be as follows:

$$Pd(m) = Rd(m) - Cd(m) = \sum_{i \in Nd(m)} \sum_{t \in T} \left[\overline{\lambda}(P^0_{d,it} - P_{IL,it}) - C_{dg,it}(P_{dg,it}) - \lambda_{it}(P^0_{d,it} - P_{IL,it} - P_{dg,it}) \right] \tag{12.4}$$

Therefore, the upper level of the proposed model will be as follows:

$$\text{Max} : \sum_{m=1}^{N_m} Pd(m) \tag{12.5}$$

Subject to:

$$P^{\min}_{dg,it} \leq P_{dg,it} \leq P^{\max}_{dg,it}, \; i \in Ndg, t \in T \tag{12.6}$$

where, constraint (6) represents DGs capacity constraint.

12.2.2 Lower Level

As previously stated, the lower level of the proposed model solves the ISO's market clearing problem in which system operating costs are minimized and system constraints are satisfied. The operating costs of the system include generation costs of generators and compensation costs for ILs. It must be mentioned that we assumed that Discos do not bid their DGs into the market but schedules them according to estimated LMPs (similar to [5]), therefore DGs cost is not considered in the lower level. So, the lower level is formulated as follows:

$$\text{Min}:\left(\sum_{i\in Ng}\sum_{t\in T}\left(a_{g,it}P_{g,it}^{2}+b_{g,it}P_{g,it}\right)+\sum_{i\in Nil}\sum_{t\in T}\left(a_{IL,it}P_{IL,it}^{2}+b_{IL,it}P_{IL,it}\right)\right) \tag{12.7}$$

Subject to:

$$P_{g,it}^{\min}\leq P_{g,it}\leq P_{g,it}^{\max}:\mu_{g(i,t)}^{\max},\mu_{g(i,t)}^{\min},i\in Ng,t\in T \tag{12.8}$$

$$P_{IL,it}^{\min}\leq P_{IL,it}\leq P_{IL,it}^{\max}:\mu_{IL(i,t)}^{\max},\mu_{IL(i,t)}^{\min},i\in Nil,t\in T \tag{12.9}$$

$$-P_{ij,t}^{\max}\leq P_{ij,t}\leq P_{ij,t}^{\max}:\upsilon_{(i,j,t)}^{\max},\upsilon_{(j,i,t)}^{\min},i,j\in Nl,t\in T \tag{12.10}$$

$$P_{g,it}+P_{dg,it}+P_{IL,it}-P_{d,it}^{0}=\sum_{j}\frac{1}{x_{(i,j)}}(\delta_{(i,t)}-\delta_{(j,t)}):\lambda_{it},\ i\in Nb,t\in T,\text{j}\in Nb(i) \tag{12.11}$$

$$-\pi\leq\delta(i,t)\leq\pi:\zeta_{(i,t)}^{\min},\zeta_{(i,t)}^{\max} \tag{12.12}$$

$$\delta(i,t)=0:\zeta_{t}^{8},\ \forall(i,t)=8 \tag{12.13}$$

The objective function (12.7) has two terms, which respectively represent the generation costs and the ILs' compensation costs. Constraints (12.8)–(12.10) represent generation capacity constraints, ILs capacity constraints, and transmission lines flow constraints, respectively. Constraint (12.11) represents the load balance limit on each bus of the system and the Lagrangian multiplier of the equivalent constraint in (12.11) gives LMP at that bus. Constraints (12.12) fix angle bounds for each node, and constraints (12.13) impose n_8 to be the slack bus. Dual variables are indicated at the corresponding equations following a colon.

12.2.3 MPEC Model

The lower level problem (12.7)–(12.13) can be replaced by its KKT conditions as follows:

$$\frac{\partial L}{\partial P_{g(i,t)}}=0\Rightarrow 2a_{g,it}P_{g,it}+b_{g,it}+\mu_{g(i,t)}^{\max}-\mu_{g(i,t)}^{\min}-\lambda_{it}=0 \tag{12.14}$$

$$\frac{\partial L}{\partial P_{IL(i,t)}}=0\Rightarrow 2a_{IL,it}P_{IL,it}+b_{IL,it}+\mu^{\max}_{IL(i,t)}-\mu^{\min}_{IL(i,t)}-\lambda_{it}=0 \tag{12.15}$$

$$\begin{aligned}\frac{\partial L}{\partial \delta_{(i,t)}}=0\Rightarrow \sum_j\frac{(\lambda_{it}-\lambda_{jt})}{x_{ij}}+\sum_j\frac{1}{x_{ij}}(\upsilon^{\max}_{(i,j,t)}-\upsilon^{\max}_{(j,i,t)})+\\ \sum_j\frac{1}{x_{ij}}(\upsilon^{\min}_{(j,i,t)}-\upsilon^{\min}_{(i,j,t)})+\zeta^{\max}_{(i,t)}-\zeta^{\min}_{(i,t)}+\zeta^{8}_{t}=0\end{aligned} \tag{12.16}$$

$$P_{g,it}+P_{dg,it}+P_{IL,it}-P^0_{d,it}=\sum_j\frac{1}{x_{(i,j)}}(\delta_{(i,t)}-\delta_{(j,t)}),\ i\in Nb, t\in T, \mathrm{j}\in Nb(i) \tag{12.17}$$

$$\delta(i,t)=0, \forall i=8\quad,\quad \forall t \tag{12.18}$$

$$P^{\min}_{g(i,t)}\le P_{g(i,t)}\perp\mu^{\min}_{g(i,t)}\ge 0; \tag{12.19}$$

$$P^{\min}_{IL(i,t)}\le P_{IL(i,t)}\perp\mu^{\min}_{IL(i,t)}\ge 0; \tag{12.20}$$

$$0\le(P^{\max}_{g(i,t)}-P_{g(i,t)})\perp\mu^{\max}_{g(i,t)}\ge 0 \tag{12.21}$$

$$0\le(P^{\max}_{IL(i,t)}-P_{IL(i,t)})\perp\mu^{\max}_{IL(i,t)}\ge 0 \tag{12.22}$$

$$0\le\left[P^{\max}_{(i,j)}+\frac{1}{x_{(i,j)}}(\delta_{(t,i)}-\delta_{(t,j)})\right]\perp\upsilon^{\min}_{(i,j,t)}\ge 0 \tag{12.23}$$

$$0\le\left[P^{\max}_{(i,j)}-\frac{1}{x_{(i,j)}}(\delta_{(t,i)}-\delta_{(t,j)})\right]\perp\upsilon^{\max}_{(i,j,t)}\ge 0 \tag{12.24}$$

$$0\le(\pi-\delta_{(i,t)})\perp\zeta^{\max}_{(i,t)}\ge 0 \tag{12.25}$$

$$0 \leq (\pi + \delta_{(i,t)}) \perp \zeta_{(i,t)}^{\min} \geq 0 \tag{12.26}$$

The result of including KKT formulation of lower level (12.14)–(12.26) in the upper-level problem is an MPEC, and the formulation is:

$$Max \quad Pd(m) = Rd(m) - Cd(m) = \sum_{i \in Nd(m)} \sum_{t \in T} \left[\overline{\lambda}(P_{d,it}^{0} - P_{IL,it}) - C_{dg,it}(P_{dg,it}) - \lambda_{it}(P_{d,it}^{0} - P_{IL,it} - P_{dg,it}) \right] \tag{12.27}$$

Subject to:

$$P_{dg,it}^{\min} \leq P_{dg,it} \leq P_{dg,it}^{\max}, i \in Ndg, t \in T \tag{12.28}$$

$$(12.14) - (12.26) \tag{12.29}$$

The above MPEC model (12.27)–(12.29) includes the nonlinearity constraints (i.e., (12.19)–(12.26)) due to complementarity conditions, which can be linearized using the well-known strong duality theorem, which is proposed in [20] as follows:

1. Linearization of (12.19) and (12.20) (see [20]):

$$\mu_{g(i,t)}^{\min} \geq 0 \tag{12.30}$$

$$P_{g(i,t)} - P_{g(i,t)}^{\min} \geq 0 \tag{12.31}$$

$$\mu_{g(i,t)}^{\min} \leq \omega_{g(i,t)}^{\min}.M^{\alpha} \tag{12.32}$$

$$(P_{g(i,t)} - P_{g(i,t)}^{\min}) \leq (1 - \omega_{g(i,t)}^{\min}).M^{\alpha} \tag{12.33}$$

$$\mu_{IL(i,t)}^{\min} \geq 0 \tag{12.34}$$

$$P_{IL(i,t)} - P_{IL(i,t)}^{\min} \geq 0 \tag{12.35}$$

$$\mu_{IL(i,t)}^{\min} \leq \omega_{IL(i,t)}^{\min}.M^{\alpha\alpha} \tag{12.36}$$

$$(P_{IL(i,t)} - P_{IL(i,t)}^{\min}) \leq (1 - \omega_{IL(i,t)}^{\min}).M^{\alpha\alpha} \tag{12.37}$$

where M^{α} and $M^{\alpha\alpha}$ are large enough constants.

2. Linearization of (12.21) and (12.22) (see [20]):

$$\mu_{g(i,t)}^{\max} \geq 0 \tag{12.38}$$

$$P_{g(i,t)}^{\max} - P_{g(i,t)} \geq 0 \tag{12.39}$$

$$\mu_{g(i,t)}^{\max} \leq \omega_{g(i,t)}^{\max}.M^{\beta} \tag{12.40}$$

$$(P_{g(i,t)}^{\max} - P_{g(i,t)}) \leq (1 - \omega_{g(i,t)}^{\max}).M^{\beta} \tag{12.41}$$

$$\mu_{IL(i,t)}^{\max} \geq 0 \tag{12.42}$$

$$P_{IL(i,t)}^{\max} - P_{IL(i,t)} \geq 0 \tag{12.43}$$

$$\mu_{IL(i,t)}^{\max} \leq \omega_{IL(i,t)}^{\max}.M^{\beta\beta} \tag{12.44}$$

$$(P_{IL(i,t)}^{\max} - P_{IL(i,t)}) \leq (1 - \omega_{IL(i,t)}^{\max}).M^{\beta\beta} \tag{12.45}$$

where M^{β} and $M^{\beta\beta}$ are large enough constants.

3. Linearization of (12.23) and (12.24) (see [20]):

$$\upsilon_{(i,j,t)}^{\min} \geq 0 \tag{12.46}$$

$$P_{(i,j)}^{\max} + \frac{1}{x_{(i,j)}}(\delta_{(t,i)} - \delta_{(t,j)}) \geq 0 \tag{12.47}$$

$$\upsilon_{(i,j,t)}^{\min} \leq \omega_{\omega(i,j,t)}^{\min}.M^{\delta} \tag{12.48}$$

$$(P_{(i,j)}^{\max} + \frac{1}{x_{(i,j)}}(\delta_{(t,i)} - \delta_{(t,j)})) \leq (1 - \omega_{\omega(i,j,t)}^{\min}).M^{\delta} \tag{12.49}$$

$$\upsilon_{(i,j,t)}^{\max} \geq 0 \tag{12.50}$$

$$P_{(i,j)}^{\max} - \frac{1}{x_{(i,j)}}(\delta_{(t,i)} - \delta_{(t,j)}) \geq 0 \tag{12.51}$$

$$\upsilon_{(i,j,t)}^{\max} \leq \omega_{\omega(i,j,t)}^{\max}.M^{\delta\delta} \tag{12.52}$$

$$(P_{(i,j)}^{\max} - \frac{1}{x_{(i,j)}}(\delta_{(t,i)} - \delta_{(t,j)})) \leq (1 - \omega_{\omega(i,j,t)}^{\max}).M^{\delta\delta} \tag{12.53}$$

where M^{δ} and $M^{\delta\delta}$ are large enough constants.

4. Linearization of (12.25) and (12.26) (see [20]):

$$\zeta_{(i,t)}^{\max} \geq 0 \tag{12.54}$$

$$\pi - \delta_{(i,t)} \geq 0 \tag{12.55}$$

$$\zeta_{(i,t)}^{\max} \leq \omega_{\omega\omega(i,t)}^{\max}.M^{\varepsilon} \tag{12.56}$$

$$(\pi - \delta_{(i,t)}) \leq (1 - \omega_{\omega\omega(i,t)}^{\max}).M^{\varepsilon} \tag{12.57}$$

$$\zeta_{(i,t)}^{\min} \geq 0 \tag{12.58}$$

$$\pi + \delta_{(i,t)} \geq 0 \tag{12.59}$$

$$\zeta_{(i,t)}^{\min} \leq \omega_{\omega\omega(i,t)}^{\min}.M^{\varepsilon\varepsilon} \tag{12.60}$$

$$(\pi + \delta_{(i,t)}) \leq (1 - \omega_{\omega\omega(i,t)}^{\min}).M^{\varepsilon\varepsilon} \tag{12.61}$$

where M^{ε} and $M^{\varepsilon\varepsilon}$ are large enough constants.

Finally, based on the above discussion, the final formulation of Disco's optimal energy acquisition model, which is obtained by replacing the lower level of the primary proposed bi-level model (12.5)–(12.13) with its KKT conditions and then application of strong duality theorem, will be as follows that is a mixed-integer linear problem and, as mentioned before, is solvable using available CPLEX solvers.

$$\text{Max}: \sum_{m=1}^{N_m} \sum_{i \in Nd(m)} \sum_{t \in T} \left[\overline{\lambda}(P_{d,it}^{0} - P_{IL,it}) - C_{dg,it}(P_{dg,it}) - \lambda_{it}(P_{d,it}^{0} - P_{IL,it} - P_{dg,it}) \right] \tag{12.62}$$

Subject to:

$$P_{dg,it}^{\min} \leq P_{dg,it} \leq P_{dg,it}^{\max}, i \in Ndg, t \in T \tag{12.63}$$

$$2a_{g,it}P_{g,it} + b_{g,it} + \mu_{g(i,t)}^{\max} - \mu_{g(i,t)}^{\min} - \lambda_{it} = 0 \tag{12.64}$$

$$2a_{IL,it}P_{IL,it} + b_{IL,it} + \mu_{IL(i,t)}^{\max} - \mu_{IL(i,t)}^{\min} - \lambda_{it} = 0 \tag{12.65}$$

$$\begin{aligned}&\sum\nolimits_j \frac{(\lambda_{it} - \lambda_{jt})}{x_{ij}} + \sum_j \frac{1}{x_{ij}}(\upsilon_{(i,j,t)}^{\max} - \upsilon_{(j,i,t)}^{\max}) + \\ &\sum\nolimits_j \frac{1}{x_{ij}}(\upsilon_{(j,i,t)}^{\min} - \upsilon_{(i,j,t)}^{\min}) + \zeta_{(i,t)}^{\max} - \zeta_{(i,t)}^{\min} + \zeta_t^{8} = 0\end{aligned} \tag{12.66}$$

$$\begin{aligned}&P_{g,it} + P_{dg,it} + P_{IL,it} - P_{d,it}^{0} = \sum\nolimits_j \frac{1}{x_{(i,j)}}(\delta_{(i,t)} - \delta_{(j,t)}), \\ &i \in Nb, t \in T, \mathrm{j} \in Nb(i)\end{aligned} \tag{12.67}$$

$$\delta(i,t) = 0, \forall (i,t) = 8 \tag{12.68}$$

$$(12.30) - (12.61) \tag{12.69}$$

The proposed problem is reformulated as a mixed-integer linear program (MILP) and can be solved using CPLEX solver [21] under GAMS optimization software [22].

12.3 CASE STUDY

12.3.1 Data

The block diagram of test system is shown in Fig. 12.1. As it is seen from Fig. 12.1, this test system includes three Discos in which L_1, L_2, and DG_1 are related to Disco 1, L_3, L_4, and DG_2 are related to Disco 2, and L_5 and DG_3 are related to Disco 3. Data of generators, DGs, ILs, and transmission lines are shown in Tables 12.1 and 12.2, and Fig. 12.2 shows 24-h demand profile of the test system. Also, it is assumed that the price of selling energy to the end consumers ($\overline{\lambda}$) is constant and is 80 $/MWh.

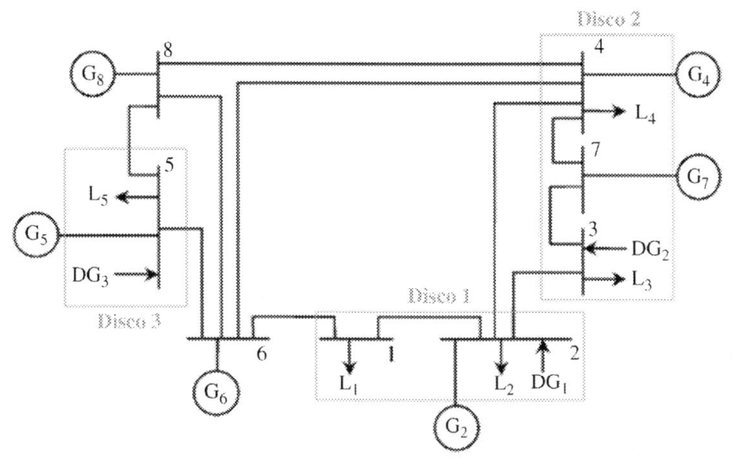

Figure 12.1 Block diagram of the test system.

Table 12.1 Data of generators, DGs, and ILs

	P_{min} (MW)	P_{max} (MW)	a ($/MW^{2}h)	b ($/MWh)
G_1	0	40	0.08	45.62
G_2	0	58	0.11	35.35
G_3	0	40	0.09	22.47
G_4	0	50	0.095	23.37
G_5	0	24	0.085	33.47
G_6	0	60	0.078	21.39
DG_1	0	5	0.09	48
DG_2	0	4	0.09	48
DG_3	0	3	0.09	48
IL_1	0	$0.1 \times L_1$	1	75
IL_2	0	$0.1 \times L_1$	1	75
IL_3	0	$0.1 \times L_1$	1	75
IL_4	0	$0.1 \times L_1$	1	75
IL_5	0	$0.1 \times L_1$	1	75

12.3.2 Simulation Results With and Without DG and IL

In this section, the simulation results of the proposed method obtained considering and neglecting DGs and ILs and its impact on the energy supply strategies and obtained profits of Discos are presented. In this section, peak-hour demand (i.e., hour 19:00) and all the system constraints

Table 12.2 Data of transmission lines

Line no.	From bus	To bus	X (p. u.)	Flow limit (MW)
1	2	1	0.011	20
2	6	1	0.030	14.2
3	3	2	0.018	30
4	4	2	0.030	20
5	7	3	0.022	40
6	6	4	0.030	30
7	7	4	0.015	38
8	8	4	0.030	40
9	6	5	0.025	38
10	8	5	0.020	20
11	8	6	0.0065	40

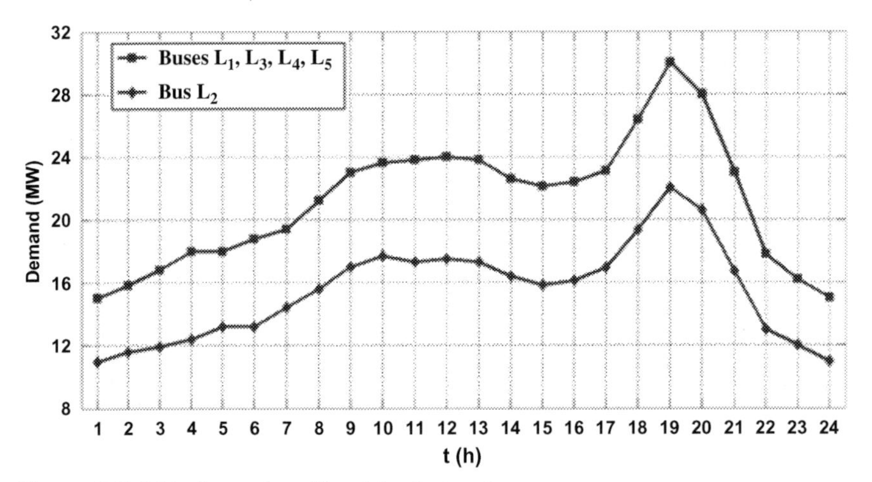

Figure 12.2 24-h demand profile of the test system.

are considered. The following four cases are investigated and the related simulation results are presented in Table 12.3 and Fig. 12.3.

Case a. Discos have no permission to use DG or IL.

Case b. Discos have permission to use only IL.

Case c. Discos have permission to use only DG.

Case d. Discos have permission to use both DG and IL.

As it is seen from Table 12.3, in case b, in which Discos have had permission to use only IL, Disco 1 has used 2.58 MW of its IL. In case c, in which Discos have had permission to use only DG, Discos 1 and 2 have

Table 12.3 Strategies of Discos to purchase energy and their obtained profits in various cases of section 12.3.2

		a	b	c	d
Disco 1	Load	52	52	52	52
	DAM[a]	52	49.42	47	46.80
	DG	0	0	5	5
	IL	0	2.58	0	0.20
	Profit	536.28	774.73	1495.40	1508.39
Disco 2	Load	60	60	60	60
	DAM	60	60	56	56
	DG	0	0	4	4
	IL	0	0	0	0
	Profit	1880.88	2034.57	2396.18	2406.55
Disco 3	Load	30	30	30	30
	DAM	30	30	30	30
	DG	0	0	0	0
	IL	0	0	0	0
	Profit	1642.27	1633.43	1625.60	1624.91
Total	Load	142	142	142	142
	DAM	142	139.42	133	132.8
	DG	0	0	9	9
	IL	0	2.58	0	0.20
	Profit	4059.42	4442.73	5517.18	5539.86

[a]Day-Ahead Market.

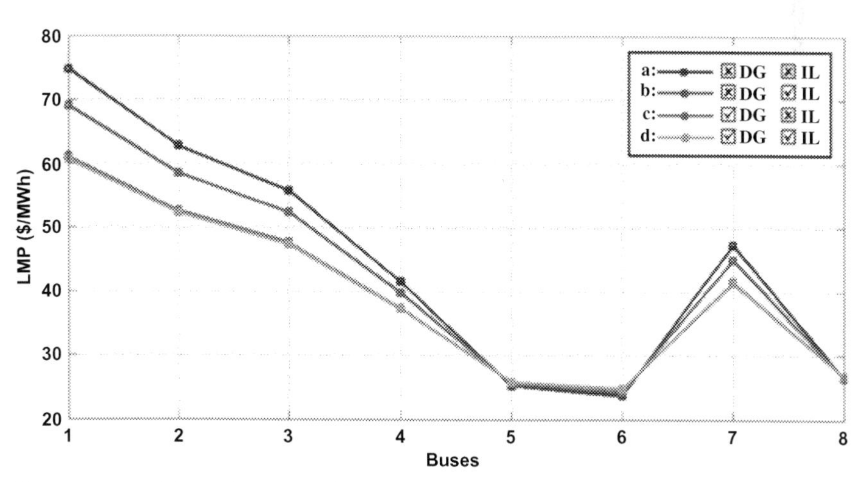

Figure 12.3 LMPs at the system buses in various cases of Section 12.3.2.

respectively used 5 MW and 4 MW of their DGs. Finally in case d, in which Discos have had permission to use both DG and IL, Disco 1 has used 0.2 MW of its IL and 5 MW of its DG, and Disco 2 has used 4 MW of its DG.

The mentioned use of ILs and DGs as local energy resources by Discos 1 and 2 in cases b, c, and d, in comparison with case a has led to decrease of congestion in the transmission lines related to Discos 1 and 2 that has in turn resulted in decrease of the LMPs in the buses related to Discos 1 and 2 (namely buses 1, 2, 3, 4, and 7) that can be seen in Fig. 12.3. It is evident from this figure that use of IL in case b, DG in case c, and both in case d has led to decrease of LMPs that the maximum reduction of the LMPs is related to case d in which Discos have had permission to use both DG and IL.

The mentioned reduction of LMPs at buses 1 and 2, and at buses 3 and 4 that are respectively related to Discos 1 and 2 has in turn increased profits of these two Discos in cases b, c, and d in comparison with case a (based on Eq. (12.4) and can be seen in Table 12.3) that the maximum increase of their profit is related to case d in which the LMPs have had the maximum reduction.

On the other hand, because Disco 3 has always been able to purchase energy from the market at a low price, it has not used any of its DG and IL in cases b, c, and d, and in comparison with case a, its profit has decreased slightly (based on Eq. (12.4) and can be seen in Table 12.3) due to small increase of LMPs at the related buses, namely at buses 5, 6, and 8 (can be seen in Fig. 12.3).

In addition, it is seen from Table 12.3 that because the costs of ILs have been higher than those of generators and DGs, Discos have used little amount of IL.

12.3.3 Simulation Results With and Without Transmission Limits

In this section, the simulation results of the proposed method obtained considering and neglecting transmission lines limits and its impact on the energy supply strategies and obtained profits of Discos are presented. To brief results and to reduce computing time, only three continuous hours from 16:00 to 18:00 are considered. The related simulation results are presented in Table 12.4 and Fig. 12.4.

According to Table 12.4, only the cheapest generators (namely, G_3, G_4, and G_6) have committed to meet the demand with not considering

Table 12.4 Strategies of Discos to purchase energy and their obtained profits considering and neglecting lines limits

		Neglecting transmission limits			**Considering transmission limits**		
		16:00	**17:00**	**18:00**	**16:00**	**17:00**	**18:00**
Disco 1	Load	39	45	48	39	45	48
	DAM	34	40	43	34	40	43
	DG	5	5	5	5	5	5
	IL	0	0	0	0	0	0
	Profit	2063.162	2344.155	2477.425	1329.088	1433.196	1469.019
	Tot. Profit		6884.741			4231.3	
Disco 2	Load	46	52	56	46	52	56
	DAM	42	48	52	42	48	52
	DG	4	4	4	4	4	4
	IL	0	0	0	0	0	0
	Profit	2426.537	2701.435	2881.567	1997.035	2195.825	2325.753
	Tot. Profit		8009.539			6518.613	
Disco 3	Load	23	26	28	23	26	28
	DAM	23	26	28	23	26	28
	DG	0	0	0	0	0	0
	IL	0	0	0	0	0	0
	Profit	1193.947	1327.043	1414.497	1253.499	1413.287	1519.105
	Tot. Profit		3935.488			4185.891	
Generators	G_1	0	0	0	19.666	28.926	34.215
	G_2	0	0	0	1.076	4.363	6.297
	G_3	31.218	36.055	38.957	16.834	17.627	18.201
	G_4	24.838	29.420	32.170	6.061	6.515	6.895
	G_5	0	0	0	24.000	24.000	24.000
	G_6	42.944	48.525	51.873	31.365	32.569	33.392

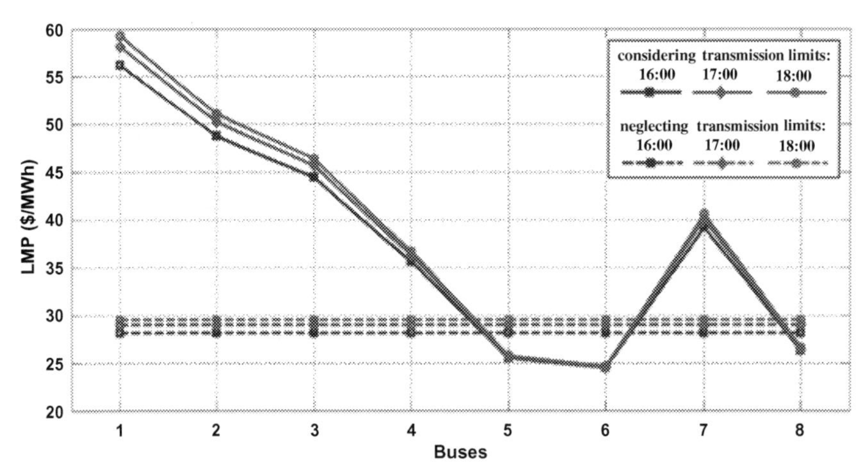

Figure 12.4 LMPs at the system buses considering and neglecting lines limits.

transmission lines limits, while with considering lines limits, all the generators have committed due to occurrence of congestion in the transmission lines.

As it is seen from Fig. 12.4, LMPs are the same at all the system buses for each hour without taking the lines limits into account. When the lines limits have been taken into account, due to occurrence of congestion in transmission lines, LMPs have increased at buses 1, 2, 3, 4, and 7, and have decreased at the other three buses.

As seen from Table 12.4, considering lines limits, has led Disco 1 and 2 to earn less profits and Disco 3 to earn more profit compared with the case in which lines limits are neglected. The reason is that, as mentioned in the previous paragraph, inclusion of transmission lines flow limits in the model has caused higher LMPs at buses related to Discos 1 and 2, and lower LMPs at buses related to Disco 3 (due to occurrence of congestion in the transmission lines). Therefore, based on Eq. (12.4), the profits of Discos 1 and 2 have reduced but the profit of Disco 3 has increased.

Furthermore, it is seen from Table 12.4 that to decrease LMPs and to achieve higher profits, Discos 1 and 2 have used the maximum capacity of their DGs in both cases considering the transmission limits and not considering them, while, due to being able to purchase energy from the market at a low price, Disco 3 has not used any of its DG and IL. In addition, because the costs of ILs have been higher than those of generators and DGs, none of the Discos have used their ILs.

12.4 CONCLUSION

An optimal energy acquisition model for Disco in a competitive day-ahead electricity market with two new resources, DG and IL, was proposed in this paper. The energy acquisition scheme was modeled as a bi-level optimization problem in which Disco's profits are maximized in the upper level and system operating costs are minimized in the lower level. To solve the proposed bi-level model with conventional optimization techniques, at first, the model was converted to a problem with nonlinear complementary constraints by replacing its lower level with KKT conditions, then, using strong duality theorem, the resulting problem with nonlinear complementary constraints was converted to a problem with mixed integer linear constrains which is easily solvable with conventional optimization approaches. An eight-bus example demonstrated that the proposed model and solution algorithm can help Discos make optimal energy purchasing plans. Also, the roles of DG and IL were studied in case of transmission congestion. Studies in this paper showed that DG and IL can alleviate congestion, reduce the market price, and reduce Gencos' market power. Based on these results, Discos are encouraged to make full use of the two useful resources by investing in DGs and signing IL contracts with the end customers. The renewable DGs such as wind turbine and photovoltaic systems in the presence of an energy storage system such as batteries and electric vehicles can be used as new options in smart grids environment and can be studied in future works.

REFERENCES

[1] M.B. Jorge, V.O. Héctor, L.G. Miguel, P.D. Héctor, Multi-fault service restoration in distribution networks considering the operating mode of distributed generation, Elec Pow Sys Res 116 (2014) 67–76.
[2] A.M. Niaki, S. Afsharnia, A new passive islanding detection method and its performance evaluation for multi-DG systems, Elec Pow Sys Res 110 (2014) 180–187.
[3] M. Sedighizadeh, M. Esmaili, M. Esmaeili, Application of the hybrid Big Bang-Big Crunch algorithm to optimal reconfiguration and distributed generation power allocation in distribution systems, Energy 76 (2014) 920–930.
[4] J. Aghaei, M. Muttaqi, A. Azizivahed, M. Gitizadeh, Distribution expansion planning considering reliability and security of energy using modified PSO (Particle Swarm Optimization) algorithm, Energy 65 (2014) 398–411.
[5] M. Ilic, J.W. Black, M. Prica, Distributed electric power systems of the future: institutional and technological drivers for near optimal performance, Elec Pow Sys Res 77 (9) (2007) 1160–1177.

[6] R. Palma-Behnke, J. Luis, A. Cerda, L.S. Vargas, A. Jofrev, A distribution company energy acquisition market model with integration of distributed generation and load curtailment options, IEEE Trans Pow Sys 20 (4) (2005) 1718–1727.
[7] R. Herranz, A.M. San Roque, J. Villar, F.A. Campos, Optimal demand-side bidding strategies in electricity spot markets, IEEE Trans Pow Sys 27 (3) (2012) 1204–1213.
[8] M. Mashhour, M.A. Golkar, S.M. Moghaddas-Tafreshi, Extending market activities for a distribution company in hourly-ahead energy and reserve markets–Part I: problem formulation, Ener Conv Manage 52 (1) (2011) 477–486.
[9] M. Mashhour, M.A. Golkar, S.M. Moghaddas-Tafreshi, Extending market activities for a distribution company in hourly-ahead energy and reserve markets—Part II: numerical results, Ener Conv Manage 52 (1) (2011) 569–580.
[10] H. Li, Y. Li, Z. Li, A multi-period energy acquisition model for a distribution company with distributed generation and interruptible load, IEEE Trans Pow Syst 22 (2) (2007) 588–596.
[11] I. Taheri, M. Rashidinejad, A. Badri, A. Rahimi-Kian, Analytical approach in computing nash equilibrium for oligopolistic competition of transmission-Constrained GENCOs, IEEE Sys J 99 (2014) 1–11.
[12] R. Baldick, Electricity market equilibrium models: the effect of parameterization, IEEE Trans Pow Syst 17 (Nov. 2002) 1170–1176.
[13] M. Latorre, S. Granville, The stacklberg equilibrium applied to AC power system—a noninterior point algorithm, IEEE Trans Pow Syst 18 (May 2003) 611–618.
[14] G. Bautista, V.H. Quintana, J.A. Aguado, An oligopolistic model of an integrated market for energy and spinning reserve, IEEE Trans Pow Syst 20 (Feb. 2006) 132–142.
[15] Y. Chen, B. Hobbs, An oligopolistic power market model with tradable NOx permits, IEEE Trans Pow Syst 20 (1) (Feb. 2005) 119–129.
[16] X. Wang, Y. Li, S. Zhang, Oligopolistic equilibrium analysis for electricity markets: a nonlinear complementarity approach, IEEE Trans Pow Syst vol. 19 (no. 3) (Nov. 2004) 1348–1355.
[17] S.H. Gorgizadeh, A. Akbari-Foroud, M. Amirahmadi, Strategic bidding in a pool-based electricity market under load forecast uncertainty, Iran J Electric Electronic Eng 8 (2) (2012) 164–176.
[18] H. Haghighat, S.W. Kennedy, A bi-level approach to operational decision making of a distribution company in competitive environments, IEEE Trans Pow Syst 27 (4) (2012) 1797–1807.
[19] A.S. Algarni, K. Bhattacharya, A generic operations framework for Discos in retail electricity markets, IEEE Trans Pow Syst 24 (1) (2009) 356–367.
[20] J. Fortuny-Amat, B. McCarl, A representation and economic interpretation of a two-level programming problem, J Oper Res Soc 32 (9) (Sep. 1981) 783–792.
[21] The GAMS Software Website, 2017. [Online]. Available: http://www.gams.com/dd/docs/solvers/cplex.pdf.
[22] A. Brooke, D. Kendrick, A. Meeraus, GAMS user's guide, The Scientific Press, Redwood City, CA, 1990. Available from: http:// www.gams.com/docs/gams/GAMSUsersGuide.pdf.

CHAPTER 13

Energy Management Systems for Hybrid AC/DC Microgrids: Challenges and Opportunities

Moein Manbachi
The University of British Columbia, Vancouver, BC, Canada

13.1 INTRODUCTION

With the advent and the expansion of smart grid new functionalities, design and development of the optimal management of energy generation-consumption in modern distribution grids has gained attention of many electric power utilities. As a smart grid technology, energy management systems (EMS) play a substantial role in "smartening up" AC power grids. Recently, DC grids have been resurging due to the development and deployment of renewable DC power sources, and to their inherent benefits for DC loads in commercial, industrial, and residential applications. To avoid unnecessary DC to AC conversions, a favorable solution is to design a hybrid DC and AC grid, to couple DC sources with DC loads and AC sources with AC loads, i.e., hybrid AC/DC microgrids. According to the fact that hybrid AC/DC microgrids should perform in different operating conditions and as they have different modes of operations, it is essential to find advanced solutions for the energy management of such grids utilizing smart microgrid functionalities and/or components such as advanced metering infrastructure (AMI) and distributed energy resources (DERs). According to CIGRÉ WG6.22 definition, microgrids are comprised of DERs and loads that can be operated in a coordinated and controlled system either while connected to the main grid, i.e., grid-connected mode, or while islanded, i.e., off-grid mode [1].

Many people around the world are living in remote communities far away from the main interconnected electricity grids. Although the global electrification rate increased from 76%in 1996 to 85%in 2012, about 1.2 billion people still do not have access to the electricity [2]. Generally,

Operation of Distributed Energy Resources in Smart Distribution Networks
DOI: https://doi.org/10.1016/B978-0-12-814891-4.00013-8

87% of people without electricity live in rural areas and remote communities with sparse population where grid extension is costly and inefficient. In recent years, refugee crisis has also forced many people to immigrate to regions close to country borders and far from the electricity grids. Statistics show that 80%of the 8.7 million people who were displaced due to the war, live in camps with minimal access to energy with high dependency on conventional biomass cooking. Moreover, most refugee campuses are equipped with poorly planned and maintained diesel generators as their main source of electricity generation [3]. The largest refugee campus in Jordan (Zaatari) serves about 80,000 refugees but it only connects to the main grid 11 h per day [4]. Hence, supplying reliable and cost-effective electricity to remote communities has become one of the main challenges of human right organizations, governments, and utilities. Regarding the remote grid market opportunity, Navigant Research predicts that the value of assets and services of the remote community market will grow from $10.9 billion in 2015 to $196 billion in 2024 [5].

Typically, many electricity grids of remote communities are operating "off-grid." In other words, they are not connected to the main grid. Natural Resources Canada defines off-grid communities as a community that is not connected to the North American main grid and not to the natural gas pipeline network. In Canada, there is about 292 remote communities with a total population of 194,281 [6]. Aboriginals are 65%of this population who live in about 170 remote communities [6]. About 79%of remote communities are using oiled-based fuel as their main electricity generation source [7].

Many remote communities around the globe are electrified by diesel generators. In Canada, more than 175 communities use diesel and 138 communities are diesel-only source communities that serve about 88,000 people [6]. In most of these communities, diesel has to be delivered by air, water, and/or winter roads. This has made people who live in these communities susceptible to diesel and other fossil fuels for the energy supply. Diesel cost volatility, rising in the electricity demand and the environmental issues using diesel, e.g., increasing greenhouse gas emissions, could significantly impact the quality of life of these communities as it brings economic, social, and environmental barriers for them. As such, many efforts have been recently made to employ alternative energy solutions to decrease diesel consumption and to provide efficient, reliable, and sustainable solutions for remote off-grid communities. One of the main solutions that most researchers have been proposed is using off-grid AC/DC microgrids.

It is possible to classify off-grid AC/DC microgrids systems in four groups based on their pre-defined tasks: off-grid system supplying community residential buildings, off-grid systems for the industry section such as mining, off-grid systems for locations that need high reliability such as military campuses or hospitals, and mobile off-grid systems that could serve small-scale consumption locations. Table 13.1 presents an off-grid system classification based on their applications.

With developments in smart microgrid technologies, increasing price or high price fluctuations of fossil fuels, especially diesels, decreasing costs of installation, operation and maintenance of renewable generations, it is more conceivable to use distributed energy resources widely in off-grid communities. Typically, supplying power for remote communities includes many challenges. These challenges could be categorized based on geographical, cultural, technical, and economic conditions. For instance, it is very hard to deliver diesel in many regions in Northern Canada due to the geographical and weather conditions. Some roads are not accessible during winter times and it is very costly to deliver diesel by air. For instance, Fort Seven, Kasabonika [8] and deer lake settlements in Ontario, Canada are accessible by air. In Ramea Island in Newfoundland and Labrador, Canada, the access is by seaway [9]. These limited ways of delivering fuel could increase diesel delivery price and impose financial pressure to remote communities to supply power for their people. As such, governments as well as electric power utilities need to invest more on sustainable and cost-effective microgrid solutions using clean energy

Table 13.1 Off-grid system classification

Grid Type	Grid Name	Grid Sub-Name	User/Application
Off-grid	Mini Grid (10 kW–10 MW)	Remote microgrid	Communities, settlements, domestic, commercial, business, industrial, institutional
		Isolated grid	Domestics, commercial, business, industrial, institutional
	Stand-alone	Pico grid	Lighting and appliances
		Home system	Households (communities)
		Productive system	Commercial (factory, clinic, hotel, etc.)

resources such as solar panels (PV), wind turbines, etc. Tables 13.2 and 13.3 represent successful off-grid systems based on AC/DC microgrid technologies in Canada and other countries respectively.

As it can be seen in Table 13.2, wind turbines have been used for the Diavink mine at Northwest Territories. The off-grid system includes four 2.3 MW wind turbines (with total generation of 9.2 MW) as well as

Table 13.2 Canadian successful remote off-grid projects using microgrid technologies [10]

Remote Microgrid Name	Technology	Generation (kW)	Objective
Nemiah Valley, BC	PV-diesel	PV mini-grid: 27.36 kW Diesel: 30. 95 kW	Build a remote small-scale microgrid as the nearest grid is about 100 km away
Ramea Island, NFL	Wind-diesel	Wind: 6*65 kW Diesel: 3*925 kW	Install first medium penetration wind installation integrated to a diesel generator based power supply in Canada
Kasabonika Lake, ON	Wind-diesel	Wind: 4*30 kW Diesel: 1000 kW, 600 kW, 400 kW	Evaluate technologies for implementation of a wind-diesel-storage system for remote communities of Northern Ontario
Hartley Bay, BC	Hydro-diesel	Hydro: 900 kW Diesel: 2*420 kW, 210 kW	Improve generation efficiency; reduce community's electrical demand, energy consumption, GHG emissions, and costs
Kluane, Yukon	Wind-diesel	Wind: 300 kW	Displace 25%of diesel and develop other energy opportunities
Diavik, NWT	Wind-diesel	Wind: 4*2.3 MW	Reduce diesel consumption by 10%(> 2 million litre)
Bella Coola, BC	Hydro-diesel	Hydro:1420 kW, 700 kW Diesel: 7*7200 kW	Hybrid renewable energy systems microgrid

Table 13.3 Successful remote off-grid projects around the world using microgrids with DERs

Name	Country	Population	Total Capacity (kW)	Peak Demand	Remote Microgrid Type	Percentage
Bonaire	Netherland	14500	25000	11000	Wind-Diesel-Battery	44-56-(100kWh)
Kodiak Alaska	USA	13000	75000	27800	Hydro-Wind-Diesel	82.8-16.9-0.3
El Hierro	Spain	11000	35000	7600	Hydro-Wind-Diesel	32-32-36
Falkland Island	United Kingdom	2500	8580	3200	Wind-Diesel	33-67
King Island	Australia	1800	8840	2500	Wind-Solar-Diesel	63-2-35
ISLE of IGG	Scotland	100	250	60	(Wind-Solar-Hydro)-Diesel	87-13
Necker Island	British Virgin Island	60	2160	400	Wind-Solar-Diesel	60-20-20
Annobon Island	Guinea	5232	5000	--	Solar-Diesel	100-0
Kimprana	Mali	3000	247	--	Solar-Diesel	30-70

diesel. Hence, it is called a wind-diesel microgrid. The main target of this project is to reduce diesel generation by 10%which conserves about 8 million liters of diesel. In the Nemiah Valley project in British Columbia, solar panels are employed to build a small-scale PV-diesel microgrid comprising of a 27.36 kW PV and 30.95 kW diesel [11].

The reason of choosing PV could have technical and economic reasons. From the economic point of view, the levelized cost of electricity (LCOE) of wind and PV are less than LCOE of other fossil fuel resources such as diesel generators.

For example, if we consider the diesel price in Canada shown in Fig. 13.1, it can be observed that price fluctuations, e.g., January 2015 to July 2015, could negatively impact the remote community power generation economy [12].

On the other hand, the price of diesel transportation would also be another major challenge for such communities especially during winter. As the price of PV and wind turbine generations are decreasing and as their related technologies are more available now, employing such DERs have become more doable in remote community grids. The method of selecting proper DER type, size, and penetration level depends on four related key factors:

1. Geographical and Climate Conditions: As an example, Kluane Wind project in Yukon, Canada aims to install 300 kW wind turbine to displace 25%of diesel and develop other energy opportunities such as advanced energy management. The reason for choosing wind turbine lies in the fact that the Yukon Territory is one of the best territories

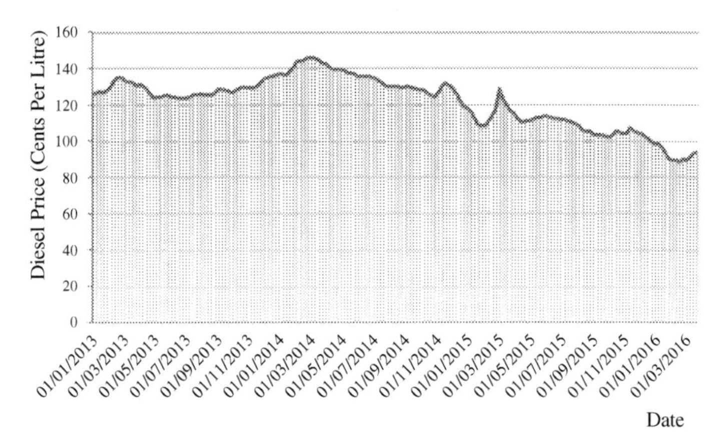

Figure 13.1 Diesel price in Canada [12].

with great wind blowing potential. Table 13.4 classifies isolated communities based on geographical and weather conditions and presents well-known examples for each category.

2. Environmental and Social Conditions: In a country like Canada, 65% of remote communities are aboriginals. Hence, selection of the DER type should be based on environmental and cultural conditions. Electric power utilities respect aboriginal rights, cultures and typically, they are eager to listen to the ideas coming from remote community people. In the case of using diesels, they are noisy, disruptive, and

Table 13.4 Remote community classification based on geographical and weather conditions

Remote area category	Remote communities
Remote areas with long winters	Small communities throughout rural Canada (CAN), small Greenland communities (DK), Iqaluit, Nunavut (CAN), Ramea Island (CAN), Kluane, Yukon (CAN), Norwegian islands (NOR), remote Alaskan communities (US), Baker Lake, Nunavut (CAN), Kodiak Island, Alaska (US)
Remote areas with temperate climates	Faroe Islands, (DEN), Japanese outer islands (JPN), King Island, Tasmania (AUS), Stuart Island (CAN), Utsira Island (NOR), Shetland Islands (UK), Falkland Islands (UK), Fair Isle (UK), Isle of Wright (UK), Chiloe Archipelago (CHILE), Isle of Eigg, Scotland, UK
Small remote areas with warm climates	Floreana Island, Galapagos, Ecuador, Coral Bay, Western Australia (AUS), Sint Eustatius (NL), Saba (NL), Saint Martin (FRA), Tokelau (NZ), Pitcairn Island (UK), Peter Island (BVI)
Large remote areas with warm climate	Bonaire, Netherlands, Anguilla (UK), American Samoa (US), San Andres (COL), Saint Barthelemy (FR), Wallis and Futuna (FR), Cook Islands (NZ), Montserrat (UK), El Hierro, Canary Islands, (SPA), Miyakojima, (JPN) Reunion Island, (FR)
Remote Institutional Stations ––National Parks	Resolute Station, Nunavut (CAN), Research stations in Antarctica, Island of Osmussaare (Estonia), Zackenberg Research Station, Greenland (DK), Rothera, British Antarctic Survey (UK), Scott Base & McMurdo Station, Antarctica
Remote Areas in Developing Regions	Akkan, Morocco, Tanjung Batu Laut Malaysia

inefficient especially in quiet regions such as remote communities. Moreover, if the generator fails, the system will face with black-out and this can be very dangerous in remote communities with long and cold winters. In many remote communities in Canada, diesel must be transported through long distances by plane, truck, or barge, that could lead to a high risk of fuel spills. On winter road, the scenario is even worth it as transportation vehicles also produce GHG emissions.

3. Technical Requirements: Technical needs define based on main objectives of the grid, required load of the community and type of it, economic and geographical conditions. The size, type, and mix of DER resources highly depend on financial budget, governmental policies and regulations and other executive capabilities of the region. For instance, Northwest Territories in Canada (NWT) aim to deploy PV systems up to 20% of the average load in diesel communities. As such, utilities may need to adapt their targets according to provincial/territorial policies if they intend to participate in remote microgrid development.
4. Economy: One of the key factor in selecting proper DER type, size and mix is financial capability based on the available budget considering future expansion plans. Present and future policies, remote community population and mix, as well as present and future load type have a significant impact on defining a remote microgrid budget. In Canada, northern locations have a high demand of diesel and heating fuel with high energy expenditures during a cold winter. Diesel fuel must be flown in, shipped in, or driven in on slippery winter roads which could lead to high fuel transportation costs.

Another important point that deserves some words here is the fact that based on a research held by Navigant, more than half of the microgrids in the world are now remote microgrids. Therefore, although selecting DER type and mix need be taken into account; designing proper control topology as well as developing efficient energy management system (EMS) need to be fully investigated.

Thus, this book chapter aims to review energy management solutions for the AC/DC microgrids using smart grid advanced functionalities such as the AMI. It primarily explains control topologies of hybrid AC/DC microgrids. It then explains key objectives and constraints of an advanced energy management system solution for hybrid AC/DC microgrids. Energy management system's for remote microgrids are fully discussed in the next section of this chapter. As a case study, a remote 33-node hybrid

AC/DC microgrid is tested under a novel energy management system. Finally, the performance and the applicability of proposed EMS are investigated in detail. In brief, the result of this book chapter could guide grid operators/planners on how to use advanced smart grid-based solutions for designing effective EMS for different hybrid AC/DC microgrids with different topologies and applications.

13.2 CONTROL TOPOLOGIES FOR HYBRID AC/DC MICROGRIDS

In general, it is possible to classify hybrid AC/DC microgrids into two main types:

- Grid-connected AC/DC microgrid: an AC/DC microgrid that is connected to the main AC grid using a breaker;
- Isolated AC/DC microgrid: an AC/DC microgrid that is not connected to the main AC grid. Isolated AC/DC microgrids can be:
    - Islanded AC/DC microgrid: that is not far from the main grid. Hence, it is possible to connect this microgrid to the main grid if it is required.
    - Remote AC/DC microgrid: that is far from the main grid. Hence, it is not possible to connect this microgrid to the main grid.

As explained above, microgrids can be classified into two major groups: isolated and non-isolated, a.k.a. grid-connected. Remote microgrids are typically categorized as isolated microgrids as most of the remote communities have long distances from the main grids. Hence, it would not be cost-effective for these remote communities to connect their grids to the main grids. The most important isolated microgrids are "renewable + fossil" type which use renewable resources along with the fossil fuel system which is typically diesel. As explained before, most of Canadian remote communities are supplied by diesel which causes environmental issues. Moreover, using renewable resources as a portion of the total supply could significantly improve system efficiency and lower the overall costs.

Within a remote microgrid, penetration of renewable resources plays an important role. In a remote microgrid with low penetration of renewables and/or DERs, a system relies more on the fossil fuel-based generation. Although fossil fuel generation would cover most of the supply, using renewable resources could improve system efficiency during specific time intervals such as peak times. Integration of renewables with battery energy storage system (BESS) could help the system to optimally dispatch

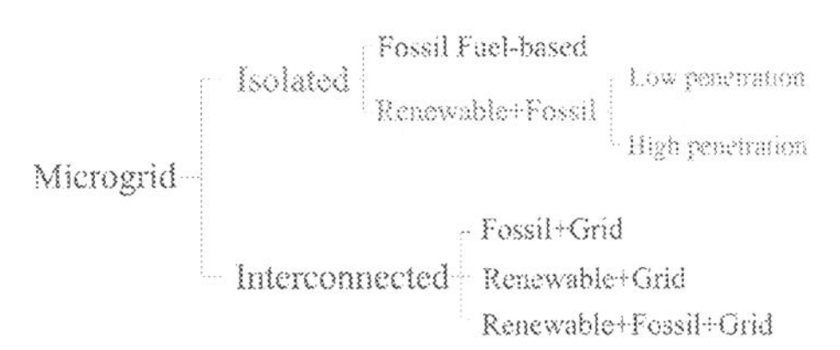

Figure 13.2 Microgrid overall classification.

itself. For instance, the system could use renewable resource generation saved in batteries during the times that the diesel cost is at its highest rate. In remote microgrids with high penetration of renewables and/or DERs, the system could rely more on clean energy rather than the fossil fuel generation. In these types of microgrids, dynamics and the stability of the microgrid would be challenging as there should be a base generation for the system to keep the frequency within standard limit all the time. Fig. 13.2 presents microgrid classification.

Using energy storage systems would be more required for these types of microgrids as the system may hardly manage intermittent generation of renewables. Briefly speaking, it is possible to classify renewable and/or DER integration to three groups: low, medium, and high integration. In low integration of renewables, typically less than 10% to 15%of total generation comes from renewables. In most of these microgrids, as generation intermittency level is low and controllable, the system could operate without using any type of EMS, i.e., local control could be sufficient. For the medium integration, EMS is required. In this case, EMS has to control renewable and DER's power, control all generating sources, provide load shedding plans as well as spinning reserve. In microgrids with high renewable/DER integration, EMS needs to be more sophisticated as it should control microgrid stability, perform demand side management, as well as base generating units for covering the base load. Fig. 13.3 summarizes renewable/DER integration classification. Moreover, Table 13.5 represents the main characteristics of isolated microgrids with different penetration levels of renewables/DERs.

For both grid-connected and isolated AC/DC microgrids, it is conceivable to consider two main control topologies: centralized and decentralized. In centralized control, the processing system is placed in a central controller unit such as energy management system (EMS) in the "utility back-office" [13]. The EMS uses relevant measurements taken from termination points, i.e., utility subscribers, supplied to it from either field

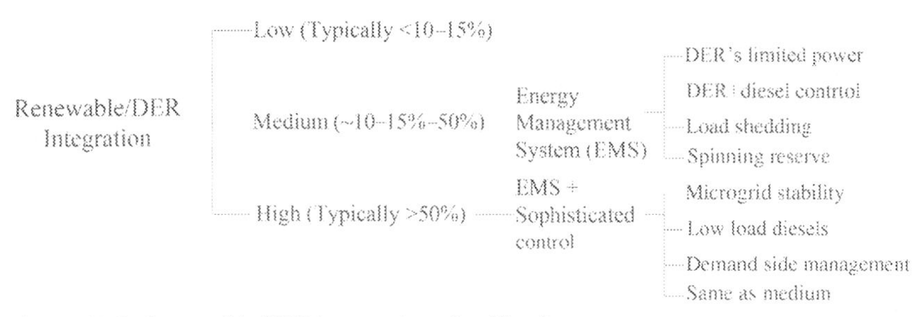

Figure 13.3 Renewable/DER integration classification.

Table 13.5 Main features of isolated microgrids with different penetration of renewables/DERs

	Low Penetration	Medium Penetration	High Penetration
Component control level	Low	Normal	High
EMS	Very simple/ local control	Normal EMS	Comprehensive EMS
Stability control	Normally not necessary	Necessary	Extremely necessary
Energy storage need	Not necessary/ optional	Optional	Necessary
Renewable operational cost	Low	Normal	More than medium
Fossil fuel operational cost	High	Normal	Low
Overall EMS cost	Low	Normal	More than medium (can design a low cost-effective EMS)

collectors or directly from measuring data management system (MDMS), to determine the best possible settings for field-bound assets to achieve the desired operation targets [13]. These settings are then off-loaded to such assets through existing downstream pipes, such as SCADA network [13]. The control system reliability is an important concern of this method that has to be considered. Failure in central control server could

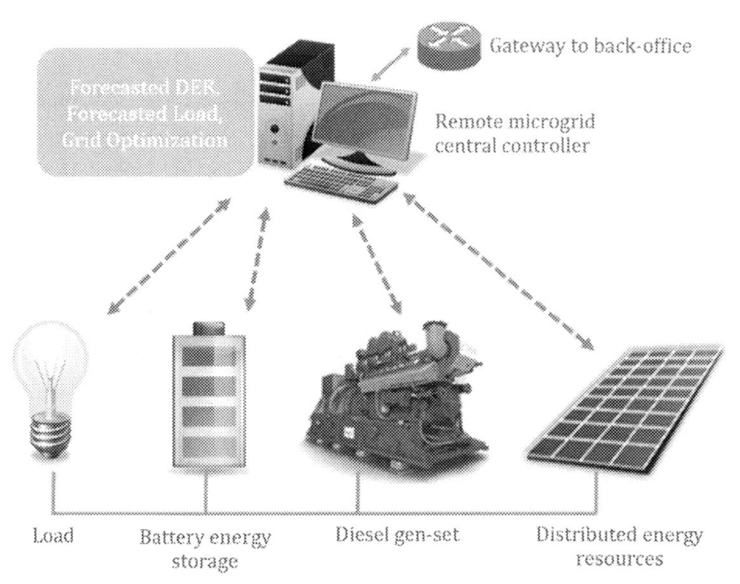

Figure 13.4 Centralized control topology of remote microgrids.

shut-down the whole system. Moreover, centralized VVO uses geographic information system (GIS) and network topology as the basis to determine the targets for each tributary. However, due to the challenges regarding access to real-time downstream sensory inputs, such functions rely on statistical load profile, rather than real-time load profiles. Fig. 13.4 shows centralize control topology for remote microgrids.

On the contrary, decentralized control technique employs local control to optimize the operation of remote microgrid components. That is why it is possible to call this approach "substation or feeder-based" approach. In this technique, local controllers receive required data from termination points to optimize specific remote microgrid feeders. Different methods can be used for sending local data from advanced metering infrastructure (AMI) to local controllers. In decentralized approach, controller engines can be located in the field, and in close-proximity to the relevant assets to optimize system needs according to local attributes of the remote microgrid [13]. Fig. 13.5 represents decentralized control topology for a typical remote microgrid. In decentralized approach, real-time local measurements of AMI do not need to travel from the field to the back-office and the new settings for remote microgrid assets are determined locally, rather than by a centralized controller. This could lead to a cheaper AMI and communication costs.

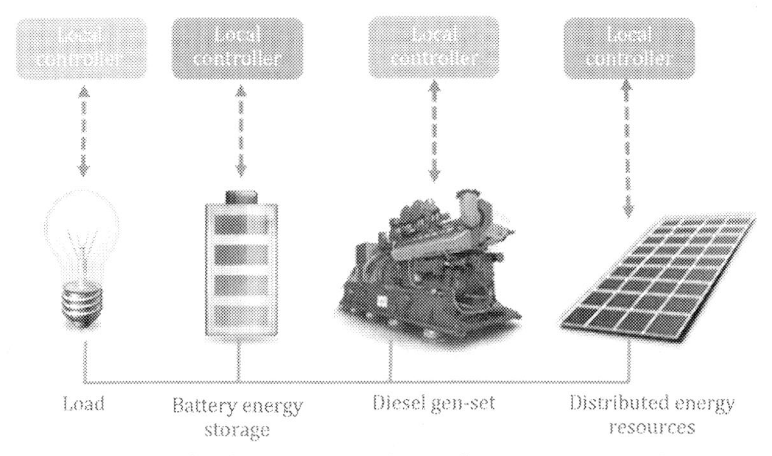

Figure 13.5 Decentralized control topology of remote microgrids using local controllers.

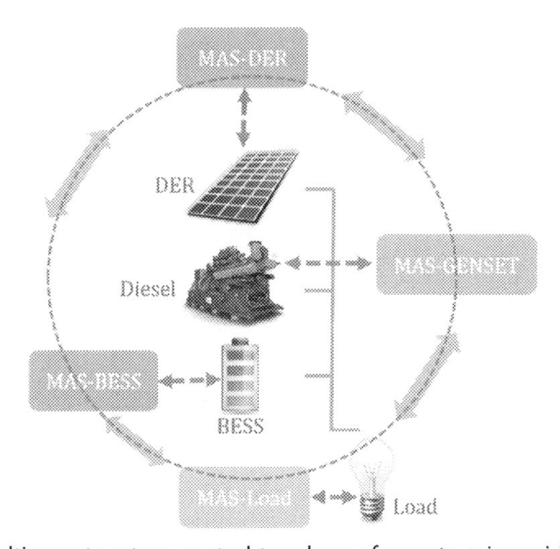

Figure 13.6 Multi-agent system control topology of remote microgrids.

As in decentralized remote microgrids control topology is not centralized anymore; controlling assets are done locally. This local control approach can be performed by a new generation of intelligent systems called multi-agent systems (MAS). In such decentralized control method, each MAS could perform its tasks and it can communicate with other multi- agents simultaneously in order to minimize the risk of failure and increase system response time. Fig. 13.6 shows a remote microgrid that is controlled by MAS topology.

13.3 ISOLATED MICROGRID ARCHITECTURES

Typically, there are three types of architectures for isolated microgrids. Each of these architectures have their own specific features. Typically, choosing proper architecture depends on system existing topology, system load types, budget, and the reliability level that the system requires.

13.3.1 Isolated Microgrid with AC Bus

In this type of microgrid architecture, all generating sources are connected to an AC bus through AC/AC converters for hydro and/or wind turbines and through DC/AC inverters for PVs and batteries. The AC bus/feeder would supply the load, which typically is AC load. As it is seen in Fig. 13.7, this topology needs to use several converters and inverters for different DER sources.

13.3.2 Isolated Microgrid with DC Bus

In this remote microgrid topology, all generating units are connected to a DC bus. Hence, the system is using AC/DC converters for generation sets and hydro as well as wind turbines. AC loads can be supplied by using a DC/AC inverter which is connected to the DC bus. This topology shown in Fig. 13.8 is more applicable when there is a significant amount of DC loads available in the system.

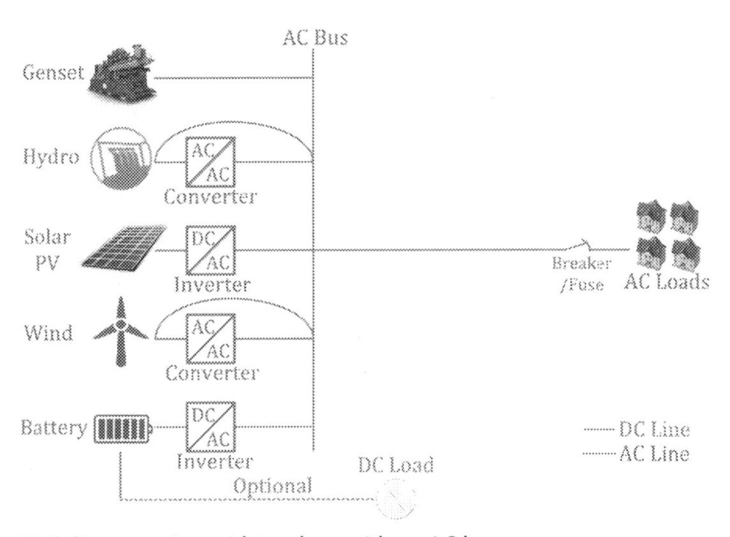

Figure 13.7 Remote microgrid topology with an AC bus.

13.3.3 Isolated Microgrid with DC Bus and Bi-Directional Inverter

In this remote microgrid topology, all generating units are connected to a DC bus. Again, the system is using AC/DC converters for hydro as well as wind turbines. AC loads can be supplied by using a DC/AC bi-directional inverter which is connected to the DC bus. Bi-directional inverter could help the system to balance active and reactive power when the system consists of both AC and DC loads. In this topology, diesel gen-set is located at the AC feeder. As topology shown in Fig. 13.9 uses bi-directional inverter, it has more flexibility in terms of power generation than other types of topologies. Moreover, this topology is used when

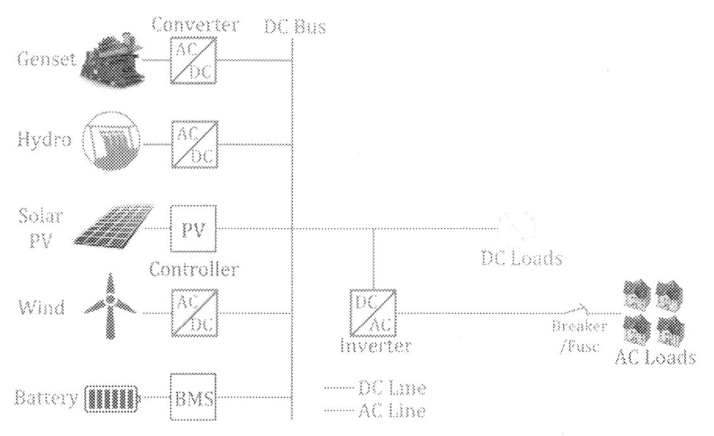

Figure 13.8 Remote microgrid topology with a DC bus.

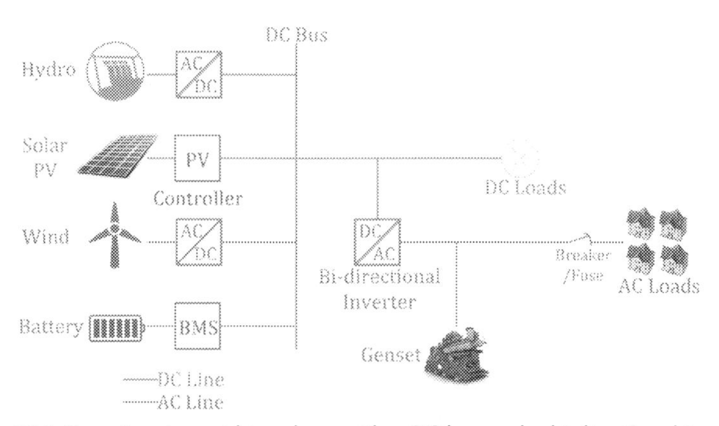

Figure 13.9 Remote microgrid topology with a DC bus and a bi-directional inverter.

a conventional gen-set is already in AC bus but the system needs to be expanded through using DC generating sources such as PV or energy storage system (ESS).

In brief, remote microgrid topology depends on several key factors that have to be taken into account in order to choose which topology would fit remote microgrid aims:

- renewable/DER source types (DC, AC);
- generation types and mix;
- number of customers;
- load types and mix;
- penetration level of renewables/DERs;
- capability of using dispatch-able sources for supplying the base load, e.g., ESS;
- heat recovery need;
- economic factors: budget, O&M cost of renewable/DER technology compared with conventional diesel generation;
- environmental factor (reducing GHG emission);
- expandability of the grid based on system planning.

From a communication network point of view, centralized approach needs a real-time bilateral communication network between remote microgrid and the back-office, but in decentralized control approach, the grid is operating locally and the data can be stored locally as well. If we study control topology from microgrid point of view, we could observe that the trend is moving from centralized control to decentralized control in many microgrid cases as decentralized control is more in-line with distributed command and control topology of microgrids. Table 13.6 illustrates microgrid control topology historical trend.

An important operating scenario has also been made by the presence of local generations, is involving consumers in electricity buy and sell process and giving the right to consumers to choose the type of their own electricity. Customers are now prosumers as they are not only purchasing electricity, but they can also supply a part of microgrid electricity as well. The important degree of this point in remote communities depends on the customers in these communities. It has been seen in some remote communities that consumers only expect the power supply from their utility and they are not eager to involve themselves in the selection process.

In summary, the method of selecting control topology of remote microgrids in remote communities is based on technical factors, e.g., objective function, constraints, number of loads, load type, type and age

Table 13.6 Microgrid control topology historical trend

Attributes	1st Generation (1980–1990)	2nd Generation (1990–1998)	3rd Generation (1998–2008)	4th Generation (2008–Now)	Near Future
Load profile	Static	Static	Static	Dynamic, source: aggregated AMI data	Dynamic, source: disaggregated AMI data
Topology	Local	Local	Centralized thru SCADA	Distributed through Local Control	Distributed through intelligent agents (IAs)
Control assets	Substation-based	Substation-based	Substation-based	Feeder-based	Feeder-based + customer assets
Off-grid application	Yes	Yes	Yes (not improved)	Yes	Yes

of existing generation sources, assets and other existing infrastructures such as the communication network, as well as economic costs such as operation and maintenance, control, communication network infrastructure, AMI, and EMS costs. As each control topology includes an EMS, having a reliable remote microgrid EMS for remote communities is one of the most important management tools of an advanced remote microgrid.

13.4 HYBRID AC/DC MICROGRID ENERGY MANAGEMENT SYSTEM SOLUTIONS

Nowadays, EMS as one of the main elements of a microgrid plays a key role in the optimized operation of microgrids. This role can be very critical in isolated AC/DC microgrids as these types of microgrids are not connected to the main grid that could cause stability and real-time dispatch issues, especially in the presence of high penetration of DERs. Since each and every microgrid is different in terms of control topology, DER sources, components, jurisdictions, etc., and according to operational and economic needs, a microgrid EMSs need to bring benefits to such grids in terms of energy management and optimization. For designing an efficient AC/DC microgrid EMS, grid main objectives, and constraints need to be addressed.

13.4.1 AC/DC Microgrid Main Objectives

Typically, it is possible to count five main objectives for an isolated AC/DC microgrid EMS:

13.4.1.1 Energy Cost Reduction

It is possible to minimize the cost of energy by precise dispatching of DERs within an isolated microgrid in specified operational time intervals. Economic dispatch could be vital for such microgrids. This objective would be more applicable in which the time-of-use (TOU) rate is used. If there is a penalty for consuming energy during peak demand times, energy cost minimization objective would be very critical. In few cases in which remote microgrids are able to connect to the main grid through a normally open tie-breaker, economic dispatch could be done based on real-time price in the presence of market participants. However, most remote communities are not dealing with market dispatch. The main challenge of EMS for most isolated microgrids would be the way EMS balances generation and load based on diesel cost and DER cost. Another challenge for isolated AC/DC microgrid EMS would be providing a real-time dispatch platform based on selected control topology of microgrid. In addition, in isolated microgrids which use a combination of DERs and conventional diesel generator, minimization of the energy cost would be an optimization problem that needs to be solved through a proper optimization technique. The objective of using DERs as an alternative source of diesel generation is in-line with other EMS objectives such as reducing greenhouse gas (GHG) emission.

Hence, the main objective function for proposed AC/DC microgrid EMS in this chapter can be defined as a multi-objective optimization problem that minimizes five different cost functions (F) that are multiplied by their correspondent weighting factors (w) shown in Eq. (13.1):

$$F = \min\left\{\sum_{r=1}^{r=5} w_r.F_r\right\} \tag{13.1}$$

As explained, the first cost function can be written as Eq. (13.2):

$$F_1 = \sum_{t=1}^{t=T}\sum_{i=1}^{i=I} \pi_{i,t} \times P_{i,t} \times \Delta t \tag{13.2}$$

where,

$\pi_{i,t}$ is the electricity price at time-t for node-i (\$/kWh),

$P_{i,t}$is the active power generation of node-i at time-t (kW),

Δt: is the operating time interval (h).

13.4.1.2 Greenhouse Gas (GHG) Emission Reduction

The overall emissions could be reduced by utilizing local generation to supply load in a remote microgrid by DER productions with less GHG emission, i.e., optimally relies more on DER generations with less GHG emission rather than diesel generation. Reducing GHG is one of the main policies of many countries. Although using DERs could decrease GHG emission, high penetration of DERs could increase system operational costs or reduce isolated microgrid reliability. GHG cost function can be written as Eq. (13.3):

$$F_2 = \left(\sum_{t=1}^{t=T}\sum_{i=1}^{i=I} G_{i,t} \times P_{i,t} \times \Delta t\right) \times \sum_{t=1}^{t=T} K_{GHC,t} \tag{13.3}$$

where,

$G_{i,t}$ is the GHG emission generation at time-t for node-i (CO2-ton/kWh),

$P_{i,t}$is the generation of node-i at time-t (kW),

Δt: is the operating time interval (h),

$K_{GHG,t}$: is the emission factor at time-t ($/CO2-tone).

13.4.1.3 Reliability Service Improvement

As typical microgrids enable supplying power locally, they could increase the system level of reliability in terms of local generation. However, developing a comprehensive EMS for isolated microgrids in a way that the operator could select the system level of reliability based on its operational and economic costs seems necessary. In remote communities such as military, mining regions, or remote hospitals, existing high reliability grids are essential. As such, minimizing the outages are being done through techniques such as using backup generation, using BESS, and avoid using high penetration of intermittent sources. Reliability cost function for isolated microgrids can be written as Eq. (13.4):

$$F_3 = \left(\sum_{t=1}^{t=T}\sum_{i=1}^{i=I}\pi_{ENS,i,t}\right) \times \sum_{t=1}^{t=T} ENS_t \tag{13.4}$$

where,

$\pi_{ENS,i,t}$ is the energy not supplied (ENS) cost at time-t for node-i ($/kWh),

ENS_t: is the total energy not supplied at time-t (kWh).

13.4.1.4 Power Fluctuation Reduction/Improve Power Quality

It is necessary to control volatile DER resources in a remote microgrid using different control layers that could be managed through EMS. Power fluctuations have to be minimized in a way that the remote microgrid can be perceived as a controllable system. Recently, using more intermittent sources have made grids more vulnerable to poor power quality. Using BESS in conjunction with DER sources, as well as using other advanced technologies such as smart inverters could improve remote microgrids in terms of power quality, quality of service, voltage fluctuations, voltage and reactive power optimization. This function can be written as Eq. (13.5):

$$F_4 = \sum_{n=1}^{n=N} K_{fluc,n} \times \left| \frac{E(t)_{total,n} - E(t-\Delta t)_{total,n}}{\Delta t} \right| \tag{13.5}$$

where,

$E(t)_{total,n}$: is total energy at time-t for node-n (kWh),
$E(t-\Delta t)_{total,n}$: is total energy at time-($t-\Delta t$) for node-n (kWh),
Δt: is the operating time interval (h),
$K_{fluc,n}$: is the power fluctuation cost for node-n (\$/kW).

13.4.2 Peak Load/Loss Reduction

Typical microgrids are able to shift system loads to minimize consumption during peak times by either curtailing a part of loads or increasing generation. This could significantly help grid planners to defer investment in equipment upgrades as they are typically rated for peak power. Reducing consumption will decrease system losses and improve the overall efficiency of the grid as well. It has to be mentioned that an efficient EMS for isolated microgrid could perform loss reduction based on generation capacity and mix. Using BESS to store PV in PV-diesel microgrids could significantly help remote communities to avoid peak shaving. In advanced EMS solutions, a volt-var optimization(VVO) function shown in Eq. (13.6) can minimize peak losses in isolated microgrids:

$$F_5 = F_{VVO} = \sum_{t=1}^{t=T} \sum_{j=1}^{j=J} \pi_{loss,j,t} \times P_{loss,j,t} \times \Delta t_j \tag{13.6}$$

where,

$\pi_{loss,j,t}$ is grid loss cost at time-t for node-j (\$/kWh),
$P_{i,t}$: is the active power loss of node-j at time-t (kW),
Δt_j: is the operating time interval (h).

Table 13.7 presents objective priorities of each remote microgrid type. In brief, as explained objectives could be represented as an optimization algorithm, the main core of a remote microgrid EMS must be a reliable and efficient optimization engine.

It has to be mentioned that [14] explained the main advantages of using hybrid AC/DC microgrid compared with single AC or single DC

Table 13.7 Main objectives and drivers for different remote microgrids

Microgrid Applications		**Key Drivers(***Main Driver, **Secondary Driver, *Tertiary Driver)**				
		Social	**Environmental**	**Operational**		**Economic**
Application	Typical customer	Easy access to kW	Reduce GHG emission	Self-supply	Reliability	Fuel and cost savings
Remote communities	Local settlements, local utility, government, IPP	***	***	***	*	***
Industrial/commercial	IPP, Mining Co., Oil & Gas Co., hotels, resorts, factories	*	**	***	***	***
Defense	Government	*	**	***	***	**
University campuses	Research institutions	*	***	*	**	***

grids. To avoid unnecessary DC to AC conversions, an effective solution is building a hybrid DC and AC grid, to couple DC sources with DC loads and AC sources with AC loads [14]. This approach can improve microgrid performance, as it can reduce power conversion losses as well as grid costs [14]. DC loads supply with DC sources directly to avoid costly and inefficient conversion from DC/AC and then AC/DC to charge a battery pack [14]. There are no higher costs, as the system does not use more cables [14].

13.4.3 Technical/Economic Constraints of AC/DC Microgrids [14]

An AC/DC microgrid needs to have the following technical/economic constraints:

13.4.3.1 Node Voltage Magnitude for AC and DC Microgrids

AC and DC node voltages should be within minimum and maximum standard ranges.

$$0.95 = v_{n,t}^{AC-\min} \leq v_{n,t}^{AC} \leq v_{n,t}^{AC-\max} = 1.05P.U \tag{13.7}$$

$$0.95 = v_{n,t}^{DC-\min} \leq v_{n,t}^{DC} \leq v_{n,t}^{DC-\max} = 1.05P.U \tag{13.8}$$

13.4.3.2 Active and Reactive Power Outputs of AC Generating Units

Active and reactive power outputs of generating sources need to be between minimum and maximum limits.

$$P_{g,t}^{AC-\min} \leq P_{g,t}^{AC} \leq P_{g,t}^{AC-\max} \tag{13.9}$$

$$Q_{g,t}^{AC-\min} \leq Q_{g,t}^{AC} \leq Q_{g,t}^{AC-\max} \tag{13.10}$$

13.4.3.3 Active Power Outputs of PV and Wind Generation at DC Microgrid

$$P_{PV,t}^{DC-\min} \leq P_{PV,t}^{DC} \leq P_{PV,t}^{DC-\max} \tag{13.11}$$

$$P_{WG,t}^{DC-\min} \leq P_{WG,t}^{DC} \leq P_{WG,t}^{DC-\max} \tag{13.12}$$

In a DC microgrid, the active power outputs of PV and/or wind power generation units should be between their minimum and maximum ranges.

13.4.3.4 Active and Reactive Power Conversions of Biirectional Inverter

$$P_{Inv,t}^{AC2DC-\min} \leq P_{Inv,t}^{AC2DC} \leq P_{Inv,t}^{AC2DC-\max} \tag{13.13}$$

$$P_{Inv,t}^{DC2AC-\min} \leq P_{Inv,t}^{DC2AC} \leq P_{Inv,t}^{DC2AC-\max} \tag{13.14}$$

$$Q_{Inv,t}^{AC2DC-\min} \leq Q_{Inv,t}^{AC2DC} \leq Q_{Inv,t}^{AC2DC-\max} \tag{13.15}$$

$$Q_{Inv,t}^{DC2AC-\min} \leq Q_{Inv,t}^{DC2AC} \leq Q_{Inv,t}^{DC2AC-\max} \tag{13.16}$$

The above constraints elucidate that the active/reactive power conversions of the smart bidirectional inverter should be between its minimum and maximum limits.

13.4.3.5 Battery Energy Storage System (BESS) Output Power During Charge and Discharge

$$\left| P_{BESS,t}^{AC} - P_{BESS,(t-1)}^{AC} \right| \leq \Delta P_{BESS}^{AC} \tag{13.17}$$

$$\left| P_{BESS,t}^{DC} - P_{BESS,(t-1)}^{DC} \right| \leq \Delta P_{BESS}^{DC} \tag{13.18}$$

$$\sum_{t=1}^{t=e} P_{BESS,t}^{Ch} \leq (\partial_1 \times P_{Cap-BESS,t}^{Ch-\max}) \tag{13.19}$$

$$\sum_{t=1}^{t=r} P_{BESS,t}^{Disch} \leq (\partial_2 \times P_{Cap-BESS,t}^{Disch-\max}) \tag{13.20}$$

Both AC and DC BESSs should have limited active power charge or discharge rates at each time interval. In addition, a BESS charge or discharge must not exceed its maximum charge or discharge capacity.

13.4.3.6 State of Charge (SOC) of AC or DC BESS

$$SOC_{BESS,t}^{AC-\min} \leq SOC_{BESS,t}^{AC} \leq SOC_{BESS,t}^{AC-\max} \tag{13.21}$$

$$SOC_{BESS,t}^{DC-\min} \leq SOC_{BESS,t}^{DC} \leq SOC_{BESS,t}^{DC-\max} \tag{13.22}$$

$$SOC_{BESS,t} = SOC_{BESS,(t-1)} \left(\frac{SOC_{BESS,(t-1)}^{Ch} - SOC_{BESS,(t-1)}^{Disch}}{P_{Cap-BESS}} \right) \tag{13.23}$$

State of charge of BESS in AC and DC grids should be limited.

13.4.3.7 AC and DC Power Flow Balances

$$P_{n,t}^{AC} = P_{n,t}^{AC-G} - P_{n,t}^{AC-Ld} \tag{13.24}$$

$$Q_{n,t}^{AC} = Q_{n,t}^{AC-G} - Q_{n,t}^{AC-Ld} \tag{13.25}$$

$$P_{n,t}^{DC} = P_{n,t}^{DC-G} - P_{n,t}^{DC-Ld} \tag{13.26}$$

The active and reactive power of node-n can be calculated by subtracting load on that load from generation of that node.

13.4.3.8 AC and DC Line Capacity Limits

$$S_{l,t}^{AC} \leq S_{l,t}^{AC\max} \tag{13.27}$$

$$P_{l,t}^{DC} \leq P_{l,t}^{DC\max} \tag{13.28}$$

In AC microgrids, the apparent power of a line (feeder branch) should be less than or equal to the capacity of that line. In DC microgrids, the active power of a line should be less than or equal to the maximum capacity of that line.

13.4.3.9 AC Microgrid Capacitor Bank (CB) limit

$$\sum_{k=1}^{k=K} Q_{CB,t}^{AC-k} \leq \sum_{n=1}^{N} Q_{req,t}^{AC-n} = Q_{req}^{\max} \tag{13.29}$$

Typically, total reactive power compensation by capacitor banks in distribution grids need to be less than or equal to the maximum required reactive power of the grid.

13.5 ENERGY MANAGEMENT SYSTEMS FOR REMOTE MICROGRIDS

The main core of an EMS is an optimization algorithm. This algorithm could either be simple or complex depends on the applications and tasks an EMS has to cover. Nowadays, many electric power utilities present their own EMS solutions for power systems in the market. They are trying non-stop to upgrade their EMS capabilities based on the most recent needs of power system market. Most of these EMS solutions could support thousands of nodes of a grid and employ complicated advanced algorithms. These capabilities in-line with the usage of recent advanced communication standards through the SCADA, increased the power of EMS although it made it more complex. For instance, an EMS in a centralized control topology has to receive a huge amount of data from the AMI and/or smart meters, classify them, optimize the grid, and send control commands to grid components in real-time or very close to real time. This bilateral communication requires a reliable communication network with proper bandwidth.

In isolated microgrid applications, there is no such number of nodes and a huge amount of transferred data. Therefore, most of the existing EMS solutions in the market are over-scaled or over-sized for such microgrids. In addition, the objective function applied in these EMS are less focused on remote microgrid main objectives. Interconnection between local EMS and remote control-center could also be costly in some remote microgrids. As stated, most EMS solutions use complicated algorithms which are not necessarily suitable for remote microgrids. On the other hand, the high purchasing price of these EMS besides limited budgets of most remote communities have become not cost-effective for remote microgrids. It is clear that each remote microgrid plan needs technical and financial studies before final approval. Factors such as internal rate of return (IRR) and payback period are some of the key factors on choosing a remote microgrid plan. In mentioned factors, the costs can be summarized as fuel cost, cost of installation of new generation source, operation and maintenance costs of existing assets, operation and maintenance costs of new assets, communication platform costs, EMS cost, controller and actuator costs, and GHG emission cost. On the contrary, the benefits can be written based on remote microgrid objective function priorities and weighted based on that. The benefits can be summarized as benefit of using DER instead of conventional fuel-based source, benefit of increasing grid level of reliability, benefit of reducing GHG emission, and optimized dispatch benefits. In general, cost-benefit analysis can be performed by comparing the abovementioned costs and benefits. The analysis should show how an isolated microgrid can get benefit from using new remote microgrid components compared with conventional off-grid system.

From technical points of view, answering to questions such as how much power needs to be generated from DER sources and how much storage the system needs could create various scenarios for choosing a microgrid plan. Finalizing an isolated AC/DC microgrid design plan depends on key factors such as economic, technical, environmental, cultural, governmental policies, and future expansion plan factors. After choosing a proper plan, minimizing EMS cost could assist the expansion of advanced microgrids. Moreover, it can definitely help the remote community economy. Now, the main question is although the market and technology exist, why the market is not serving the needs of remote microgrid EMS. Table 13.8 summarizes the main features of a remote microgrid EMS cost with different DER penetration levels. Other than cost features depicted in Table 13.8, remote microgrids from business perspective have to be studied.

13.5.1 Remote Microgrid EMS Main Features

It is understandable that remote microgrid systems are not the main prior target for most electric power utilities as they intend to produce EMS that could simultaneously serve both large grids and small-scale grids.

However, EMS solutions in the market forced remote communities to use EMS that is not typically designed for remote communities. Mostly, they are over-scaled and their objective function algorithms are not necessarily optimized for such grids. Moreover, they impose high EMS costs to the remote communities as they need to purchase a license and in some cases, upgrade their license annually. It seems that electric power utilities have not yet attracted the remote community market although recent researches explained in this chapter show the market the potential of remote microgrids in terms of renewable generation installation,

Table 13.8 Main features of remote microgrids with different penetration of DERs

	Low Penetration	Medium Penetration	High Penetration
Component control level	Low	Normal	High
EMS	Very simple/local control	Normal EMS	Comprehensive EMS
Stability control	Normally not necessary	Necessary	Extremely necessary
Energy storage need	Not necessary/ Optional	Optional	Necessary
Renewable operational cost	Low	Normal	More than medium
Fossil fuel operational Cost	High	Normal	Low
Inertia issue	Low (rely on conventional gen-sets)	Can be harmful (rely on conventional gen-sets and battery)	Very risky (should rely on sources like batteries)
Overall EMS cost	Low	Normal	More than medium (can design a low cost-effective EMS)

infrastructural construction, and EMS production. Here, the missing puzzle of EMS is design and developing specific EMS for isolated microgrids. This EMS needs to have the following main features:

13.5.1.1 Efficiency

Remote microgrid EMS has to perform efficiently in real-time. Moreover, its objective function has to be in compliance with remote microgrid objectives. Each objective function sub-part of isolated microgrid EMS could be weighted based on remote microgrid objective priorities. As the objectives of an isolated microgrid is typically simpler than a large-scale network, EMS that is going to be designed for isolated microgrids would have a simple but fast core engine, with less computational expenses using advanced optimization algorithms that are suitable for a few number of nodes.

13.5.1.2 Flexibility

According to the main objective function of this EMS, this system has to be flexible to be used in various types of isolated microgrids such as military, mining companies, and residential regions. Moreover, using recent advanced protocols such as IEC 61850 that is extended to be used for DERs could support the development of such EMS.

13.5.1.3 Selectivity

Isolated microgrid EMS has to be able to give the operator the chance of regulating its system based on technical, environmental, and economic conditions at any time. With the advancement in artificial intelligence technologies, it is possible to design an EMS for remote communities to perform autonomously by learning from the system.

13.5.1.4 Cost-Effectiveness

One of the most important features of remote microgrid EMS besides being efficient is being cost-effective. Typically, the main optimization of a remote microgrid EMS is with fewer data inputs, lower number of nodes, and smaller objective function and constraints. As such, it is conceivable to design a simple but complete cost-effective EMS for remote communities.

13.5.1.5 Availability and Reliability

The availability of EMS with reasonable price, i.e., affordable, besides capability of using this EMS for different types of isolated microgrids could significantly help remote communities' economy. In few operational cases, the EMS could be an open source to have a very cost-effective solution but important issues such as security and cyber-security could avoid usage of open source approach for the EMS design. Microgrid EMS has to operate automatically and manage the whole microgrid in real-time. In many remote communities, the operators or consumers do not intend to be involved in operation and maintenance (O&M) of such systems. The reason lies in a fact that it is more economical for remote communities to have the system with minimum O&M costs such as crew and maintenance costs. Therefore, a proper EMS system could help such a microgrid to perform autonomously with minimum outages and reliability costs.

Table 13.9 summarizes the main features of a cost-effective user friendly isolated microgrid EMS compared with current EMS solutions in market. It has to be mentioned that the communication between

Table 13.9 Main features of current utility EMS compared with isolated microgrid EMS

System Functionality	Current Utility EMS	Isolated/Remote Microgrid EMS
Control and command architecture	Centralized	Distributed/ local control
Network topology	Independent	Configurable
Optimization engine	Multi-objective	Adaptive/predictive, multi-objective
Planning	Slow	Fast
Data capturing	Close to real-time	Close to real-time
Nodes to control	More than thousands	Less than few hundreds
Control parameters	QoS, reliability, etc.	Different cost, GHG, etc.
Monitoring and control platform	Hard to understand	User-friendly
Operational instruction	Hard to understand	Easy to learn
Overall EMS cost	Expensive	Cheap

remote microgrid EMS and other grid components would also be critical. With most remote microgrids with centralized control topology, the communication is between EMS with AMI, EMS and back-office (or sometimes EMS sits in back-office), and EMS with other power generation and power quality components. On the contrary, in remote communities with decentralized control, communication of EMS with the AMI, local controllers and/or MAS are important. In general, reliability of such communication platform is based on the type of the application as well as the type of control topology. In conclusion, it is possible to reach a new generation of EMS specifically designed for remote communities with great capabilities such as being small-scale, cost-effective, use-friendly, reliable, flexible, accessible, and efficient according to the importance and market potential of remote microgrids in the near future.

13.6 CHALLENGES AND OPPORTUNITIES

13.6.1 Monitoring and Control

Providing a simple user-friendly monitoring and control platform that could monitor and control generation, loads and grid operation are one of the most important objectives of this project. The proposed EMS has to be able to visualize grid events in a simple and user-friendly manner. It has to store and recall historical data and analyze energy and power flows in real-time. Providing a graphical interface could help system operators and even non-professional users to understand a general overview of the grid as well as grid snapshot at the same time. Power system parameters related to power generation, energy consumption, storage system operation, breaker status, costs, etc. as well as communication parameters have to be recorded and graphically shown. The graphical display has to be based on grid topology and color coding which assist users to track different system parameters easily.

13.6.2 Planning

As the first level of operation of proposed EMS, reliable planning of generation and loads is necessary. Planning needs extensive knowledge of different operational and economic parameters with some that are fixed and some variable within short-term, mid-term, or long-term horizons. To provide a short-term planning, proposed EMS has to be able to track generation and loads in short-term horizons and optimally find the best planning solution that has to be in-line with mid-term and long-term plans as well. Studying DER capacity, expansion planning, and expected demand growth besides monitoring microgrids asset conditions, are some of the main tasks of proposed EMS regarding its planning level. Providing an accurate load forecasting engine for different time horizons (short, mid, and long terms) would be one of the main objectives of this project. The proposed EMS should be able to plan a microgrid in advance. This could be done for a short term (e.g., a week) through using short time intervals such as 15min time intervals, mid-term (e.g., month to 1 or 2 years) or long term (typically more than 2 years) through using hourly time intervals. Precise modeling of generating units, loads, and dispatchable resources such as ESS and electric vehicles (EV) as well as accurate data capturing close to real-time have to be done within proposed EMS optimization and planning engine. Regarding load and generation forecasting, configurable forecast periods could increase the system's level of flexibility in a way that different forecast periods can be chosen for different planning and optimization aims.

13.6.3 Dispatch

The next level of an effective EMS to guarantee that the energy supplied would be equal to the energy demanded is dispatch. A dispatch problem should be solved based on EMS;s objective function and grid operational and economic constraints according to the latest data that could be captured from different microgrid components. Adequate power flow technique has to be chosen in order to be sure that the system is balanced in terms of generation and consumption. Other constraints such as thermal overload of lines, frequency and voltage fluctuations, have to be avoided. Operational issues may occur while using DERs such as PV and wind turbine from the volatile nature of DER power generation as inconsistent and intermittent sources of energy may lead to a system with excess generation, or with excess load. Although such renewable resources can be curtailed in order to generate lower levels of power, they cannot produce more power than what is being provided by their nature. Hence, load following capabilities have to be done by the proposed EMS to lower the peak power flow of the grid when there is a low correlation between peak load time and the peak of the intermittent generation. As such, optimal dispatch of microgrids within proposed EMS has to significantly improve power quality of the grid.

13.6.4 Real-time Optimization and Control

It is crucial for isolated/remote microgrids to operate autonomously without or by optimal interruptions to secure a reliable grid. Hence, proposed EMS has to operate with different controllers of components within a microgrid. Power conversion systems, battery management systems, protection intelligent electronic device manager, and substation control systems are some of the controllers that proposed EMS has to work with reliably. Moreover, any deviations from the energy demand that is calculated in a previous time interval should be smoothen in a cost-effective approach by re-adjustment of DER generations, storage, or controllable loads within microgrid.

13.6.5 Typical Energy Management System

Fig. 13.10 presents a typical EMS architecture. In general, an EMS has to communicate with supply, e.g., DERs, diesel, etc., to control power generation. Moreover, it has to operate with demand side/customers in order to collect required data from customers through systems such as AMI, and to communicate with other systems such as battery/EV energy management system (BEMS, EV-EMS). Moreover, performing demand side management can be achieved through a reliable communication between EMS and customers' dynamic load profiles. If a remote microgrid can be connected to the main grid through a breaker, the EMS has to operate

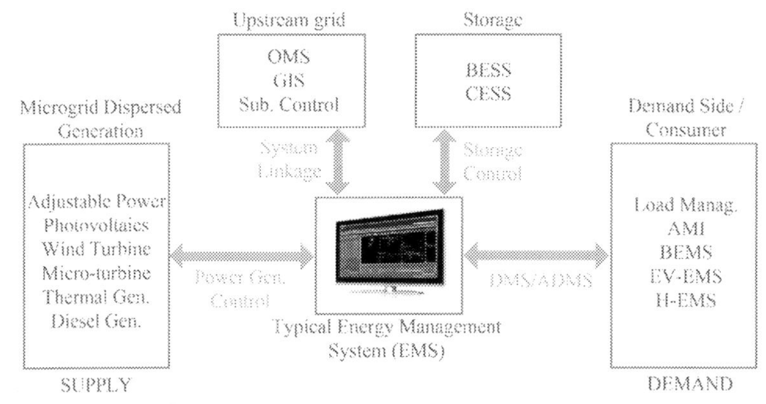

Figure 13.10 Typical energy management system.

with upstream grid systems such as operation management systems, GIS as well as substation control. As stated before, other recent important components of a remote microgrid could be storage. EMS has to work with local energy storage systems through a reliable communication platform. In brief, typical EMS tasks can be classified into controlling power generation, control and monitoring demand side, system linkage to the main grid and storage control.

13.7 REMOTE MICROGRID EMS OPTIMIZATION ENGINE

As explained before, remote microgrid EMS follows a multi-objective optimization approach since it is comprised of different objectives with different weights. However, the main optimization model should have the following capabilities:

- Ability to capture component's live data: this could help the optimization engine to perform faster. Some applications such as energy conservation and smart grid adaptive VVO [15] can use quasi real-time stages based on AMI data capturing time intervals which are typically 15min in North America.
- Ability to receive main settings: remote microgrid EMS optimization engine has to receive system topology, system configuration changes, live fuel costs, start-up/shut down costs, etc. The main settings can change in different operating time intervals. Hence, it is necessary for the EMS to receive the main settings at the beginning or before each optimization start time.

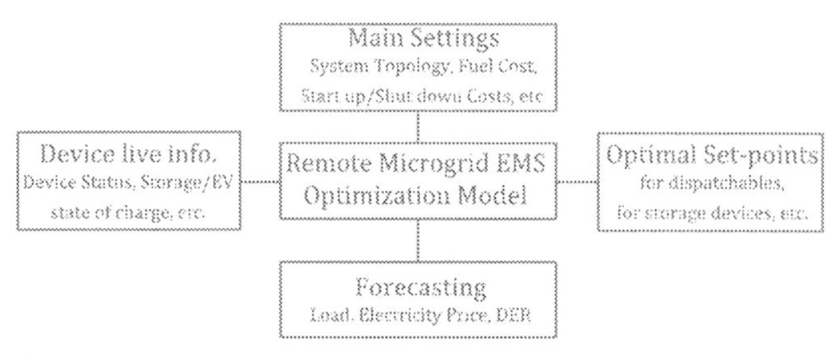

Figure 13.11 Remote microgrid EMS optimization model.

- Close to real-time operation capability: in order to have more accurate EMS, forecasting of consumption, generation, electricity and fuel prices could be very helpful. It is possible to provide the EMS optimization engine with a higher level of accuracy and efficiency using advanced optimization algorithm.
- Optimal set-points: for dispatchable resources such as storage and EV, providing optimal set-points for the optimization model can be very helpful. Moreover, in the main objective function of the EMS, if the operator determines maximum and minimum desired set-points of each objective function subparts, it would be much easier for the optimization engine to solve the optimization problem faster and in less iterative steps. Fig. 13.11 depicts remote microgrid EMS optimization model

13.8 REMOTE MICROGRID PROPOSED EMS

A user-friendly cost-effective remote microgrid EMS has to operate with the following systems:

- remote microgrid generation: PV, wind, diesel, gas engines, hydro, and other technologies;
- remote microgrid loads: typical loads, and dispatch-able loads such as EVs;
- optimization engine: it is possible to create an optimization engine that balances generation and consumption using proper power flow;
- microgrid controllers, power conversion systems: remote microgrid EMS has to communicate with other microgrid controllers and power conversion systems of renewables and DERs to accurately control the system's stability and power balance;

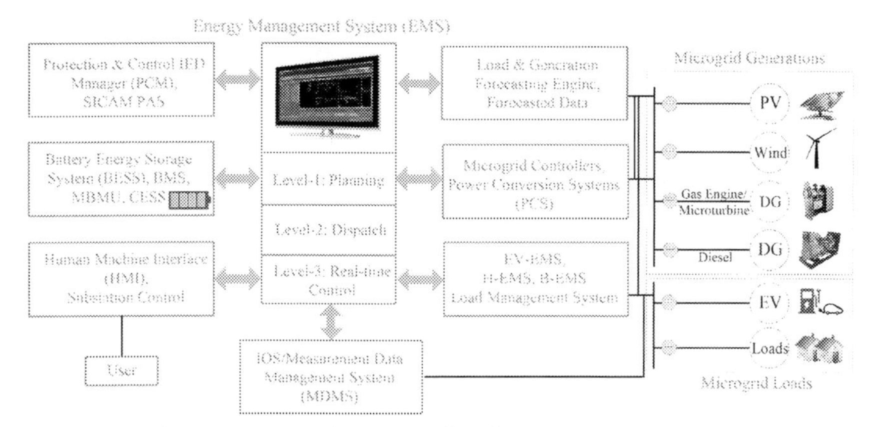

Figure 13.12 Remote microgrid EMS overall architecture.

- demand side systems: remote microgrid EMS requires operating with demand side systems such as local EV-EMS, load management systems, and Home EMS (H-EMS) as well. All these local management systems are able to locally control microgrid components. Hence, reliable links have to provide between these systems and the main EMS to avoid operational interferences;
- measurement aggregation system: remote microgrid EMS has to receive its required data from measurement aggregation systems such as MDMS, AMI or directly from the meters;
- protection: remote microgrid EMS is in direct communication with protection and control IEDs within remote microgrid in order to guarantee system reliability;
- BESS: remote microgrid EMS has extensive data exchange with BESS different sections;
- human machine interface: as one of the main control parts of EMS, remote microgrid EMS has to link with HMI in order to fully monitor the system;s level of operation.

Fig. 13.12, presents overall remote microgrid EMS architecture based on abovementioned communication/grid factors.

13.9 REMOTE MICROGRID PROPOSED EMS TOPOLOGY

Fig. 13.13 presents proposed remote microgrid EMS topology. Remote microgrid EMS communicates bidirectionally with local controllers of generating units and loads, conversion management systems of

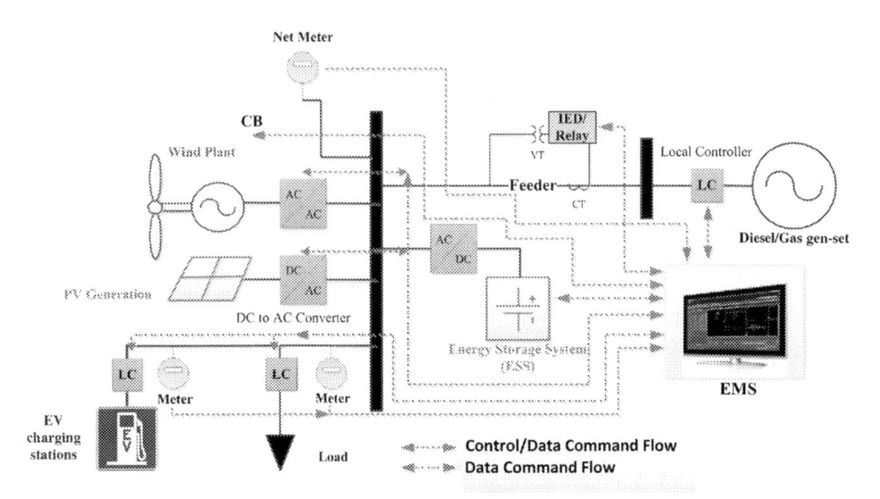

Figure 13.13 Remote microgrid EMS overall topology.

renewables, protection system, ESS and grid connection breaker (if needed). Data command flow shows by green lines in Fig. 13.13. Hence, required data can be collected from meters. As the main loads of remote communities are AC, a topology with an AC bus is proposed here. However, a hybrid AC/DC topology is also preferable in cases in which DC generations can supply DC loads to some extent.

As shown in Fig. 13.13, a user-friendly cost-effective remote microgrid EMS has to communicate with several systems and components within remote microgrid. These communication interactions highly depend on several factors that can be summarized as:

- generation Technology: to design an efficient remote microgrid, the generating units have to be selected carefully based on several factors:
  - geographical situation: close to water, easy access to the main grid, availability of road, availability of solar radiation, availability of wind;
  - remote microgrid main targets: fuel cost saving, emission reduction, O&M cost reduction, reliability improvement, etc.;
  - budget: availability of budget for using different generating sources, availability of budget for putting storage, budget to level-up system reliability.
- load considerations: number of customers as well as the type of customers have significant impacts on remote microgrid topology;

- other required systems such as protection: feeder protection is necessary for the system. The expected reliability level can change the topology of EMS;
- planning: future planning of remote microgrid project could also impact remote microgrid EMS topology as well.

In conclusion, it is necessary to provide a comprehensive cost-benefit analysis in order to find out which generation, load, protection, and communication technologies are suited for a remote microgrid.

13.10 ISOLATED AC/DC MICROGRID CASE STUDY [14]

In this part, a modified 33-node distribution system [16] is applied for the case study. This AC/DC microgrid includes a DC microgrid at one of its feeders and a bidirectional inverter which is located between the AC microgrid and DC microgrid (node-2 and node-19). It is possible to consider this grid as an isolated microgrid or an AC/DC microgrid in an islanded mode considering the fact that typical isolated microgrids have sufficient generation to supply loads during peaks. Fig. 13.14 presents case study SLD and Table 13.10 gives AC/DC microgrid general information. Fig. 13.14 illustrates locations of control actuators such as onload tap changer of transformer (OLTC), voltage regulator (VR), BESS, capacitor banks (CBs), EV charging stations, bidirectional inverter, and AC/DC generating sources. The AC microgrid in this system is comprised of 3 MW natural gas engine generator that is connected to a low voltage grid by a 3750 kVA MV/LV transformer. The AC grid includes various types of components, such as electric vehicle level 1 and level 2 chargers, a BESS, an OLTC with 32 tap steps, a VR with 16 tap steps, and CBs. The DC

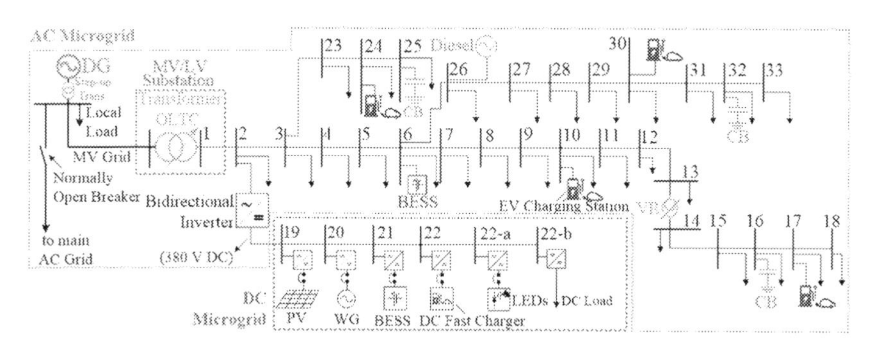

Figure 13.14 Case study: 33-node islanded AC/DC microgrid [14].

Table 13.10 General information of hybrid AC/DC microgrid [14]

AC Microgrid		DC Microgrid		Bidirectional Inverter/ DC-DC Power Converters	
DG + diesel	3000 kW + 1000 kW	PV	250 kW	Capacity	280 kW
BESS	4000 kWh	Wind	100 kW	Efficiency	97 (%)
Average load	2553 kW	BESS	500 kWh	PV	700/380 VDC
Charging stations	Node: 10-17-24-30	Fast charger	0-50 kW	BESS	2*125 kW DC/DC
CBs	0-250 kVAr	LED	3.8728 kW	Charger/ DC Load	380/380 VDC
OLTC/ VR	32/16 Tap 0.95-1.05 P.U	DC Load	89.5562 kW	LEDs	24/380 VDC

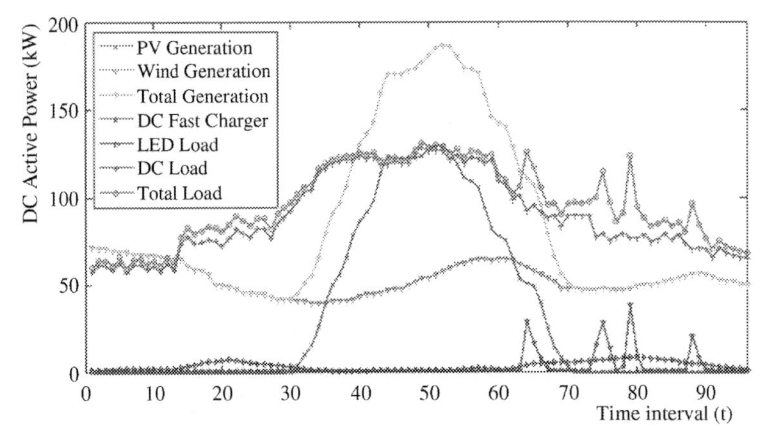

Figure 13.15 DC microgrid active power generations and consumptions for 96-time intervals [14].

microgrid has different generating units and loads such as a PV, a wind turbine, a DC fast charger, a BESS, an integrated LED system that supplies lighting for a parking lot and a neighbor building, and a DC load [14]. Fig. 13.15 shows DC microgrid's generating power and loads captured from smart meters in quasi real-times (every 15mins) for a whole day, and Fig. 13.16 gives maximum generation limit and daily load profile of the AC grid for 96-time intervals of the targeted day collected by the AMI.

The reason for using 96-time intervals a day lies in a fact that most AMI data, especially in North America, are collected every 15 min from the smart meters using data collectors.

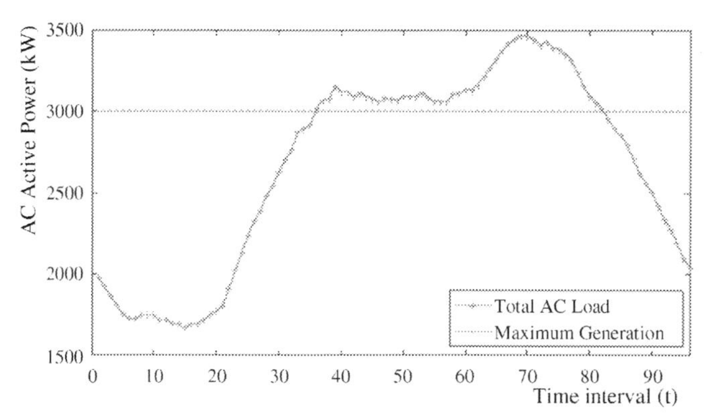

Figure 13.16 AC microgrid maximum generation limit and load profile for 96 operating time intervals [14].

Table 13.11 ZIP coefficients of AC/DC Microgrid [14,20,22]

Coefficients:	*Z*	*I*	*P*
AC microgrid loads	0.418	0.135	0.447
AC EV charging stations	0.16	0.26	0.58
DC microgrid loads	0.1	0.65	0.25

PV generation, DC fast charging station data and DC load data are captured by the AMI [17] for a whole day. For wind power generation, data of the same day reported in [18] is used in this study. ZIP coefficients from [19] are applied for LED aggregated load. Various types of ZIP coefficients are introduced and explained in [20] for various types of electric vehicles in different scenarios. To find ZIP coefficients for level 1 and 2 charging stations, this study calculates the average ZIP values of all four scenarios explained in [20]. For normal operating condition of the AC loads, the ZIP coefficients in [21] are applied. Table 13.11 represents ZIP coefficients used in this case study. Moreover, Table 13.12 depicts hourly electricity prices of the studied day taken from [14].

13.10.1 Islanded AC/DC Microgrid Case Study Results

In this part, the proposed energy management solution is utilized to obtain AC/DC microgrid optimal operation in islanded mode using explained objective functions and constraints. Proposed energy management engine programmed in MATLAB and AC/DC backward-forward sweep (BFS) technique is used for the power flow. Optimization algorithm setting parameters and final results of the optimization engine can

Table 13.12 Hourly electricity price of AC/DC microgrid [14]

Hour	Price (¢/kWh)	Hour	Price (¢/kWh)	Hour	Price (¢/kWh)
1	6.61	9	24.48	17	26.14
2	5.53	10	24.58	18	43.39
3	6.06	11	24.18	19	31.27
4	9.5	12	27.54	20	24
5	14.84	13	25.04	21	22.41
6	18.58	14	27.45	22	21.49
7	23.16	15	25.01	23	21.47
8	24.72	16	25.02	24	20.54

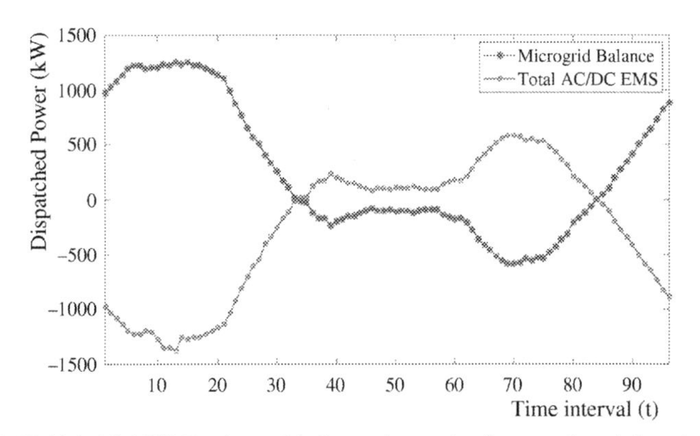

Figure 13.17 Hybrid AC/DC microgrid dispatch results for quasi real-time intervals.

be found in [14]. Fig. 13.17 shows hybrid AC/DC microgrid dispatch results for different time intervals of the studied day. Accordingly, Fig. 13.18 demonstrates how EMS performed to charge/discharge BESS in AC and DC grids and Fig. 13.19 presents AC to DC and DC to AC conversions using proposed EMS.

Fig. 13.20 depicts how performing VVO in the AC/DC microgrid decreased power grid losses. Fig. 13.21 proves the fact that all AC/DC microgrid nodes are within the ANSI-band (0.95-1.05 P.U.). Finally, Table 13.13 gives islanded AC/DC microgrid case study results.

13.10.2 Result Analysis and Discussions

In order to assess how proposed energy management solution could optimize AC/DC microgrid performance, it is necessary to classify operating time intervals into four main scenarios: total generation in both AC and DC microgrids is greater than total AC and DC loads (case1), total DC

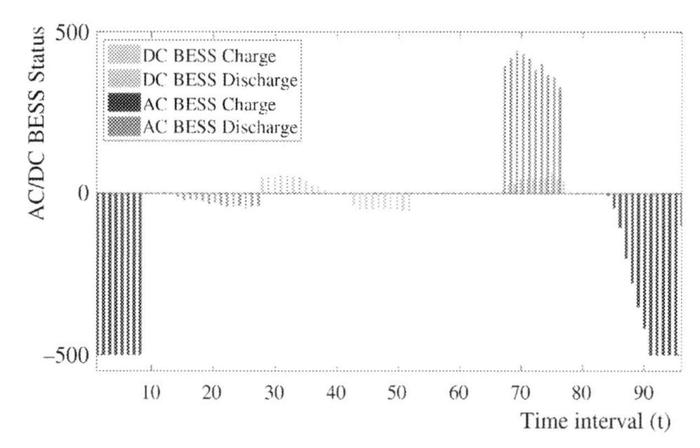

Figure 13.18 Charge/discharge of batteries in AC and DC microgrids.

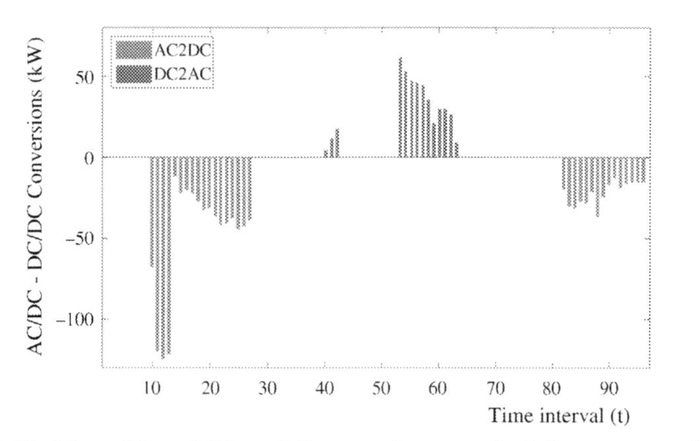

Figure 13.19 AC to DC and DC to AC conversions, resulted from proposed hybrid AC/DC EMS.

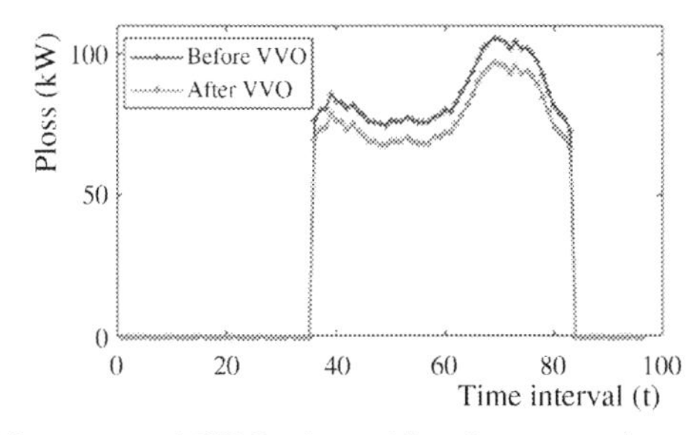

Figure 13.20 VVO impact on AC/DC microgrid active power loss minimization.

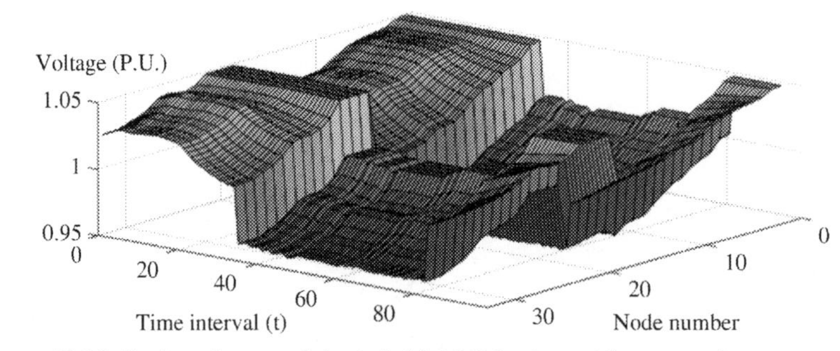

Figure 13.21 Node voltages of the hybrid AC/DC microgrid case study.

Table 13.13 Proposed energy management engine result summary

Summary of Results	Before EMS	After EMS
Average losses	2735.59 kW	2528.12 kW
Average AC grid VVO	0 kW	55.638 kW
Average DC grid VVO	0 kW	1.407 kW
GHG reduction (%)	0	3.23(%)
Total AC2DC	0 kW	1201.66 kW
Total DC2AC	0 kW	412.19 kW
ENS AC	18258.91	2667.41 (KWh*24, i.e., for a whole day)
ENS DC	911.64	213.81 (KWh*24, i.e., for a whole day)
Reliability improvement	0	16287.78
Objective function improvement (%)	(Non-optimized)	18. 05 (Optimized)

generation is greater than total DC load, but total AC generation is lower than total AC load (case2), total AC generation is greater than total AC load, but total DC generation is lower than total DC load (case 3) and both total AC and DC generations are lower than total AC and DC loads (case4).

Now, it is conceivable to analyze proposed EMS solution results in grid operating time intervals:

1. From time interval-1 to time interval-13 (case 1):

   During this time interval, both AC and DC microgrids have extra generation. Hence, both microgrids charge their BESS that provides the system with lesser cost. The energy management engine primarily charges the BESS at the AC side. AC grid can charge the BESS at the DC side as it has extra power generation.

2. From time interval14 to time interval35 (case 3):

   In this time interval, BESS in DC microgrid should be discharged to supply demand. The AC grid has extra power generation to supply the first part of this period. So, proposed EMS discharged DC BESS at the end of this time interval.
3. From time interval 36 to time interval 39 (case4):

   Here, the AC grid cannot be discharged (because of high operating cost). The reason lies in the fact that the system operator already knows that the grid needs BESS power for peak time intervals. The BESS at DC microgrid discharges in this period but, it uses VVO to save the energy consumption and keep BESS charge a bit more.
4. From time interval 40 to time interval 63 (case 2):

   Here, the AC microgrid cannot discharge as well (because of the same reason). VVO at the AC microgrid decreases load curtailment. Moreover, DC with extra generation enables supplying a part of AC consumption. Consequently, BESS at DC microgrid should be charged. Thus, proposed algorithm charges DC BESS in times with cheaper electricity price and lower grid operating costs.
5. From time interval 64 to time interval 83 (case 4):

   AC microgrid peak takes place during this period. Therefore, the BESS of the AC microgrid discharges during peak when power price is high but, in order to decrease the amount of discharge and extend discharge time intervals, VVO performs at both AC/DC sides. DC BESS should be discharged as well. As such, proposed EMS discharges DC BESS when overall DC microgrid cost is at its highest rate. To reduce the amount of discharge and reduce DC load shedding VVO at DC side is performed as well.

   In brief, proposed EMS method has a lower operating cost compared with conventional EMS solutions, as it creates its objective function using precise cost functions.
6. From time interval 84 to time interval 96 (case 3):

   At this time period, the AC microgrid is able to charge its BESS and transfer its extra generation to the DC microgrid if required. BESS in DC side cannot discharge, as it has discharged during peak time intervals. Thus, DC grid requires AC microgrid power. Hence, proposed EMS performs VVO at DC side first and then try to supply the rest of needed power from the AC grid. The operating cost of performing VVO at DC side is low enough so that proposed EMS performs VVO to primarily minimize losses and then supplies required power from the AC grid.

As a result, proposed energy management system solution could optimize AC/DC microgrid operation using an effective AMI-based method. By evaluating case study results (Table 13.13), it can be concluded that the proposed EMS solution could optimize the grid 18% more than using conventional EMS. The greatest impact of performing such an energy management solution that utilizes loss reduction is that it minimizes energy not supplied (ENS) and GHG emissions. In addition, by using precise load models and receiving AMI data, the proposed solution could elevate system precision, especially in control commands that the AC/DC microgrid should enforce to control components every 15 min. In short, the main impacts of the proposed EMS solution include but are not limited to optimizing the AC/DC power transfer, minimizing operating costs of the grid, minimizing GHG emission, minimizing energy not supplied, minimizing grid losses, and saving consumers' energy.

13.11 CONCLUSION

This chapter showed that although the technology exists, the market has not yet served remote microgrids with a reliable, flexible, and cost-effective EMS. It discussed the main reasons such as unawareness of electric power utilities on remote microgrid future market potential, and low priority level of remote microgrids compared with other power system applications and/or networks for electric power utilities. As such, this chapter primarily investigated the necessity of designing an EMS specifically for remote microgrids based on remote community factors such as technical, economic, cultural ,and environmental factors. Then, it elucidates the main issues of presented EMS in the market and numerated the main features of a cost-effective user-friendly EMS for remote microgrids in near future. To test the precision and the applicability of proposed EMS, an islanded 33-node AC/DC microgrid tested for 96 time-intervals for a whole day.

The results of this chapter ensured that by designing smart grid adaptive EMS solutions for islanded or isolated AC/DC microgrids, it is possible to reduce the amount and time of charge/discharge of batteries in both AC and DC grids. Even in some operating time-intervals, using proposed EMS which includes energy conservation subpart, could supply AC or DC demand in full without doing any other operating actions.

The most valuable impact of utilizing such an EMS engine was the minimization of AC/DC microgrid energy not supplied (ENS). Minimizing ENS would be very important for AC/DC microgrids that are operating in isolated/islanded mode, in which load shedding actions are critical, and the optimal load shedding performance is needed. The results of the case study also presented how applying sophisticated EMS objective function can lead energy management systems, as well as AC/DC microgrids, to achieve higher levels of operation, GHG reduction and reliability.

REFERENCES

[1] Microgrids Engineering, Economics, & experience, CIGRE working group C6. 22, Oct. 2015.
[2] Energy Access Database, International energy agency, Available for: <http://www.worldenergyoutlook.org/resources/energydevelopment/energyaccessdatabase/>
[3] Renewable Energies for Remote Areas and Islands (Remote), IEA-Renewable Energy Technology Development (RETD), Final Report, Apr. 2012.
[4] Zaatari Refugee Camp Fact Sheet, April 2016, Available for: <http://data.unhcr.org/syrianrefugees/download.php?id = 10812>
[5] Market Data: Remote Microgrids and Nanogrids, Commodity extraction, physical island, village electrification, military, commercial, and residential remote microgrids and nanogrids: global market analysis and forecasts, navigant research, 2015. Available from: <https://www.navigantresearch.com/research/market-data-remote-microgrids-and-nanogrids>
[6] Status of Remote/Off-Grid Communities in Canada, Natural Resources Canada, (2011).
[7] Enabling a Clean Energy Future for Canada's remote communities, Advanced Energy Centre MaRS Discovery District, Ontario, Canada, Dec. 2015.
[8] D.A. Johnson, Wind-diesel-storage project at Kasabonika Lake First Nation, Wind Energy Group Department of Mechanical and Mechatronics Engineering University of Waterloo, 2009.
[9] F. Katiraei, C. Abbey, Diesel plant sizing and performance analysis of a remote wind-diesel microgrid, IEEE PES General Meeting, Tampa, FL, USA, (2007).
[10] Remote communicates database, Available from: <https://www2.nrcanrncan.gc.ca/eneene/sources/rcdbce/index.cfm?fuseaction = admin.home1>.
[11] G. Colgate, A. Swingler, Remote 'micro-grids' in Nemiah Valley, British Columbia Canada: past, present and future potentials, International Micro-Grid Symposium, Vancouver, Canada, July 2010.
[12] Diesel Price in Canada, Available from: <http://www2.nrcan.gc.ca/eneene/sources/pripri/prices_byyear_e.cfm?ProductID = 5>.
[13] M. Manbachi, M. Nasri, B. Shahabi, H. Farhangi, A. Palizban, S. Arzanpour, et al., Real-time adaptive VVO/CVR topology using multi-agent system and IEC 61850-based communication protocol, IEEE Trans Sust Ener 5 (2) (April 2014) 587–597. Apr. 2014.
[14] M. Manbachi, M. Ordonez, AMI based energy management for islanded ac/dc microgrid utilizing energy conservation and optimization, IEEE Trans Smart Grid (Aug. 2017). Available from: <https://doi.org/10.1109/TSG.2017.2737946>.

[15] M. Manbachi, A. Sadu, H. Farhangi, A. Monti, A. Palizban, F. Ponci, et al., Real-time co-simulation platform for smart grid volt-var optimization using IEC 61850, IEEE Trans Indus Inform 12 (4) (Aug. 2016) 1392−1402.
[16] M.J.E. Alam, K.M. Muttaqi, D. Sutanto, An approach for online assessment of rooftop solar PV impacts on low-voltage distribution networks, IEEE Trans Sust Ener 5 (2) (Apr. 2014) 663−672.
[17] M.J.E. Alam, K.M. Muttaqi, D. Sutanto, A SAX-based advanced computational tool for assessment of clustered rooftop solar PV impacts on LV and MV networks in smart grid, IEEE Trans Smart Grid 4 (1) (Mar. 2013) 577−585.
[18] R. Tonkoski, D. Turcotte, T.H.M. El-Fouly, Impact of high PV penetration on voltage profiles in residential neighborhoods, IEEE Trans Sust Ener 3 (3) (Jul. 2012) 518−527.
[19] M. Tasdighi, H. Ghasemi, A. Rahimi-Kian, Residential microgrid scheduling based on smart meters data and temperature dependent thermal load modeling, IEEE Trans Smart Grid 5 (1) (Jan. 2014) 349−357.
[20] D. Infield, F. Li, Integrating micro-generation into distribution systems - a review of recent research, Proceedings of the IEEE Power and Energy Society General Meeting, Pittsburg, PA, Jul. 2008.
[21] M. Thomson, D.G. Infield, Network power-flow analysis for a high penetration of distributed generation, IEEE Trans Pow Syst 22 (3) (Aug. 2007) 1157−1162.
[22] B. Asare-Bediako, W.L. Kling, P.F. Ribeiro, Integrated agent-based home energy management system for smart grid applications, Proceedings of the IEEE PES innovative smart grid technologies Europe, Copenhagen, Denmark, Oct. 2013.

FURTHER READING

F. Mahdloo, M. Manbachi, M.S. Ghazizadeh, R. Vasigh, New efficient approach for optimal sizing and placement of micro combined heat and power systems on low voltage grids, Proceedings of the CIGRE Canada Conf., Toronto, ON, Canada, Sept. 2012.

CHAPTER 14

Integration of Distributed Energy Resources Under the Transactive Energy Structure in the Future Smart Distribution Networks

Mohammadreza Daneshvar, Behnam Mohammadi-ivatloo and Kazem Zare
University of Tabriz, Tabriz, Iran

14.1 INTRODUCTION

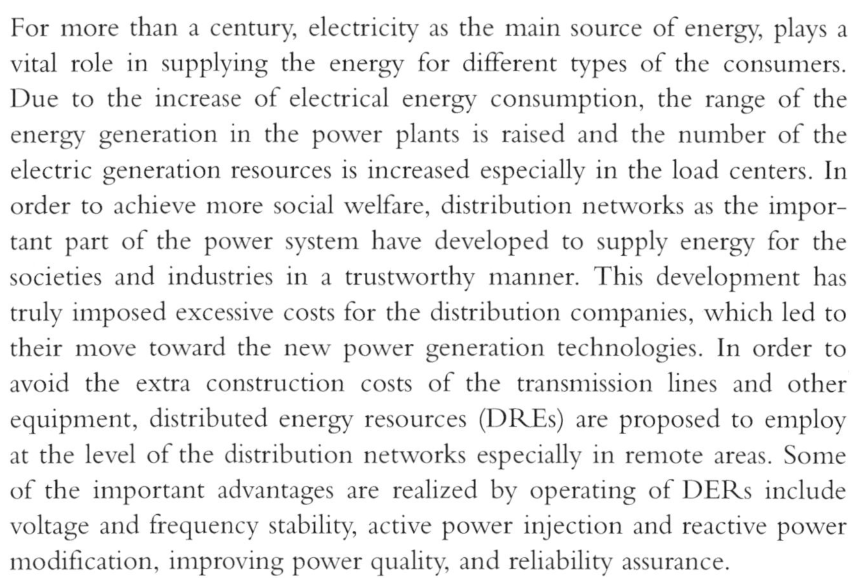

For more than a century, electricity as the main source of energy, plays a vital role in supplying the energy for different types of the consumers. Due to the increase of electrical energy consumption, the range of the energy generation in the power plants is raised and the number of the electric generation resources is increased especially in the load centers. In order to achieve more social welfare, distribution networks as the important part of the power system have developed to supply energy for the societies and industries in a trustworthy manner. This development has truly imposed excessive costs for the distribution companies, which led to their move toward the new power generation technologies. In order to avoid the extra construction costs of the transmission lines and other equipment, distributed energy resources (DREs) are proposed to employ at the level of the distribution networks especially in remote areas. Some of the important advantages are realized by operating of DERs include voltage and frequency stability, active power injection and reactive power modification, improving power quality, and reliability assurance.

Although a part of convenience (as a significant factor of the social welfare) is provided by the electricity, but because of the use of fossil fuels such as gas, coal, and oil in the traditional power plants, environmental factors (as an other important factor of the social welfare) have been violated and greenhouse gas emissions are increased, too. These issues caused the power system engineers to use new energy resources, which

Operation of Distributed Energy Resources in Smart Distribution Networks
DOI: https://doi.org/10.1016/B978-0-12-814891-4.00014-X

environmental issues are considered by them. These resources, which is called renewable energy resources (RERs) can reduce the cost of the electricity generation and environmental problems are also solved by operation of them. Among different types of RERs, using the energy of the wind and solar for electricity generation is conventional and most promising for mankind [1]. The energy of the wind and solar, which can naturally be found in the environment are used to convert to the electrical energy by the wind turbine and solar PV systems, respectively. With mentioned testimonials, RERs are expected to reduce the emissions and total cost of the energy generation and facilitate the operation of them that will cause they have a widespread presence in the future smart distribution networks [2].

Although using the RERs has facilitated the energy supply in the distribution networks and it has also provided significant advantages for the power system, some key challenges are created by applying them for the power system planners. One of the important challenges is that the electrical energy production of RERs depends directly on climate conditions. The uncertainty in the weather forecasting has caused uncertainty in the production of RERs, too. Because of this uncertainty in the electricity generation by RERs, the scheduling of the RERs for optimal presence in the network electrification and the network professional analysis are difficult to realizing the smart structure of the future networks [3]. Due to the existing large amount of t renewable resources in the universe, it is predicted that over the next two decades, RERs will account for approximately one fifth of the total energy consumption in future [4].

In the smart future distribution networks, operation of the various DERs will not be inevitable to respond the amount of the energy demand of consumers. This issue caused that the researches in the electrical field moved to the network assessment with high penetration of the DERs. Currently, these researches have indicated that the power system is faced with basic challenges in the operation of the DERs, which one of the important of them is establishing a dynamic balance between the demand and supply that needs more attention in the scheduling of the electrical generations. Integration of the DERs has been proposed as the main solution to respond to these challenges. Reconfiguration of the existing smart energy systems would be needed for the integration process of the numerous DERs into the power system [5]. The power grid with intelligent devices or smart grid is vital to this transformation. In the future power system, smart grids will be composed of several devices such as all types of the DERs, an intelligent power control system, and a flexible consumption [2]. In this regards, RERs (e.g., PV panels, wind turbine, biomass, fuel cell, smart house, etc.) and energy storage systems (e.g., battery, superconducting magnetic energy storage, electric double layer capacitor, etc.) are anticipated to have a widespread presence in the future smart grid and to support the future electricity demand [6,7]. On the other hand, integration of the DERs need the sustainable technology, which can be applied successfully on the future smart networks. In this regard, transactive energy (TE) is introduced as a reliable and sustainable manner to integrate the numerous DERs, which it would be more efficient technique for the future smart distribution networks. For these goals, the advocator of TE provide reasonable arguments to application of TE in creating dynamic balance in both demand and supply side. The specialized researches in this regard illustrated that this technology can reliably integrate the DERs in the future smart grid with high penetration of the RERs. Therefore, TE as a robust contender is presented to consider the concerns of the both consumers and distribution companies in the future distribution networks with smart structure. TE is consisted of the numerous agent-based nodes, which are continuously distributed in the level of the network. Each of the advanced nodes has key duty on the TE market, which correct performances of them will realize the comprehensive smart control of the system. Indeed, TE nodes are composed of some intelligent multipurpose optimization software, which can be run at any time of the smart system periods. This feature abled the system control center to be informed of the amount of the energy consumption and generation at the any time of periods, which this information would be required to make the optimal decisions for the operation of all intelligent devices in the next periods of the system. Actually, these agents receive the communication signals from each intelligent device, which each of them contains basic information such as energy consumption. Telecommunication platforms are created for transmitting signals at the highest speed allows the all TE nodes to exchange information among themselves and TE control center, which this system capability provides key information about all the smart device activity to make the suitable decisions by the power system planners. This process of the TE market allows for all types of the DERs to the presence in the smart grid and establish the interaction between all intelligent devices in a sustainable manner. Therefore, TE as a modern technology can be applied to the electricity market area with high penetration of the smart systems and by employing the TE technology not only integration of the large amount of the DERs can be realized, but also interoperability between the intelligent systems can be provided.

14.2 DISTRIBUTED ENERGY RESOURCES (DERs)

Wind turbines, PV panels, and other distributed generations are usually located in remote areas, which are fully needed to integrate into transmissions and especially distribution networks for the system effective operation. Reducing greenhouse gas emissions and total costs are two main goals of the integration of all types of power plants in the smart grids. This is done by employing the comprehensive technology and using the intelligent microgrid controllers and devices, which are considered to be executed by distribution companies in the future smart networks. Distributed generation, on-site generation, and local generation are three conventional terms of the distributed energy, which produce energy from the small energy resources. Different types of the DERs are illustrated in Fig. 14.1 [8].

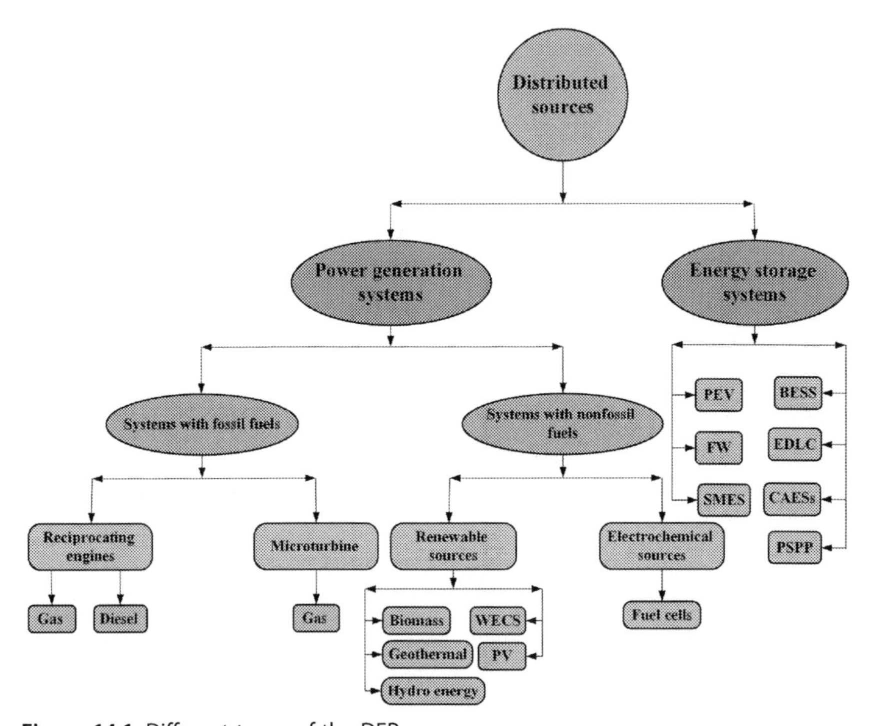

Figure 14.1 Different types of the DERs.

According to Fig. 14.1, there are different types of energy storage, which are used depending on their application in different parts of the network. Some of these storage systems are: battery energy storage systems (BESS), compressed air energy storages (CAESs), pump storage power plant (PSPP), superconducting magnetic energy storage (SMES), flywheel (FW), electric double layer capacitor (EDLC), plug in electric vehicle (PEV), etc. In addition, a wind energy conversion system (WECS) and photovoltaic (PV) systems are two special types of renewable energy, which are widely used for energy production process in smart grids [9,10].

14.3 CLASSIFICATION OF DERs

Characteristically, DERs consist of the small part of the electric power generation, which typically the range of them is from less than a kW to several MW; that is not a large amount of energy for providing the all consumers energy requirements, but they can at least help in establishing energy balance between demand and supply, especially in the peak times. DERs can commonly be operated in two modes, which includes connected to the grid usually with distribution systems and separated from the network or independent operation, which is used effectively for remote areas [11].

In addition to conventional power plants, electrical energy can be generated by the two typical DERs, which includes conventional generation systems such as gas and diesel generators, and nonconventional generation systems such as fuel cells, RERs, and various types of energy storage. The various types of the renewable and non-renewable energy resources are illustrated in Fig. 14.2. Moreover, the features of both renewable and non-renewable energy resources are described in the following subsections.

14.3.1 Renewable Energy Resources

In recent years, greenhouse gas emissions caused by power plants with fossil fuels have increased, which has led to growing concerns about the adverse effects of burning them on the health of living creatures and the environment. On the other hand, the geopolitical climate associated with energy generation from fossil fuel and high energy prices of them leads to the presence of RERs as a vital component in the energy world with mix consumption [12]. Although fusion energy faces with technical challenges

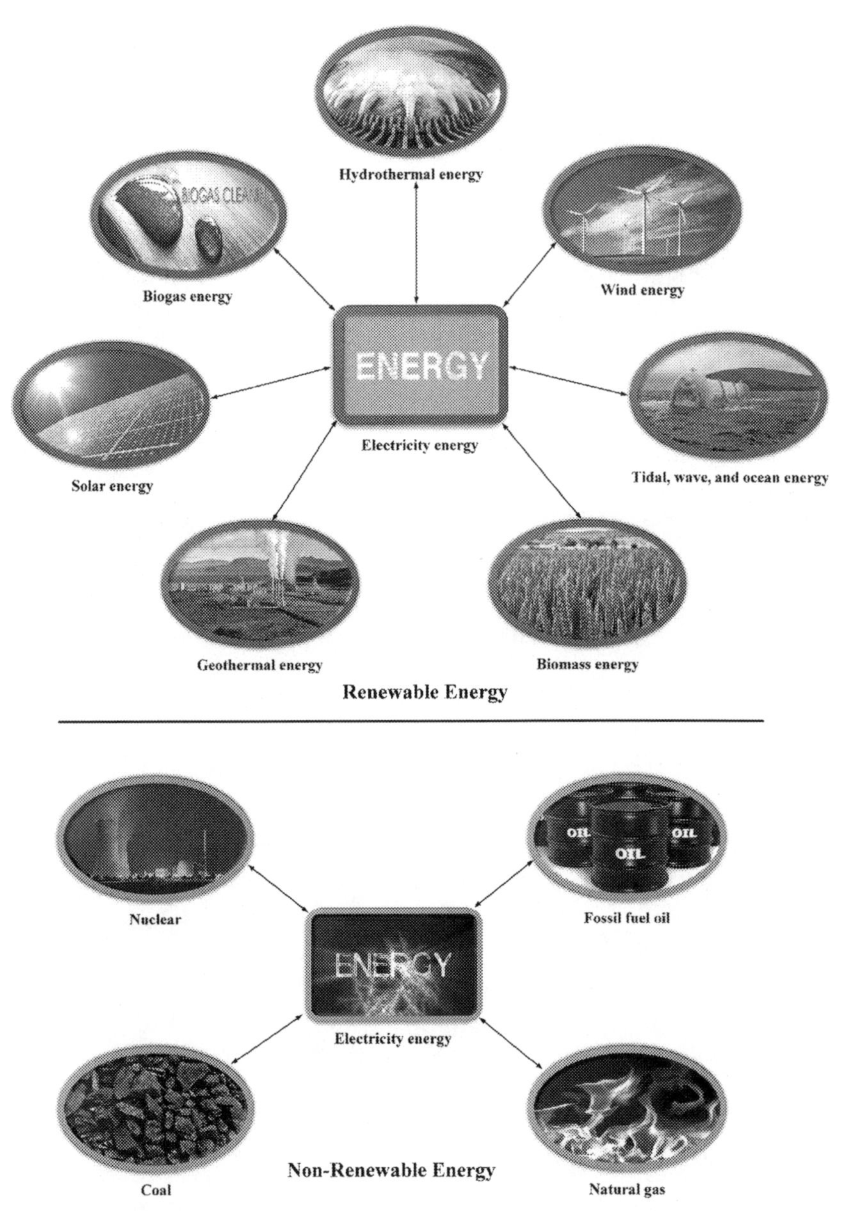

Figure 14.2 Various types of renewable and non-renewable energy resources.

and it takes undoubtedly into account as a fatal and important option, but achieving the advances in the RERs technologies are much easier. RERs include wind, solar, biogas, hydro, biomass, wave and ocean energy, and geothermal are typically non-problematic and clean sources of energy;

these features and their technical simplicity have certainly caused them as an interesting option for mankind [13].

Therefore, because of the zero greenhouse gas emissions of RERs and low electricity generation costs, these resources are converted to the key part of the smart grid, which not only help more to network modernization and facilitate electrification in the future smart distribution networks, but also these features of RERs would be very effective to the comprehensive presence of them in the future bulk power system.

14.3.2 Non-renewable Energy Resources

In past years, conventional or non-renewable energy resources (non-RERs) were the only energy resources that were used for electricity generation in thermal power plants. However, increasing the energy consumption and its popularity among people caused construction of a large number of power plants in the load centers. This development in the field of electricity not only has increased electricity generation costs, but also it has led to the release of a large amount of emissions in the environment. These features of the non-RERs have led researchers to find better energy resources, thus RERs are proposed to distribution networks.

14.4 FEATURES OF DERs

Since electricity energy was invented, it quickly became more popular among people because it has been able to provide more convenience for people with lower social welfare costs. In this regard, distribution networks as the last part of the power system are responsible to supply quality and reliable energy for consumers. Due to the presence of some loads in the remote area, electricity must be transmitted to these regions by the transmission lines. High cost of energy transmission from conventional power plants to remote loads, absence of sufficient land for the construction of large power plants on the periphery of cities, greenhouse gas emissions from large energy generation resources with fossil fuels, and other important reasons certainly caused power system engineers to employ DERs as the best solution for the distribution networks to better cover the mentioned reasons. In general, using the distributed generation resources has some important advantages for consumers which are listed in Table 14.1.

Due to the vital role of electricity companies, especially distribution companies in delivering quality energy to the consumers, distributed

Table 14.1 Some important advantages of distributed generation resources for consumers

Advantages of distributed generation resources for consumers:
1. Enhancing the electrical reliability of distributed generation,
2. Provide appropriate energy source at the suitable place,
3. Preparation of desired electrical energy,
4. Increasing system efficiency for local application with simultaneous use of electricity and heat,
5. The possibility of reducing the cost of paying for electric energy using dedicated units,
6. Providing the possibility of using electric energy for remote areas where the construction of electric networks is impossible or costly,
7. Possibility of exchanging demand response programs.

generation resources have provided some advantages for electricity companies, which help them to achieve their goals. A summary of these benefits is explained as follows:

- avoiding excessive costs of construction and development of transmission and distribution systems by installing the distributed generation units in the vicinity of consumers;
- limiting the level of risk and threats due to size, duration of installation and commissioning, flexibility, environmental compatibility and flexible fuel systems of these production resources;
- providing relatively low cost paths for competitive electricity markets;
- avoiding the costs of uncertainty in predicting available load and capacity by increasing installed capacity in accordance with the load growth;
- providing markets in remote areas lacking transmission and distribution systems, and areas lacking electrical energy due to geographical considerations.

Nowadays, government policies are moving to widespread usage of the DERs, especially RERs in power network. Therefore, some of the considerable advantages of employing distributed generation resources, which convinces the governments to apply them, are tabulated in Table 14.2.

In addition to mentioned advantages for distributed generation resources, applying these resources has some significant disadvantages, which need to be carefully analyzed before the employing them in the power grid. These disadvantages are briefly explained as follows:

- System protection becomes more complex.

Table 14.2 Some effective national benefits of distributed generation resources

National benefits of distributed generation resources
1. Reducing greenhouse gases due to the use of distributed generation resources based on the RERs,
2. Providing thousands of direct and indirect employment opportunities,
3. Increasing the return on various investment due to reliability improvement and improving the quality of electricity delivered to consumers,
4. Creating competitive markets and increasing direct and indirect trades,
5. Maintaining national capital through postponing the construction of new power plants,
6. Creating diversity in energy sources and increasing energy security.

The rate of effectiveness of distributed generation on the coordination of protective elements such as fuses, reclosers, and relays depends on the size of the capacity and type of these resources and their location. In addition, to prevent the power supply by DERs for the sectors of the system that their electricity has been interrupted, appropriate protective measures should be taken.

- In some cases, it causes power quality and reliability problems.
- Because of the intermittent nature of the production of RERs and large number of them in the power grid, these issues have negative impacts on the power quality; and the system's reliability becomes complex, too.
- Operation and network control are more difficult in distribution networks with high penetration of DERs. For example, applying the DERs should not create technical problems for the consumers and networks.
- Telecommunication systems needed for exchanging information between all intelligent devices and their control centers must be have advanced technology and a platform to handle the high volume of exchanging information.

14.4.1 Operation and Control

Generally, operation and control of DERs are a main part of the power grid, which should follow a number of specific guidelines. In the conventional power plants, automatic voltage regulator (AVR) and governor are used to control and regulate voltage and frequency in standard magnitude of them, respectively, which their amounts should not be exceeded from the acceptable range. In addition, the large number of complex equipment in such power plants have reduced the operation speed of them. Today, with a tendency towards using DERs, operation of these resources not only becomes easier, but also the speed of employing them is increased, which these features of DERs have helped more for the distribution system to deliver electrical energy for the consumers in any geographic location. Therefore, applying DERs has followed important advantages from the different aspects such as control, operation, environmental effects, and other benefits, which encourages engineers to use them in the future smart distribution networks. However, in addition to the mentioned advantages for DERs, some considerable concerns are discovered about the coordination and integration of the DERs and other challenges, which many researches are focused on them to find the best solution that can be satisfied network constraints.

14.4.2 Protection and Coordination

The main goal of the power grid protection is preparation of reliable energy supply to consumers. In the power grid protection, control and coordination of the all electrical devices are compulsory. Future smart power grid with a large number of DERs will require the advanced and comprehensive protection systems that can cover all smart devices in terms of control, coordination, and management without any irregularities and inconsistencies in the system. This is done by applying the modern merger technology, which is be able to receive information about the operation of devices at any moment and can integrate them with essential data from the other part of the system to ordain the best program for equipment operation. On the other hand, the protection system will depend on the mode of distribution networks that is ring or radial. However, protection and coordination of numerous DERs are a very complex action, which are converted to the basic concern for the future smart distribution network designers [14].

14.4.3 Stability and Power Quality

Generally, to continue the system performances, some key factors will be needed, which one of the important of them is stability. All of the control system mechanisms try to have interaction with each other to keep the power system in the stable mode. Voltage and frequency stability are two conventional states of stability, which establishment of them is vital for power grid.

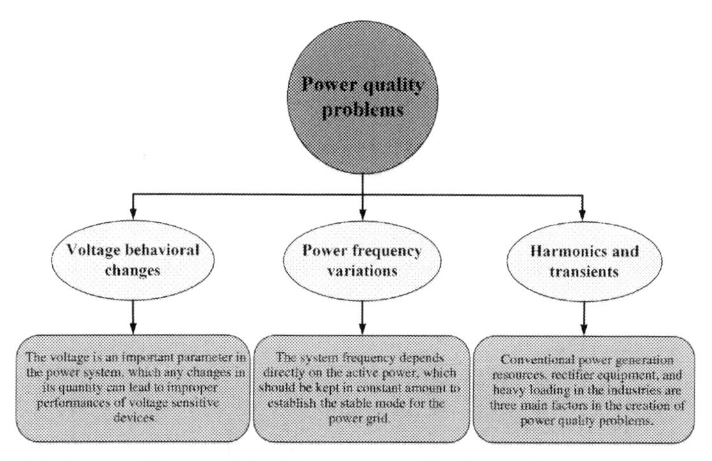

Figure 14.3 Some of the considerable power quality issues.

Recently, growing concerns about the lack of energy resources along with liberalization of the electricity market are two main reasons, which has led to increasing the use of RERs. In addition, the move to exploitation of smart homes with all electrical equipment to get the economic benefits caused to increase the electricity demand, which has resulted to more power grid capacity utilization in severe situations [15]. More electricity consumption increases power exchanged on the transmission lines, which can lead to the operation of the power grid near the system stability limits. Under these conditions, the possibility of voltage collapse can increase more than ever in the power system [16]. Therefore, system stability analysis will be needed as a substantial step of a network modernization in the future smart grids. In this regard, another considerable issue is power quality. The lifetime of the electrical equipment increases if the power output of energy generation resources has a standard quality. Some of the power quality issues are presented with their description in Fig. 14.3.

14.4.4 Problems and Challenges

In general, all main parts of the power system, namely power generation, transmission, and distribution are controlled and managed independently. Due to the development in the technology of network management, power system is derived to the modernization to provide both high level of convenience for the consumers and widespread smart control of devices. This is needed for the intelligent infrastructure, which can be handled by all smart platforms with any range of software. However, conventional infrastructure of the existing network is converted to the main problem in the network modernization process, which has also created some key challenges about the power grid planning and operation for the power system engineers. On the other hand, distributed generations (DGs) will be an inseparable part of the smart distribution network, and implementation and integration of them on the conventional networks will create some other significant challenges that need more analysis and attention to avoid basic problems in the network with a smart structure [17]. In this regard, some important issues about the future networks with a large amount of the DGs are briefly explained as follows, which are required for more assessments to apply them in smart grid.

- Power losses:

  Future network is expected to consist of a large number of DG units. Considering that some DG units such as PV and wind farms are located far from the load centers, the long transmission lines will be needed to deliver electric power to the consumers. Increasing the line length has a direct correlation with the increase in losses that caused the power losses and system costs increase.

- Power balancing:

  The production of some DGs such as wind turbine and solar PV is variable at the all-time, which causes their power production to fluctuate. Such power fluctuations from renewable energies will require the advanced system controller, and establishing the energy balance between supply and demand will also be complex under these conditions. Therefore, providing the advanced power-balancing mechanism is a basic challenge to power system, which will need energy storage systems and conventional generation to cover and fill these energy gaps.

- High renewable energy penetration:

  In the smart grids, all plans for network design are modified to consider the maximum use of clean energy resources and minimize dependency on conventional energy generation. Therefore, RERs are expected to have significant presence in the future distribution networks. On the other hand, production of the RERs has a direct dependency on the climate changes, which does not guarantee a deterministic amount of energy. This problem is solved with the integration of numerous RERs in the grid, which this performance requires the advanced and sustainable technology that achieving to this technology caused the creation of new challenges regarding the costs, implementation and other important aspects of it in the system. Some serious concerns about the fluctuation power from RERs, which needed more attention by power system engineers are illustrated in Fig. 14.4 [2,18].

Researches have been conducted to solve mentioned concerns, indicated that employing the energy storage systems can usefully cover the deviation of energy production from RERs, but applying them in the power grid not only increases the total system costs; but also if the power fluctuation of RERs increases, a large capacity of storage systems will be required, which installation of large size of energy storage is very difficult in some cases [19].

- Impacts of real-time power market:

  In some cases, electrical energy is generated by independent power producers. Presence in the electricity market with this condition leads to the competition in energy trades between claimant participants in the all power sectors [20]. In order to maximize the profit and reduce the operation costs, electrical energy producers should employ power units with fossil fuels efficiently. Therefore, optimal operation of fossil fuels based power units integrated with DGs is converted to the key issue for the smart grid systems, which unit commitment schemes are proposed as a suitable method to determine the optimal operation of the smart power system [21–23]. Because of the deviation in the output power of DGs units, correct power forecasting of the DGs units has become a substantial challenge for the real time power market, which may increase the price of power.

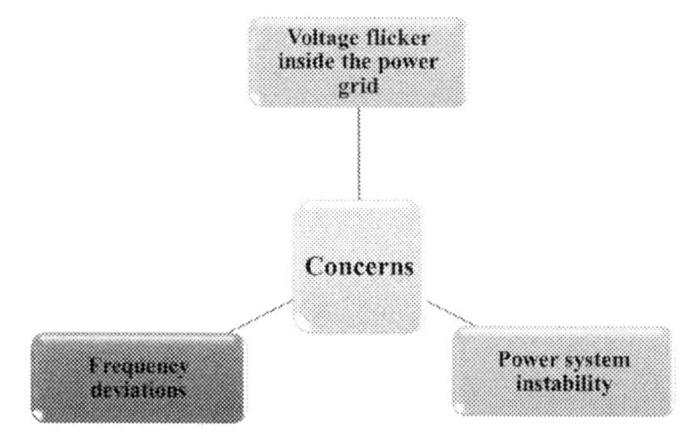

Figure 14.4 Some serious concerns about the fluctuation power from RERs.

14.5 DERs IN THE FUTURE SMART DISTRIBUTION NETWORKS

Recently, DERs are more and more promoted by distribution networks to cover electricity loads in total regions, especially remote areas, which the cost of the transmission electricity to them is uneconomical. Because of the mentioned advantages about the RERs, power system designers try to reduce dependency of the non-RERs to the minimum amount of own and, as far as possible, replace them with RERs. However, they are faced with some key problems in their planning, which without solving them, the dream of a smart grid with a modern structure will not be realized in the future. One of the most fundamental of these problems is how to intelligently manage and coordinate the large number of DERs. This is done by integration of all the DERs in the network, but this integration requires modern and mighty technology, which this duty can be performed without any problems.

14.6 INTEGRATION OF DERs

The concept of interconnection or integration of the DERs for the formation of the smart grid will be very basic in the near future. The issue of integration of DERs is evolved as an emerging challenge for three main parts of electric system, i.e. power generation, transmission, and distribution. This is done based on substantial topics such as deregulation of electric utilities, reduction of fossil fuel resources, widespread extension of advanced DERs technologies, and increasing consumer awareness about the environmental impacts of conventional electrical energy generation. These issues are opening up significant challenges and changing the concept of electricity generation in the distribution markets. Due to the connection of the large number of the RERs such as large and small PV, hydro power plant, and wind parks, as well as other types of DERs, such as industrial and micro combined cooling, heat and power to the medium and low voltage distribution networks, employing integration technology is more and more key for distribution companies [24]. In this regard, transactive energy (TE) is proposed as a reliable and sustainable technology for the smart grid, which has been able to meet the expectations of the power system experts. Consequently, TE technology is vital for network modernization, which is widely considered by the power system engineers in the future smart distribution network process.

14.7 TRANSACTIVE ENERGY

Nowadays, many institutes are formed in the field of energy to evaluate and respond to emerging challenges of the power grid and they are trying to provide suitable conditions for providing better services to various consumers in any geographic location. United States Department of Energy (DOE) is one of the most important energy centers in the world, which uses connoisseur experts. DOE is divided into different parts of energy, which can support all sections of the energy to have comprehensive researches for attention to the problems and challenges in the energy world [25]. In order to realize the power grid with smart structure, DOE is starting to form some special councils to solve key emerging problems, which one of significant of them is Grid Wise Architecture Council (GWAC). Indeed, GWAC is created by the DOE to enable all intelligent components to have interaction with each other in all parts of the power system especially distribution sector. By establishing interaction between smart elements of the grid, some capabilities required to continue the system performances can be realized such as interoperability between system elements and integration of all sectors of the system especially DERs [26]. GWAC is consisted of several research fields, which some of the most important of them are: information technology and advanced telecommunications, commercial and industrial systems coordination and control, economic policy and regulatory, building automation, and energy generation and supply [25,27].

Today, the challenge of optimal management of both the energy supply and demand sides has transformed it into one of the important and hot topics in the field of energy. In addition, the coordination issue of all energy resources with various loads and intelligent devices should also be added to a number of significant contemporary issues. Indeed, from the perspective of power grid operators, coordination and control of demand side resources with a large number of smart elements through the communicative price signals is the feasible technique to take benefit of their flexibility, which has defined to the TE concept [28].

14.8 TRANSACTIVE ENERGY DESCRIPTION

In order to consider the challenges ahead of the integration of large numbers of DERs in the TE market, GWAC held specialized workshops to familiarize specialists with emerging challenges and to take advantage of

Table 14.3 Some of the important features of the TE technology

Transactive energy descriptions
1. It provides personal services for consumers on the demand side.
2. It is a market-based architecture with widespread synergies, which is provided to the all platform providers and participants.
3. It has comprehensive structure, which eliminates all intermediaries.
4. It follows a new ecosystem, which based on it, integration of suppliers and buyers, to establish interaction between them are accomplished using the information technology in the market area.

top ideas to solve the challenges. The output of meetings of the GWAC experts led to the creation of a market of practice, which can integrate numerous DERs based on the TE structure [29]. According to the GWAC group conclusion, GWAC defines the TE as "a set of economic and control mechanisms that allows the dynamic balance of supply and demand across the entire electrical infrastructure using value as a key operational parameter" [30].

In addition to the mentioned definition of the TE market, TE technology can be described with some basic features, which are tabulated in Table 14.3 [31].

14.9 TRANSACTIVE ENERGY ATTRIBUTES

With recent advances in grid modernization technology, the need for a given basic framework about an activity of all parts of the market was felt more than ever. This tendency in the development of the electricity market has led the GWAC to codify a comprehensive TE framework, which the performances of each part of the TE market will be guided based on this framework. The aim of the TE framework is to provide a proper area for useful discussions about the common features of the specific part of the smart market process, including researches, models, evaluation and designs, and implementation of the TE systems at the conceptual level [32]. Indeed, the document of TE framework is intended to cover requirements for the all elements involved in TE market, including asset owners, developers, vendors, regulators, utility decision makers, system integrators, and researchers. On the other hand, the community of practice built by GWAC will be needed to enable various approaches and methodologies to establish interoperability between intelligent elements of the TE market. This is done by designing the TE system with

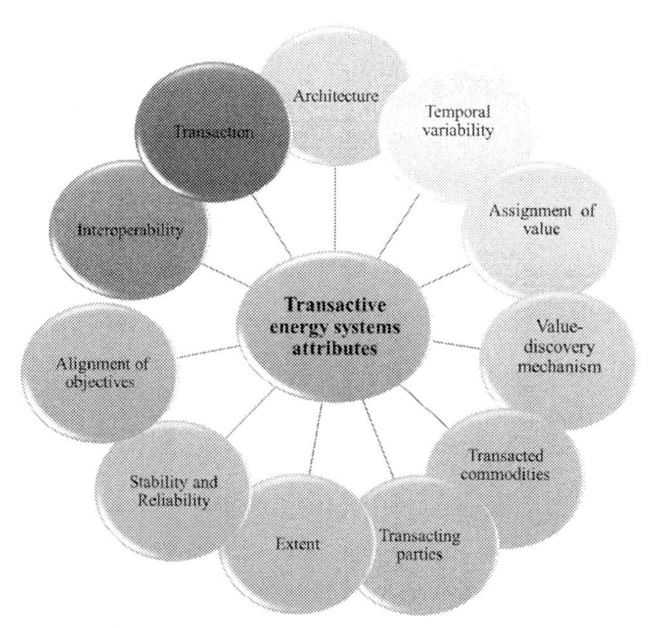

Figure 14.5 Some of the special attributes of the TE systems.

significant attributes. Therefore, TE framework is considered to define practical attributes for the TE systems, some of the special of attributes are summarized in Fig. 14.5 [25].

14.9.1 Purposes

GWAC builds the TE market for key purposes and develops it to realize these purposes with high efficiency. TE framework is one of the effective steps to achieve TE goals, which is provided by GWAC experts to help more for system development. Indeed, the main purpose of TE market is the organization of intelligent devices with information technology and other paradigms to realize the community of practice to integrate automation systems and different types of the DERs in the TE market area with interoperability property between all market components. In addition, the other considerable purpose of the TE framework is that it allows all shareholders and other participants in the market to negotiate with each other about the challenges and concerns of the power grid. This feature not only helps to find the suitable solutions for system challenges, but also it provides a good level of knowledge background for people about the grid modernization, which can increase popularity among people to use smart appliances in their lives and helps more and more the future distribution networks to have full smart structure.

14.9.2 Principles

In general, the electric infrastructure is expected to become more reliable, secure, and efficient in the future smart grids. Indeed, in the network with numerous intelligent components, which all of them must have communication with each other, providing mentioned factors would be very complex and difficult without integration of them with information technology by the sustainable technology based on the influential platforms. Under these conditions, all of the smart elements require to act based on some guiding principles. Therefore, GWAC has organized some basic principles for the relationship between the smart devices and information technology, which can ultimately lead to the effective operation of them in the smart grid area [26,33]. Some of the basic principles, which designed by GWAC are briefly defined in Table 14.4.

14.9.3 Challenges

Nowadays, because of increasing penetration of the DERs and self-controlled resources in the electricity generation process, power system control, coordination, and management have been more and more complexed, which the need for influential technology seems to be more than necessary. TE is introduced as a sustainable technology to the smart grids with high penetration of the DERs especially RERs and intelligent

Table 14.4 Some of the basic principles for the TE market

Transactive energy principles
1. Coordinated self-optimization is expected to be provided by TE systems.
2. It should integrate numerous DERs effectively, while system reliability should not be violated.
3. Participants in the TE market are responsible for compliance with standards of performance.
4. Extensible, adaptable, and scalable are key three properties, which should be satisfied by TE systems across a number of smart devices, participants, and geographic area.
5. TE systems should be able to provide fair TE market opportunities for all qualified participants.
6. TE systems should be able to provide auditability and observability at interfaces.

devices, which transformed the structure of network management and control with its clever mechanisms. In addition to the mentioned advantages of the TE technology, it is faced with some significant challenges, which consideration to them can be very useful for all the participants and shareholders to determine the optimum of their investment. The basis of these challenges is divided into the four parts that are explained as follows:

- System management:

  Currently, the power grid is only equipped with a medium number of DERs and they are covered with the low percentage of electricity generation in the network. Future smart distribution networks have been targeted to use as much as possible RERs as an especial type of DERs, which leads to the production of a large amount of energy from clean sources and helps to reduce the greenhouse gas emissions. However, management of such systems with the presence of a large number of energy resources with uncertainty is converted to the basic challenges of the power system planners. Meanwhile, power system experts along with many researchers believe that interoperability among all intelligent elements and system management status can be improved by converting the system control from the centralized mode to the decentralized one, which is done in the TE technology [34,35].
- Technology:

  Nowadays, electrical energy consumption in our life has grown more than ever and it is also expected to grow dramatically in the future as well. Given these predictions about the future power consumption, the exact planning for the future smart grid is essential. The existence of the intelligent controllable devices would be required to realize comprehensive system control needed for integration smart elements to establish interaction effectively. The realization of all of these goals depends on the mighty technology that can handle all system necessities. Therefore, does existence technology provide integration of all intelligent components of the network under the unit and reliable manager? And, is there such technology with mentioned properties in reality? These are two basic questions, which power system engineers try to respond to them and help to continue the process of modernizing the network [34,35].
- Consumer behavior:

  In general, the electrical consumers have a different behavior in their daily electricity consumption. Indeed, this uncertainty in various consumer performances has led to the system probabilistic analysis. The power system should be provided practical conditions, which do not impose any restrictions on the electricity consumption to consumers.

- Scalability:

  Because of the future smart grid system is expected to have large scale, the power system will have a large number of controller nodes, which makes the power system management more difficult. Optimal control of such systems with large scale will require the trusted platform at the distribution level to improve electrification level and make the energy supply to consumers easier. However, designing such a platform for larger scale systems is difficult and needs complex technology. Therefore, this issue should be taken into account by power system planners in their plans [34–36].

14.10 TRANSACTIVE ENERGY SYSTEMS

About a few years ago, the Pacific Northwest National Laboratory (PNNL) was supported and obligated by the DOE to conduct one of the significant fields of the grid electrification that is called transactive energy system (TES). This conduction includes all part of research studies such as standardization of theoretical principles, simulation, appraisement, validation and implementation [30]. Indeed, TES is promoted to create distributed information systems to achieve advanced market based architecture for the system with different elements. Uniform signaling as a coordinating mechanism is applied by the TES to make the collaboration between system elements more interactive [37].

Due to these goals, TES is consisted of the set of controllable equipment and multifunctional agents to provide a widespread grid architecture to engage all DERs and other system components for the realization of the robustness power system structure [31]. The TES agents are formed by the intelligent software and they are also developed by powerful optimization techniques to manage system operations at any time. In the smart power grid area, several important tasks have been assigned to each of the agents and they are responsible to manage and control performances of all smart devices. This is done by agents through the telecommunication signals between the devices and agents at the various frequency levels. Some of the agents are integrated in a set that is called a TE node. Each TE nodes covers the special part of the system, which receives all operational information from all devices in their covered area and then the data is sent from the TE nodes to the TE control center to determine the optimal performance of the components in the next period [37–39].

14.11 TRANSACTIVE ENERGY SYSTEMS IMPLEMENTATIONS

In smart grid customer services, TE systems have significant advantages over traditional coordination systems such as centralized optimization and price reactive systems that have led to special focuses on it all over the world. One of the main focuses of the TE systems is on the smart agent based systems, which involves important sectors of the grid such as power technology and energy supply companies, grid operators, and regulators. This perspective of the TE systems has led to some implementation of them as practical projects across Europe and the United States. For example, DOE has contributed to three TE based mechanisms projects along with several energy organizations to coordinate and control the DERs and manage them effectively in the United States. Some of the important projects are Olympic Peninsula Demonstration (2006–2007), AEP Ohio grid SMART Real-Time Pricing Demonstration (2010–2014), and Pacific Northwest Smart Grid Demonstration (2010–2015), which the full description of the mentioned projects can be found in [26].

14.12 TRANSACTIVE ENERGY IN INTEGRATION OF DERs

Integration issue for the smart distribution networks was key challenge, which had created plenty concerns for the engineers on the path to modernizing the network. However, with the advent of TE as a modern, sustainable, and reliable technology, the integration of large numbers of DERs is accomplished effectively. TE is a market based technology, which is created by the GWAC to respond to emerging challenges of the future modern grid. It employs advanced optimization technique in making optimal decisions for the network smart elements. In addition, TE not only established interoperability between power grid components, but also interaction as a key feature for integrated systems is realized by it. Therefore, TE has enough potential that can integrate current equipment and components that are coming into the network in the future with utilization of its equipped nodes in the smart grid with any scale.

14.13 CASE STUDY

Transactive energy is one of the new emerging technology that needs more attention to be operational. In order to evaluate the TE effects on the network energy exchanges, electrical energy transactions between the three microgrids are considered. TE has been targeted to respond to the concerns of both electrical utilities and consumers, which this important aim of TE has also been scrutinized in this case study. In this research, two operational models for the relationship of the microgrids have been proposed, in which each of them illustrates one aspect of the energy transaction in the smart grid. The first model (model 1) of the case study evaluates energy transactions of microgrids in individual mode. In this model, each of the microgrids provides their energy demand only from two methods: 1-from the energy resources inside the microgrids, 2-from the power grid. In other words, this operation mode of microgrids does not allow them to have an energy exchange with each other and they can only connect to the network. The second model (model 2) is considered for assessment of microgrids' operation in a cooperative mode. Energy transaction can be accomplished among the microgrids with each other and power grid in model 2. The main goal of this model is to minimize energy costs for each microgrid via providing the possible energy trades between them. Therefore, three microgrids with different load profiles are selected to evaluate the effectiveness of the two mentioned models with their features in the TE structure.

14.13.1 Problem Formulation for the Microgrids

In this study, three microgrids are considered to appraise the usefulness of TE in the smart grid. Each of these microgrids is organized with some electrical energy resources and storage systems. For each microgrid, power generation unit (PGU) and solar PV systems are used as the two energy generation resources and electrical storage system is engaged in microgrids, too. In this research, electrical load as one type of load is considered, which this load not only can be fed from the energy resources of their own microgrids, but also it can receive energy from the power grid and other microgrids. In addition, the microgrids can sell their extra energy production to the power grid and local transaction market (other microgrids) or they can purchase energy from them when they cannot provide their energy demand through energy generated inside the microgrid. A modeling framework for the each microgrid is shown in

Fig. 14.6. In this figure, all the components of each microgrid with communication among them and power grid are also demonstrated. The energy cost of the microgrids can be reduced and reliability of them can also be increased, if the mentioned energy exchanging is done under the TE structure. Therefore, the effects of TE technology on the microgrid energy costs are covered in this study based on the following formulations.

14.13.1.1 Objective Function

In this research, minimizing energy costs is the main goal of each microgrid. In this regards, energy purchasing (selling) from (to) the grid has a cost (revenue) for each microgrids. In addition, the PGU unit consumes gas fuel for the electrical energy production, which has a cost for the

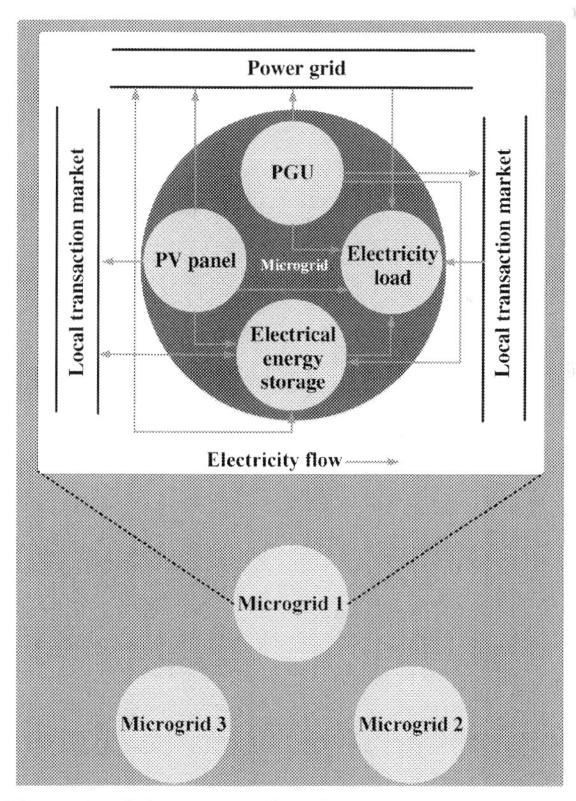

Figure 14.6 Schematic of the microgrid with its energy transaction in the power grid.

microgrids, too. Therefore, the objective function of this study is consisted of two mentioned terms, which is formulated as follows:

$$OBJ = \sum_t (Ep_{m,t}.Epp_t - Es_{m,t}.Eps_t) + \sum_t (Fpgu_{m,t}.Ppgu_t), \quad \forall\ m \quad (14.1)$$

where, $Ep_{m,t}(Es_{m,t})$ is the electricity purchased (sold back) from (to) the electrical power grid by microgrid m at time t. Epp_t and Eps_t represent, respectively, the electricity purchasing and selling price in power grid at time t. The $Fpgu_{m,t}$ is the fuel consumed by PGU unit in microgrid m at time t. $Ppgu_t$ is the fuel price for PGU unit at time t.

14.13.1.2 Constraints

Generally, in order to control energy transaction, generation, and consumption, some constraints should be satisfied by all parts of system, which are organized as the following equations.

14.13.1.2.1 Electrical Power Balance Constraint

In general, the microgrids must provide all their energy demand in a reliable manner. Therefore, their energy generation and energy coming from the power grid and other microgrids should be equivalent to energy demand and energy coming out of each microgrid, which leads to the following constraint:

$$\begin{aligned} &Ep_{m,t} + EPV_{m,t} + Epgu_{m,t} + EBd_{m,t}.\eta_{Bd} + ETin_{m,t} \\ &= EL_{m,t} + Es_{m,t} + \frac{EBc_{m,t}}{\eta_{Bc}} + ETout_{m,t} \quad \forall\ m, \ \forall\ t \end{aligned} \quad (14.2)$$

where, $EPV_{m,t}$ and $Epgu_{m,t}$ are the electrical energy generated by PV panel and PGU unit in microgrid m at time t, respectively. $EBd_{m,t}$ and $EBc_{m,t}$ are, respectively, discharged and charged ratio of battery storage system in microgrid m at time t. $EL_{m,t}$ represents the electricity demand in microgrid m at time t. η_{Bd} and η_{Bc} signify, respectively, discharging and charging efficiency of the storage system. Finally, $ETin_{m,t}(ETout_{m,t})$ is the electrical energy transmitted from (into) the local transaction market into (from) the microgrid m at time t, respectively.

14.13.1.2.2 PGU constraints

Because of the PGU unit consumes the gas fuel for electricity generation, the constraints of the PGU unit are divided into the two parts, which are defined as follows:

a. Fuel consumption constraint

The fuel consumed by the PGU unit is limited to the maximum capacity. This limitation is illustrated by the following equation:

$$Fpgu_{m,t} \leq Xpgu_{m,t}.Spgu_{m} \quad \forall\ m,\ \forall\ t \tag{14.3}$$

where, $Spgu_m$ is the size of the PGU unit in microgrid m, $Xpgu_{m,t}$ represents the state of the PGU unit in microgrid m at time t.

b. Electrical energy generation constraint

$$Epgu_{m,t} \leq (Fpgu_{m,t} - b_{pgu}.Xpgu_{m,t})/a_{pgu} \quad \forall\ m,\ \forall\ t \tag{14.4}$$

where a_{pgu} and b_{pgu} are the coefficients of fuel-to-electricity conversion of PGU unit.

14.13.1.2.3 PV Constraints

The constraint of the electrical energy generated by PV panel is formulated as follows:

$$EPV_{m,t} \leq Spv_{m}.SOL_{t}.\eta_{pv} \quad \forall\ m,\ \forall\ t \tag{14.5}$$

where, Spv_m is the size of solar PV panel in microgrid m, SOL_t and η_{pv} represent the solar radiation at time t and electricity production efficiency of PV panel, respectively.

14.13.1.2.4 Battery Storage Constraints

Battery storage system constraints are consisted of three following parts:

a. Charging and discharging constraint

The state of battery charging and discharging cannot happen at the same time according to the following constraint:

$$XBc_{m,t} + XBd_{m,t} \leq 1 \quad \forall\ m,\ \forall\ t \tag{14.6}$$

where, $\mathrm{XBc}_{m,t}$ and $\mathrm{XBd}_{m,t}$ are the charging and discharging state of the battery storage system in microgrid m at time t, respectively.

b. Battery stored electricity constraints

Electrical energy stored in the battery should not be violated from its permissible range, which is identified using the charging and discharging state of the battery. These constraints are defined as follows:

$$\mathrm{Sbs}_{m}.\alpha_{\mathrm{Bmin}} \leq Eb_{m,t} \leq \mathrm{Sbs}_{m} \quad \forall\ m,\ \forall\ t \tag{14.7}$$

$$Eb_{m,t} = Eb_{0} + (Ebc_{m,t} - Ebd_{m,t}).\Delta t \quad \forall\ m,\ t = 1 \tag{14.8}$$

$$Eb_{m,t} - Eb_{m,t-1} = (Ebc_{m,t} - Ebd_{m,t}).\Delta t \quad \forall\, m, \;\; \forall\, t \geq 2 \qquad (14.9)$$

Where, Sbs_m is the size of battery storage in microgrid m, $Eb_{m,t}$ represents the electricity stored in battery in microgrid m at time t, $Ebc_{m,1}$ and $Ebd_{m,1}$ are the charging and discharging rate of the battery storage in microgrid m at time t, respectively, Δt is the time interval with setting 1 h for this research, and $\alpha_{B\min}$ is the coefficient of battery considering the minimum storage limit of battery.

c. Electricity charging and discharging power constraints

The magnitude of charging and discharging power must be kept in the allowable range, which are formulated as the following equations:

$$Sbs_m.\alpha_{Bc\min}.XBc_{m,t} \leq Ebc_{m,t} \leq Sbs_m.\alpha_{Bc\max}.XBc_{m,t} \qquad (14.10)$$

$$Sbs_m.\alpha_{Bd\min}.XBd_{m,t} \leq Ebd_{m,t} \leq Sbs_m.\alpha_{Bd\max}.XBd_{m,t} \qquad (14.11)$$

where, $\alpha_{Bc\max}$ and $\alpha_{Bc\min}$ are coefficients of battery denote the maximum and minimum charging of battery storage, respectively, $\alpha_{Bd\min}$ and $\alpha_{Bd\max}$ are coefficients of the maximum and minimum discharging of battery, respectively.

14.13.1.2.5 Local Transaction Market Constraints

In general, the electrical energy cannot be sent and received between microgrids and local transaction market at the same time, which is considered in Eq. (14.12). The mutual transmission indicator condition affects the energy transfer between local transaction market and microgrids. Eqs. (14.13) and (14.14) are formulated to show this. In the local transaction market, establishing balance between energy supply and demand is essential, thus Eq. (14.14) is defined to consider this.

$$XEin_{m,t} + XEout_{m,t} \leq 1 \quad \forall\, m, \;\; \forall\, t \qquad (14.12)$$

$$ETin_{m,t} \leq M.XEin_{m,t} \quad \forall\, m, \;\; \forall\, t \qquad (14.13)$$

$$ETout_{m,t} \leq M.XEout_{m,t} \quad \forall\, m, \;\; \forall\, t \qquad (14.14)$$

$$\sum_m ETin_{m,t} = \sum_m ETout_{m,t} \quad \forall\, t \qquad (14.15)$$

where, $XEin_{m,t}$ and $XEout_{m,t}$ are the electrical energy transmission and contribution state of microgrid m at time t, respectively. $ETin_{m,t}(ETout_{m,t})$ is the electricity transmitted from (into) the local transaction market into (from) the microgrid m at time t. M is a parameter, which is used in the mixed integer programming (MIP) and has a big amount.

14.13.2 Operation Models of Microgrids

In this study, two models for microgrid operation are considered based on the TE management to evaluate the positive effects of TE on the microgrids energy costs. The first model evaluates the operation of microgrids when they do not have any energy exchanging with each other. Moreover, the second model is considered to analyze the operation of microgrids in the interconnected mode, which microgrids can exchange energy with each other based on this mode.

a. Model I

In this model, the electrical energy of each microgrids will be provided from the electricity generation resources inside the microgrid and energy coming from the power grid. In other words, microgrids operate in the separated mode. The model I with mentioned features is presented as follows:

$$\min OBJ_I = \sum_m OBJ_{I,m}$$

s.t. constraints in Eqs. (14.2)–(14.15)

$$XEin_{m,t} = 0, \quad XEout_{m,t} = 0, \quad \forall\ m, \ \forall\ t$$

where, $OBJ_{I,m}$ is the electrical energy cost of the mth microgrid in the model I, which is computed by Eq. (14.1).

b. Model II

According to this model, electricity can be exchanged among the microgrids and power grid based on the TE structure. Indeed, each microgrids can freely connect to the other microgrids to maximize the collective interests. Model II is defined as follows:

$$\min OBJ_{II} = \sum_m OBJ_{II,m}$$

s.t. constraints in Eqs. (14.2)–(14.15)

where, $OBJ_{II,m}$ is the electrical energy cost of the mth microgrid in model II, which is computed by Eq. (14.1).

Table 14.5 Simulation results of the microgrids in model I and II

Microgrid index	Cost in model I ($OBJ_{I,m}$)($)	Cost in model II ($OBJ_{II,m}$)($)	Amount of cost saving ($)	Percentage of cost saving (%)
1	25.024	21.757	3.267	13.0554
2	27.811	27.857	− 0.046	− 0.1654
3	26.356	24.997	1.359	5.1563
Total cost	79.190	74.611	4.579	5.7822

14.13.3 Simulation Results

In this research, three microgrids are considered to evaluate the effectiveness of TE technology for the smart distribution networks. For this study, electricity load profile for June 1st is chosen. The load data needed for each microgrid is accessible in [40]. Summer TOU prices are considered to determine the electricity selling and purchasing prices, which can be located in [41]. The data about the solar radiation is achievable in [42]. All other parameters needed for this research are collective from [43,44]. For this study, GAMS software is used to solve this problem with applying the CPLEX solver for the MIP problem.

In order to assessment of usefulness of TE technology, three microgrid are considered with various electricity load profile. The simulation results of them are tabulated in Table 14.5.

According to the obtained results in Table 14.5, total electrical energy cost for all microgrids in models I and II is 79.190 ($) and 74.611 ($), respectively. In this research, some microgrids may be a loss when they operate in the interconnected mode. For example, the amount of cost saving for microgrid 2 is −0.046 ($) after connection to the other microgrids. However, the total cost of microgrids is improved by 5.7822%, which indicated that the energy cost is reduced in model II in comparison with the model I.

14.14 CONCLUSION

In this research, the various aspects of DERs were highlighted and the effectiveness of them for the future smart distribution networks was evaluated completely, too. The application of the DERs was explained with assessment of two types of DERs, i.e. RERs and non-RERs, which the problems and challenges of them were also described to consider them in designing the effective plans for modernizing the future distribution networks. The integration of the DERs was presented as an important topic in the distribution sector and the challenges ahead were also analyzed. The TE technology as a sustainable and reliable manner is proposed for the DERs' integration issue to the smart distribution grids with high penetration of DERs. The evaluation of TE technology indicated that it can integrate and coordinate numerous DERs through the comprehensive nodes with advanced agents, which establish the dynamic balance between energy demand and supply based on the interoperability created between themselves. In order to prove the positive effects of TE technology, three microgrids with different load profiles are considered and energy exchange between them is scrutinized in two separate and interconnected modes based on the TE management. The obtained results are demonstrated that if the microgrids have only energy exchanges with the power grid, they will have more energy cost in comparison with when they trade energy among themselves and power grid under the TE management. Consequently, the total energy cost of model II is lower than the model I, thus model II was preferred to microgrids operation in the future smart distribution networks.

REFERENCES

[1] M. Moradi-Dalvand, B. Mohammadi-Ivatloo, N. Amjady, H. Zareipour, A. Mazhab-Jafari, Self-scheduling of a wind producer based on information gap decision theory, Energy 81 (2015) 588–600.

[2] A.M. Howlader, N. Urasaki, A.Y. Saber, Control strategies for wind-farm-based smart grid system, IEEE Trans Ind App 50 (5) (2014) 3591–3601.

[3] S. Shargh, B. Khorshid ghazani, B. Mohammadi-ivatloo, H. Seyedi, M. Abapour, Probabilistic multi-objective optimal power flow considering correlated wind power and load uncertainties, Renew Ener 94 (Supplement C) (2016) 10–21. 2016/08/01/ .

[4] A.M. Howlader, N. Urasaki, A. Yona, T. Senjyu, A.Y. Saber, Design and implement a digital $H\infty$ robust controller for a MW-class PMSG-based grid-interactive wind energy conversion system, Energies 6 (4) (2013) 2084–2109.

[5] B. Khorshid-Ghazani, H. Seyedi, B. Mohammadi-ivatloo, K. Zare, S. Shargh, Reconfiguration of distribution networks considering coordination of the protective devices, IET Gen Trans Dist 11 (1) (2017) 82–92.

[6] H.O.R. Howlader, H. Matayoshi, T. Senjyu, Distributed generation incorporated with the thermal generation for optimum operation of a smart grid considering forecast error, Energy Convers Manage 96 (2015) 303–314.

[7] M.L. Ferrari, A. Traverso, M. Pascenti, A.F. Massardo, Plant management tools tested with a small-scale distributed generation laboratory, Energy Convers Manage 78 (2014) 105–113.

[8] M.F. Akorede, H. Hizam, E. Pouresmaeil, Distributed energy resources and benefits to the environment, Renew Sust Ener Rev. 14 (2) (2010) 724–734.

[9] S. Nojavan, K. Zare, B. Mohammadi-Ivatloo, Application of fuel cell and electrolyzer as hydrogen energy storage system in energy management of electricity energy retailer in the presence of the renewable energy sources and plug-in electric vehicles, Ener Convers Manage. 136 (2017) 404–417. 2017/03/15/.

[10] T. Funabashi, Integration of distributed energy resources in power systems: implementation, operation and control, Academic Press, 2016.

[11] P. Dondi, D. Bayoumi, C. Haederli, D. Julian, M. Suter, Network integration of distributed power generation, J Power Sources 106 (1) (2002) 1–9.

[12] N. Apergis, J.E. Payne, Renewable and non-renewable energy consumption-growth nexus: evidence from a panel error correction model, Ener Econ 34 (3) (2012) 733–738.

[13] A.H. Ghorashi, A. Rahimi, Renewable and non-renewable energy status in Iran: art of know-how and technology-gaps, Renew Sust Ener Rev. 15 (1) (2011) 729–736.

[14] A. Mokari-Bolhasan, H. Seyedi, B. Mohammadi-ivatloo, S. Abapour, S. Ghasemzadeh, Modified centralized ROCOF based load shedding scheme in an islanded distribution network, Int J Electric Pow Ener Syst 62 (2014) 806–815.

[15] Y.-K. Wu, A novel algorithm for ATC calculations and applications in deregulated electricity markets, Int J Electric Pow Ener Syst 29 (10) (2007) 810–821.

[16] M. Nojavan, H. Seyedi, B. Mohammadi-Ivatloo, Preventive voltage control scheme considering demand response, correlated wind and load uncertainties, J Ener Manage Technol 1 (1) (2017) 43–52.

[17] P. Järventausta, S. Repo, A. Rautiainen, J. Partanen, Smart grid power system control in distributed generation environment, Ann Rev Con 34 (2) (2010) 277–286.

[18] A.M. Shotorbani, S.G. Zadeh, B. Mohammadi-Ivatloo, S.H. Hosseini, A distributed non-Lipschitz control framework for self-organizing microgrids with uncooperative and renewable generations, Int J Electric Pow Ener Syst 90 (2017) 267–279.

[19] A.M. Shotorbani, S. Ghassem-Zadeh, B. Mohammadi-Ivatloo, S.H. Hosseini, A distributed secondary scheme with terminal sliding mode controller for energy storages in an islanded microgrid, Int J Electric Pow Ener Syst 93 (2017) 352–364.

[20] R. Kyoho, et al., Thermal units commitment with demand response to optimize battery storage capacity, Power Electronics and Drive Systems (PEDS), 2013 IEEE 10th International Conference on, IEEE, 2013, pp. 1207–1212.

[21] T. Goya, T. Senjyu, A. Yona, N. Urasaki, T. Funabashi, Optimal operation of thermal unit in smart grid considering transmission constraint, Int J Electric Pow Ener Syst 40 (1) (2012) 21–28.

[22] J. Wang, M. Shahidehpour, Z. Li, Security-constrained unit commitment with volatile wind power generation, IEEE Trans Pow Syst 23 (3) (2008) 1319–1327.

[23] S. Chakraborty, T. Senjyu, A.Y. Saber, A. Yona, T. Funabashi, A fuzzy binary clustered particle swarm optimization strategy for thermal unit commitment problem with wind power integration, IEEJ Trans Elect Elect Eng 7 (5) (2012) 478–486.

[24] F.P. Sioshansi, Distributed generation and its implications for the utility industry, Academic Press, 2014.

[25] D. Forfia, M. Knight, R. Melton, The view from the top of the mountain: building a community of practice with the GridWise transactive energy framework, IEEE Pow Ener Mag 14 (3) (2016) 25–33.

[26] K. Kok, S. Widergren, A society of devices: integrating intelligent distributed resources with transactive energy, IEEE Pow Ener Mag 14 (3) (2016) 34–45.

[27] R. Masiello, Transactive energy: the hot topic in the industry [Guest Editorial], IEEE Pow Ener Mag 14 (3) (2016) 14–16.

[28] E.A.M. Ceseña, N. Good, A.L. Syrri, P. Mancarella, Techno-economic and business case assessment of multi-energy microgrids with co-optimization of energy, reserve and reliability services, Appl. Energy (2017).

[29] M.R. Knight, Stochastic impacts of metaheuristics from the toaster to the turbine [Technology Leaders], IEEE Elect Mag 4 (4) (2016) 40–52.
[30] D.J. Hammerstrom, S.E. Widergren, C. Irwin, Evaluating transactive systems: historical and current US DOE research and development activities, IEEE Elect Mag 4 (4) (2016) 30–36.
[31] R. Melton and J. Fuller, Transactive energy: envisioning the future, 2016.
[32] F. Rahimi, A. Ipakchi, F. Fletcher, The changing electrical landscape: end-to-end power system operation under the transactive energy paradigm, IEEE Pow Ener Mag 14 (3) (2016) 52–62.
[33] M. Olken, Transactive energy: providing an enabling environment [From the Editor], IEEE Pow Ener Mag 14 (3) (2016), pp. 4-4.
[34] C. Sijie, L. Chen-Ching, From demand response to transactive energy: state of the art, J Mod Pow Syst Clean Ener 5 (1) (2017) 10–19.
[35] N. Atamturk, M. Zafar, Transactive energy: a surreal vision or a necessary and feasible solution to grid problems, California Public Utilities Commission Policy & Planning Division, Los Angeles Google Scholar, 2014.
[36] P. Mazza, The smart energy network: electricity's third great revolution, Pacific Northwest National Laboratory, Washington, 2003.
[37] R. Ambrosio, Transactive Energy systems [Viewpoint], IEEE Elect Mag 4 (4) (2016) 4–7.
[38] J. Haack, B. Akyol, N. Tenney, B. Carpenter, R. Pratt, T. Carroll, VOLTTRON™: An agent platform for integrating electric vehicles and Smart Grid, Connected Vehicles and Expo (ICCVE), 2013 International Conference on, IEEE, 2013, pp. 81–86.
[39] S. Chen, H.A. Love, C.-C. Liu, Optimal opt-in residential time-of-use contract based on principal-agent theory, IEEE Transactions on Power Systems 31 (6) (2016) 4415–4426.
[40] M. Alipour, B. Mohammadi-Ivatloo, K. Zare, Stochastic scheduling of renewable and CHP-based microgrids, IEEE Transactions on Industrial Informatics 11 (5) (2015) 1049–1058.
[41] L. Goel, P. Viswanath, P. Wang, Evaluation of probability distributions of reliability indices in a multi bilateral contracts market, Power System Technology, 2004. PowerCon 2004. 2004 International Conference on, vol. 1, IEEE, 2004, pp. 96–101.
[42] M. Matos, J. Fidalgo, L. Ribeiro, Deriving LV load diagrams for market purposes using commercial information, Intelligent Systems Application to Power Systems, 2005. Proceedings of the 13th International Conference on, IEEE, 2005, pp. 105–110.
[43] R. Dai, M. Hu, D. Yang, Y. Chen, A collaborative operation decision model for distributed building clusters, Energy 84 (2015) 759–773.
[44] Y. Chen, M. Hu, Balancing collective and individual interests in transactive energy management of interconnected micro-grid clusters, Energy 109 (2016) 1075–1085.

APPENDIX: NOMENCLATURE

CHAPTER 3

Indices/sets

$\{i, j, t\} \in T$	Indices for time
$\{m, n\} \in MG$	Indices for microgrids
l	Index for loads
u	Index for generation units
OM	Index for operation and maintenance cost
g	Index for generated power
BAT	Index for battery packs
CH	Index for the amount of battery charge
DCH	Index for the amount of battery discharge
$sell$	Index for sold power
pur	Index for purchased power
k	Index for pollutants
w, z	Indices for particles in PSO
RI	Resilience index

Parameters and constants

$E(i, i)$	Self elasticity
$E(i, j)$	Cross elasticity between ith and jth hour
λ	cost coefficient
C_{nl}	Natural gas price
L	Natural gas low-hot value kWh/m^3
UR	Ramp up rate
DR	Ramp down rate
$\in_{rec}$	Heat recovery factor
η_e^t	Electrical efficiency of MT at hour t
η_b	Boiler efficiency
$P_{BAT,CAP}$	Capacity of battery kW
P_{BAT}^{loss}	Power loss of battery kW
Υ	Price coefficient of different pollutants
r_1, r_2	Random functions in the range (0,1)
W	Inertia weight factor
c_1, c_2	Acceleration coefficients of PSO
N	Number of variables

Variables

C_{l0} Initial electricity price before DRP ($/kWh)
P_{l0} Initial demand value before DRP (kWh)
$P_{l,new}$ Power consumption after DRP (kWh)
$C_{l,new}$ Electricity energy price after DRP ($/kWh)
ΔP_l Difference between demands before and after DRPs
S Customer's benefit
B Customer's income
C Energy cost in different modes
P RER output power (kW)
η Efficiency of generation units
u Commitment status of generators
SOC State of charge of batteries
$C_{pur,mn}$ Cost of purchased power by MG-m from MG-n ($/h)
$C_{sell,mn}$ Cost of sold power by MG-m to MG-n ($/h)
$P_{pur,mn}$ Purchased power by MG-m from MG-n (kW)
$P_{sell,mn}$ Sold power by MG-m to MG-n (kW)
$P_{tran,m}$ The amount of transactive power of MG-m (kW)
OF Objective function
$Cost_{op}$ Operation cost ($/h)
$Cost_{em}$ Emission cost ($/h)
ρ Emission factor of pollutants
$Cost_{shed}$ The cost of load shedding ($/h)
P_{shed} The amount of load shedding (kW)
x Vector of uncertain input variables
Y Vector of uncertain output variables
χ Position vector of particles in PSO
ϕ Velocity vector of particles in PSO
P_{best} Best previous position of particles in PSO
g_{best} Best particle among all P_{best} in PSO

Acronyms

CHP Combined heat and power
DG Distributed generation
DNO Distribution network operator
DR Demand response
DRP Demand response program
EMS Energy management system
FC Fuel cell
MCS Monte Carlo simulation
MG Microgrid
MGCC Microgrid central controller
MT Microturbine
NMG Networked microgrid
PDF Probability distribution function
PSO Particle swarm optimization
PV Photovoltaic panel

RER	Renewable energy resources
RTP	Real time pricing
TOU	Time of use
WT	Wind turbine

CHAPTERS 6 AND 7

Indices

T	Identifier of optimization periods, $t = 1, 2, \ldots, N_T$
j	Identifier of DG units, $j = 1, 2, \ldots, N_{DG}$
I	Identifier of large loads, $i = 1, 2, \ldots, N_{LL}$
D	Identifier of demand response aggregators, $d = 1, 2, \ldots, N_{DRA}$
n, m	Identifier of distribution network buses, $n = 1, 2, \ldots, N_{Bus}$
K	Identifier of steps of bid-quantity offers, $k = 1, 2, \ldots, K$
w	Identifier of wind turbines, $w = 1, 2, \ldots, N_{wind}$
$\alpha_1, \alpha_2, \alpha_3$	Coefficients of DG units cost function

Parameters

ς_{ug}	The price of upstream grid
η_{ch}, η_{dis}	Coefficients of battery energy storage system charge/discharge efficiency
ς_{su}	Startup cost
T_{ut}, T_{DT}	Minimum up time/minimum down time of DG units
$Re(Z)$	Real part of the feeder impedance
$Im(Z)$	Imaginary part of the feeder impedance
$V_{\min}, V_{\max}$	Minimum and maximum amount of bus voltage
$I_{substation}^{\max}$	Substation maximum current limit
a, b, c	Binary variables for DG unit commitment, start-up and shut down status
$P_{DG}^{\min}, P_{DG}^{\max}$	Minimum and maximum amount of DG units power
SOC	Battery energy storage system state of charge
$SOC^{\min}, SOC^{\max}$	Minimum and maximum amount of battery energy storage system state of charge
P_{rated}	Rated power of wind turbine
v_{cut-in}	Cut in speed of wind turbine
$v_{cut-out}$	Cut out speed of wind turbine
$\hbar_{\min}, \hbar_{\max}$	Minimum and maximum amount of load decline offered by demand response aggregators

Functions and Variables

P_{ug}^{DAS}	Active power purchased from upstream grid
$cost_{DG}^{DAS,E}$	Energy cost of DG units
$cost_{DG}^{DAS,su}$	Startup cost of DG units
$cost_{DG}^{DAS,R}$	Reserve cost of DG units
$cost_{LL}^{DAS,E}$	Cost of energy reduction by large loads
$cost_{LL}^{DAS,R}$	Cost of energy reduction by large loads
$cost_{DRA}^{DAS,E}$	Cost of energy reduction by demand response aggregators

$cost_{DRA}^{DAS,R}$	Cost of reserve provided by demand response aggregators
$P_{load}^{DAS}, Q_{Load}^{DAS}$	Active/reactive load consumption of each bus
$P_{DRA}^{DAS}, Q_{DRA}^{DAS}$	Active/reactive load reduction of demand response aggregators
P_{flow}, Q_{flow}	Active/reactive power flow of feeders
ℓ, v	Auxiliary variables introduced in the AC power flow equations
$P_{DG}^{DAS}, Q_{DG}^{DAS}$	Active /reactive power of DG units
$P_{Wind}^{DAS}, Q_{Wind}^{DAS}$	Active/reactive power of wind turbines
$P_{LL}^{DAS}, Q_{LL}^{DAS}$	Active/reactive load reduction of large loads
H	Consumption decline offered by demand response aggregator
$\omega_{DRA}^{k,d}$	The bid price of DRP d in step k
H_k^d	The amount of energy reduction of DRP d in step k
ω_{DRA}	The cost of energy decline bids by demand response aggregator
Ω_{DRA}	The cost of reserve provided by demand response aggregator
ω_{LL}	The cost of energy decline provided by large loads
Ω_{LL}	The cost of reserve provided by large loads
P_{RU}, P_{RD}	The ramp up and ramp down ratio of DG units
$SOC(t, v)$	State of charge of vehicle v in time t
η_v^{ch}	Charging efficiency of vehicle v
$P_{ch}(t, v)$	Charged power of vehicle v in time t
η_v^{dis}	Discharging efficiency of vehicle v
$P_{dis}(t, v)$	Discharged power of vehicle v in time t
$P_{tra}(t, v)$	Traveling requirement of vehicle v in time t
$Uch(t, v)$	Binary variable indicating the charging state
$Udis(t, v)$	Binary variable indicating the discharging state
$\underline{P_{ch}}, \overline{P_{ch}}$	Min/max charging limits of vehicle v
$\underline{P_{dis}}, \overline{P_{dis}}$	Min/max discharging limits of vehicle v
$\Delta D(t, v)$	Traveling distance
Ω_v	Efficiency of the vehicle

CHAPTER 12

The main notation used in the paper is stated below for quick reference. Other symbols are defined as needed throughout the text.

Nomenclature

i,j	Index for buses
m	Index for Discos
t	Index for hours
T	Set of hours
Nb	Set of all buses
$Nb(i)$	Set of buses connected to bus i
Ng	Set of Gencos buses
Ndg	Set of DG buses
Nil	Set of IL buses
Nl	Set of all branches
N_m	Set of Discos
$Nd(m)$	Set of buses for Disco m
X_{ij}	Reactance of line i–j

$\overline{\lambda}$	Disco's retail energy rate
$P_{d,it}^{0}$	Maximum demand of Disco
$P_{g,it}^{\min}$, $P_{g,it}^{\max}$	Generation limits on a generator
$P_{dg,it}^{\min}$, $P_{dg,it}^{\max}$	Limits on generation of a DG
$P_{IL,it}^{\min}$, $P_{IL,it}^{\max}$	Limits on *IL* submitted by a Disco
$P_{ij,t}^{\min}$, $P_{ij,t}^{\max}$	Limits on flow of line *i*–*j*
$a_{g,it}$, $b_{g,it}$	Generation cost coefficients of a generator
$a_{dg,it}$, $b_{dg,it}$	Generation cost coefficients of a DG
$a_{IL,it}$, $b_{IL,it}$	*IL* cost coefficients of a Disco
$Rd(m)$	Revenue of Disco *m*
$Cd(m)$	Cost of Disco *m*
$Pd(m)$	Profit of Disco *m*
$C_{dg,it}(P_{dg,it})$	Generation cost of a DG
$P_{g,it}$	Generation of a generator
$P_{dg,it}$	Generation of a DG
$P_{ij,t}$	Line flow of line *i*–*j*
$P_{IL,it}$	IL granted to Disco
λ_{it}	Locational Marginal Price at bus *i*
$\mu_{g(i,t)}^{\max}$, $\mu_{g(i,t)}^{\min}$	Dual variables associated with the upper and lower bounds of generators capacity constraints
$\mu_{IL(i,t)}^{\max}$, $\mu_{IL(i,t)}^{\min}$	Dual variables associated with the upper and lower bounds of ILs capacity constraints
$\upsilon_{(i,j,t)}^{\max}$, $\upsilon_{(j,i,t)}^{\min}$	Dual variables associated with the upper and lower bounds of transmission lines constraints
$\zeta_{(i,t)}^{\min}$, $\zeta_{(i,t)}^{\max}$	Dual variables associated with the upper and lower bounds of voltage angle at bus *i*
ζ_{t}^{8}	Dual variable associated with the voltage angle at bus 8 upper and lower bounds of voltage angle at bus *i*

INDEX

Note: Page numbers followed by "*f*" and "*t*" refer to figures and tables, respectively.

E

F

R

S

T

W

Printed in the United States
By Bookmasters